CALCULUS

An Introduction Using Concrete Examples

A. Shabazz

L. Yao   F. Semwogerere

April 24, 2010

Order this book online at www.trafford.com
or email orders@trafford.com

Most Trafford titles are also available at major online book retailers.

Printed in Victoria, BC, Canada.

ISBN: 978-1-4251-9088-0 (sc)

*Our mission is to efficiently provide the world's finest, most comprehensive book publishing
service, enabling every author to experience success. To find out how to publish your book,
your way, and have it available worldwide, visit us online at www.trafford.com*

Trafford rev. 5/5/2010

 www.trafford.com

North America & international
toll-free: 1 888 232 4444 (USA & Canada)
phone: 250 383 6864 ♦ fax: 812 355 4082

CONTENTS

4

PREFACE

This book is intended to cover two semesters of Calculus. The first part aims at introducing derivatives as quickly as possible in order to give the students ample time to master techniques of differentiation. To this end, the first topic of the book is derivatives which are introduced as slopes of tangents. The student is shown how to compute these slopes for polynomial functions and, in the process, he/she is introduced to a limit of a quotient. (Formal limits are introduced in Chapter 2 after the student has had a glimpse at this particular limit.) Slopes of tangents are then used to solve optimization and approximation problems, including Newton's method. These standard applications are covered in the first 50 pages. Half-way through the book, we present the definite integral as a limit of sums in preparation for its applications to volumes of revolution, work done by a variable force, fluid forces, etc. Antiderivatives are shown to be tools one needs to evaluate limits of sums. They are actually introduced in Chapter 1 to provide additional practice in taking derivatives. The following is a summary of what is covered in the various chapters:

Chapter 1 starts off with a challenge, to the student, to determine the largest possible volume of a box that can be constructed from a rectangular metal plate of a given size. This is intended to set the tone for using calculus to solve problems. We guide the student to the conclusion that if one knows how to calculate slopes of tangents then one can easily solve the above problem. This is followed by a systematic calculation of slopes of tangents to the graphs of polynomial functions, using secant lines, and a derivation of the general power rule. Now the student has a fairly large class of functions he/she can differentiate, therefore problems that can be reduced to determining slopes of tangents, (of simple rational functions), are addressed. These include maxima/minima problems, curve sketching, linear approximations, Newton's method, and rates of change. The Mean Value Theorem for derivatives is also introduced. For additional practice in taking derivatives, the student is asked to guess simple antiderivatives.

Chapter 2 exposes the student to more general limits of real valued functions. The usual properties of limits are addressed, and the idea of a limit is used to introduce continuous functions.

Chapter 3 introduces the product and quotient rules for derivatives, followed by the chain rule as a generalization of the power rule. Higher order derivatives are introduced, and the second derivative test for relative maximum/minimum values is derived. The Taylor polynomials and L'Hopital's rule are covered, to give the student more practice in taking derivatives. Implicit differentiation and derivatives of inverse functions are addressed. The hyperbolic functions, their inverses and derivatives are also introduced in this chapter. Guessing antiderivatives, in order to strengthen differentiation, is continued.

The basic concept in Chapter 4 is the area of regions enclosed by curves. We approximate a given region with a number of rectangles. The total area of the rectangles is introduced as a Riemann sum and is used to approximate the area of the region. The required area is the limit of the Riemann sums. We then show how to use antiderivatives to calculate limits of Riemann sums. Various techniques for determining antiderivatives are then investigated. These include integration by substitution, integration by parts, integration by partial fractions, etc.

Chapter 5 exposes the student to problems that are reducible to determining limits of Riemann sums. These include the area between curves, work done by a variable force, volumes of solids of revolution, lengths of graphs of functions, and areas of surfaces of revolution.

Chapter 6 addresses the problem of approximating indefinite integrals. We briefly discuss the use of Riemann sums, the trapezoidal rule, and Simpson's rule to calculate such approximations.

Chapter 7 addresses sequences and series plus their limits. A number of tests for convergence of series are discussed. The Taylor and Maclaurin series are also mentioned.

The last chapter is an appendix that reviews function notation, and the common trigonometric and exponential functions plus their inverses. It also contains the solutions to about half of the exercises in the book.

The book is self-contained. Most of the major theorems we use are proved directly or in exercises with ample hints.

We would like to acknowledge the inclusion of many problems that were prepared by Dr. Sims. We are grateful that he gave us the permission to include them.

A. Shabazz, Ph.D.   L. Yao, Ph.D.   F. Semwogerere, Ph.D.
Grambling State Univ.   Grambling State Univ. Grambling State Univ.

Chapter 1   SLOPES OF TANGENTS

Introduction - A Maximization Problem

The first part of this course is devoted to calculating slopes of tangents, and using them to solve problems. You should have an intuitive idea of a tangent to the graph of a given function f. It is a line that "lies flat on the graph". In the figures below two tangents and a line that is not a tangent to the graph of $f(x) = x^2$ are shown.

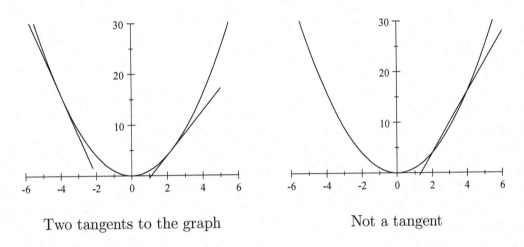

Two tangents to the graph   Not a tangent

There are many problems one can solve if one knows how to calculate slopes of tangents to graphs. Here is one example requiring you to maximize a volume:

You are the production manager of a small metal factory. You receive a shipment of 20 cm by 26 cm metal plates to be transformed into boxes without tops, by cutting off identical squares from each corner of a plate then fold along the dotted lines to get a box with no cover as shown below.

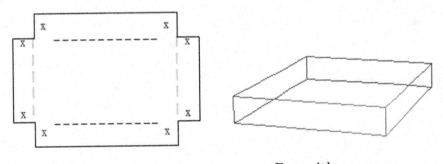

Box with no cover

What size of square will you recommend be cut off in order to form a box with the largest possible volume?

The most natural thing to think of is to try out several possible sizes. Use photocopies of the 20 by 26 rectangle on page 294 to make such boxes and calculate their volumes. It soon becomes clear that the volume of a box you make depends on the size of the square you remove. Soon, you also figure out how to calculate the volume of a box formed when a particular size of square is cut out without first constructing the box. Copy and complete the following table:

Length removed	2	3	4	5	5.5	6	6.5	7
Length of box	22							
Width of box	16							
Volume of box (in cu. units.)	352							

The length x of a square that may be cut from each corner must be less than 10 cm, (can you see why?), and it must be positive. This places x in the interval $[0, 10]$. Therefore you must recommend a length c, between 0 and 10, of a square that should be cut from each corner to form the box with the largest possible volume. It is impossible to try out all the numbers between 0 and 10 because they are infinitely many, therefore a different approach is necessary. Plotting a graph of the function that gives the volume of the box formed in terms of the length x of each square removed is one option. To obtain its formula, note that if you remove a square of length x from each of the 4 corners then the box you form has length $(20 - 2x)$ cm, width $(26 - 2x)$ cm, and height x cm., therefore its volume is

$$v(x) = x(20 - 2x)(26 - 2x) = 4x(10 - x)(13 - x), \quad 0 \le x \le 10$$

We may remove parentheses to get $v(x) = 4x^3 - 92x^2 + 520x$. Its graph is shown below. It does not provide the exact answer, but it suggests that c is between 3.5 cm and 4 cm. It also provides the following insight into what is going on: If you do not cut off anything, (which corresponds to $x = 0$), you get a zero volume. As you increase the size of the square you cut off, the volume of the box you get increases, but not for ever. When x gets to the required value c, (which we do not know at the moment), you get the maximum possible volume. Then further increases in x give smaller volumes, eventually shrinking to 0 when $x = 10$ cm.

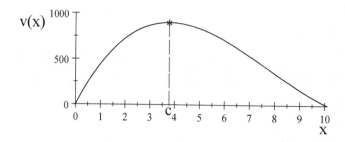

One property that distinguishes the number c we are looking for from all the other numbers between 0 and 10 is that the tangent to the graph of v at the point

$(c, v(c))$ is horizontal, (see the figure below), hence its slope is 0.

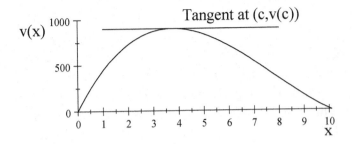

This suggests that to solve this problem, it is sufficient to devise a method of determining the slopes of tangents to graphs of functions. With such a tool at our disposal, the problem is solved by determining the number c between 0 and 10 such that the slope of the tangent to the graph of v at $(c, v(c))$ is zero. That tool has already been developed and refined. It is called **differential calculus**, and it is the main focus of the next couple of chapters. We will soon use it to show that, in the case of $v(x) = 4x^3 - 92x^2 + 520x$ the slope of the tangent at any point $(x, v(x))$ on its graph is $12x^2 - 184x + 520$. Therefore, the required number c is the solution of the equation

$$12x^2 - 184x + 520 = 0$$

which is between 0 and 10. This equation may be solved by using the quadratic formula, and the result is

$$x = \frac{184 \pm \sqrt{184^2 - 4(12)(520)}}{24}$$

Of the two solutions, $x = 3.74$ and $x = 11.64$, (rounded off to 2 decimal places), it is the first one which is between 0 and 10. It follows that the largest possible volume is approximately

$$4(3.74)^3 - 92(3.74)^2 + 520(3.14) = 867.20 \text{ cubic centimeters, (to 2 decimal places).}$$

Slope of a Tangent to a Graph of a Function

The example in the introduction should have convinced you that knowing how to calculate slopes of tangents is a powerful problem-solving tool. In this section, we systematically compute the slopes of tangents to graphs of some familiar functions.

We start with the "linear functions", (i.e. the functions whose graphs are straight lines). The simplest are the constant ones like $g(x) = -1$ and $h(x) = 2.2$ whose graphs are shown below. In general, if c is a

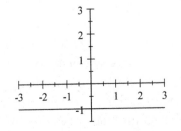

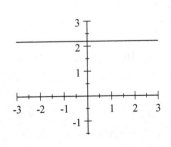

constant then the graph of $f(x) = c$ is a horizontal straight line. The tangent to its graph at any point (x, c) is the horizontal line itself. Since a horizontal line has zero slope, the slope of the tangent to the graph of $f(x) = c$ at any point (x, c) is 0.

In general, a linear function has a formula $f(x) = mx + c$ where m and c are constants, and its graph is a straight line ℓ with slope m. The tangent to ℓ at any point $(x, f(x))$ is the line ℓ itself. Therefore the slope of the tangent at any point on the graph of $f(x) = mx + c$ is m. For example, the slopes of the tangents to the graphs of $g(x) = 3x - 1$ and $h(x) = -2x + 5$, (shown below), are 3 and -2 respectively.

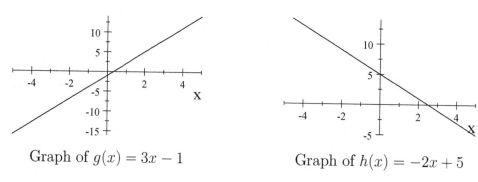

Graph of $g(x) = 3x - 1$ Graph of $h(x) = -2x + 5$

After the linear functions, the quadratic functions like $f(x) = x^2$ should be next. The graph of f is a parabola shown below. We wish to determine the slope of the tangent at an arbitrary point $A\,(a, a^2)$. The second figure shows the tangent.

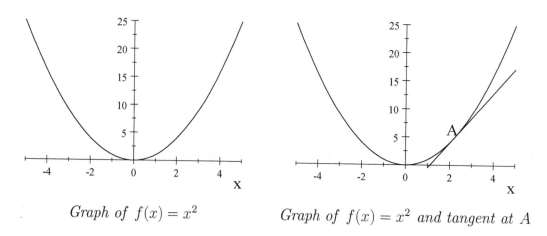

Graph of $f(x) = x^2$ *Graph of $f(x) = x^2$ and tangent at A*

Denote its slope by m. A standard method of calculating the slope of a line is to determine a "rise" and a "run" then use the formula

$$m = \frac{Rise}{Run}$$

Unfortunately, this standard approach does not work in this case because we know only one point on the tangent, namely (a, a^2), whereas we need two in order to calculate a "rise" and a "run". The Calculus approach is to determine approximate values of m and use them to deduce its exact value. The approximate values are obtained using **secant lines** through (a, a^2). By definition, a secant line through

(a, a^2) is a line segment that passes through (a, a^2) and another point on the graph of f. An example is the line segment through $A(a, a^2)$ and $B(a + 2, (a + 2)^2)$ shown in the figure below. It is drawn with the tangent at A in the second figure.

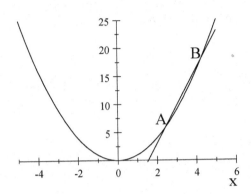 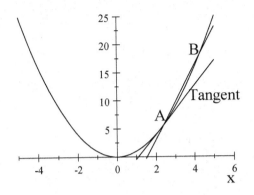

Secant line joining A and B   *The secant line and the tangent at A*

Its slope, can be calculated because we know two of its points, (namely A and B), and it is equal to

$$\frac{(a + 2)^2 - a^2}{(a + 2) - a} = 2a + 2.$$

This is an example of an approximate value of m, therefore we may write

$$m \simeq 2a + 2.$$

It is not a particularly good approximation because, (as the second figure above shows), AB does not approximate the tangent that well. We should get a better one by taking a secant line joining (a, a^2) to a closer point than $(a + 2, (a + 2)^2)$. An example is $C = (a + 0.2, f(a + 0.2)) = (a + 0.2, a^2 + 0.4a + 0.04)$, drawn in the figure below.

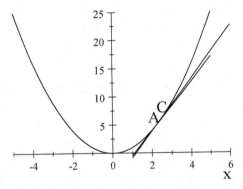

Its slope is $\dfrac{f(a + 0.2) - f(a))}{a + 0.2 - a} = \dfrac{0.4a + 0.04}{0.2} = 2a + 0.2$, therefore $m \simeq 2a + 0.2$.

For an even better one, take the line joining (a, a^2) and $(a + 0.01, f(a + 0.01))$. Its slope is

$$\frac{f(a + 0.01) - f(a)}{a + 0.01 - a} = \frac{0.02a + 0.0001}{0.01} = 2a + 0.01,$$

hence $m \simeq 2a + 0.01$.

To get the exact value of m we proceed as follows: Take a *general* point P close to (a, a^2). It is obtained by adding a small variable h, (which may be positive or negative), to a then evaluate $f(a+h)$. Thus, P has coordinates

$$(a + h, f(a+h)) = \left(a + h, a^2 + 2ah + h^2\right).$$

The slope of the secant line joining A and P is

$$\frac{f(a+h) - f(a)}{h} = \frac{a^2 + 2ah + h^2 - a^2}{h} = \frac{h(2a+h)}{h} = 2a + h \qquad (1.1)$$

Every sufficiently small number h, (positive or negative), gives a secant line that is pretty "close" to the tangent at (a, a^2). Two more are shown in the figures below.

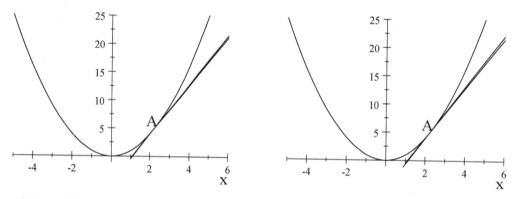

Tangent and secant line when $h = -0.1$ Tangent and secant line when $h = 0.15$

The tangent at (a, a^2) is the only line that is "close" to *all* these secant lines. No other line through $A(a, a^2)$ has that property. This points to one conclusion:

- *The slope of the tangent at (a, a^2) is the number that is close to the slopes of all these secant lines. In other words, the slope of the tangent is the number that is close to all the numbers of the form $2a + h$ when h is close to 0.*

That number is $2a$. Therefore the slope of the tangent at (a, a^2) must be $2a$.

For a quadratic function like this, it is possible to double check this result using an algebraic argument as follows: If the slope of the tangent is m, (it is unknown at this moment), then its equation is $(y - a^2) = m(x - a)$, which we may also write as $y = mx - ma + a^2$. The equation that gives the point(s) of intersection of the graph of f and the line $y = mx - ma + a^2$ is $x^2 = mx - ma + a^2$. In the standard form of a quadratic equation, it is

$$x^2 - mx + ma - a^2 = 0 \qquad (1.2)$$

Since a tangent to a parabola intersects the parabola in only one point, equation (1.2) above has only one solution. (This is the condition that enables us to determine m in the case of a parabola. It may not hold for other curves.) Recall that for a quadratic equation $\quad Ax^2 + Bx + C = 0 \quad$ to have one root, its discriminant

$B^2 - 4AC$ must be 0. In the case of (1.2), $A = 1$, $B = m$ and $C = ma - a^2$, therefore m must satisfy the equation

$$m^2 - 4(ma - a^2) = 0 \qquad (1.3)$$

When we remove the parentheses, (1.3) becomes $m^2 - 4am + 4a^2 = 0$ which factors as $(m - 2a)^2 = 0$. Therefore $m = 2a$, as expected.

Returning to the ratio

$$\frac{f(a + h) - f(a)}{h} = 2a + h,$$

we noted that if you substitute small numbers h into this expression, then $2a$ is the number that is close to the answers you get. Because of this, we say that

> **"the limit of** $\dfrac{\mathbf{f(a+h) - f(a)}}{\mathbf{h}}$ **as h approaches 0 is $2a$"**

This is written briefly as

$$\lim_{h \to 0} \frac{f(a + h) - f(a)}{h} = 2a$$

Remember that this is just a formal way of stating that if you substitute any number h close to 0 into the expression $\dfrac{f(a + h) - f(a)}{h}$, you get a value close to $2a$.

Remark 1 We have essentially established that the slope of the tangent to the graph of $f(x) = x^2$ at any point (x, x^2) is $2x$. We may use this result to answer a number of specific questions about the graph of f. For example

- What is the slope of the tangent at $(2.5, 6.25)$? And the answer is $2 \times (2.5) = 5$.

- What is the equation of the tangent at $(-3, 9)$? Answer: Its slope is $2 \times (-3) = -6$, therefore its equation is

$$(y - 9) = -6(x + 3) \qquad \text{or} \qquad y = -6x - 9.$$

- At what point on the graph of g is the slope of the tangent equal to 7? And the answer is $(3.5, (3.5)^2) = (3.5, 12.25)$ since 3.5 is the solution to the equation $2x = 7$.

- Where, on the graph of f, is the tangent parallel to the line $3y + 4x = 2$? Answer; at the point where the slope of the tangent is $-\frac{4}{3}$. The solution to the equation $2x = -\frac{4}{3}$ is $x = -\frac{2}{3}$. Therefore the point is $(-\frac{4}{3}, \frac{4}{9})$.

- For what values of x is the slope of the tangent at (x, x^2) negative? Answer; any $x < 0$

The next function to consider is $u(x) = x^3$. It turns out that the slope of the tangent at any point $A(x, x^3)$ on its graph is $3x^2$. To see this, take a point $P(x + h, (x + h)^3)$ near A. The slope of the secant line joining A and P is

$$\frac{u(x+h) - u(x)}{h} = \frac{3x^2h + 3xh^2 + h^3}{h} = \frac{h(3x^2 + 3xh + h^2)}{h} = 3x^2 + h(3x + h)$$

The slope of the tangent is the number that is close to all the values $3x^2 + h(3x+h)$ when h is close to 0. More precisely, it is the limit of $3x^2 + h(3x+h)$ as h approaches 0. That number is

$$3x^2 + 0(3x + 0) = 3x^2,$$

therefore, the slope of the tangent at (x, x^3) is $3x^2$.

Next up the ladder is $v(x) = x^4$. You have probably guessed that the slope of the tangent at $A(x, x^4)$ is $4x^3$. If YES, you guessed the right answer. To confirm it take a point $P(x + h, (x + h)^4)$ near A. The slope of the secant line joining A and P is

$$\frac{v(x+h) - v(x)}{h} = \frac{(x+h)^4 - x^4}{h} = \frac{x^4 + 4x^3h + 6x^2h^2 + 4xh^3 + h^4 - x^4}{h}$$

$$= \frac{h(4x^3 + 6x^2h + 4xh^2 + h^3)}{h} = 4x^3 + h(6x^2 + 4xh + h^2)$$

The slope of the tangent is the limit of $4x^3 + h(6x^2 + 4xh + h^2)$ as h approaches 0, which is $4x^3$

Claim 2 *If r is any real then the slope of the tangent to the graph of $f(x) = x^r$ at (x, x^r) is rx^{r-1}.*

In the next two examples, we verify the claim for the negative integer $n = -1$ and the fraction $r = \frac{1}{2}$. You are asked to prove it in problems 3, 4, 5, and 6 of Exercise 26 on page 21, when r is any rational number, positive or negative. The proof when r is not rational requires tools that are not develop in this book, therefore we will not give it.

Example 3 *The slope of the tangent to the graph of $g(x) = x^{-1} = \dfrac{1}{x}$, $x \neq 0$ at any point $\left(x, \dfrac{1}{x}\right)$ is $(-1)x^{-2} = -x^{-2} = -\dfrac{1}{x^2}$. For,*

$$\frac{g(x+h) - g(x)}{h} = \frac{\dfrac{1}{(x+h)} - \dfrac{1}{x}}{h} = \frac{x - (x+h)}{hx(x+h)} = \frac{-h}{hx(x+h)} = \frac{-1}{x(x+h)}.$$

The required slope is the limit of $\dfrac{-1}{x\,(x+h)}$ *as h approaches 0, which is*

$$\frac{-1}{x\,(x+0)} = -\frac{1}{x^2} = -x^{-2}.$$

Example 4 *The slope of the tangent to the graph of* $h(x) = x^{\frac{1}{2}} = \sqrt{x}$ *at* $(x, \sqrt{x})$
is $\frac{1}{2}x^{\frac{1}{2}-1} = \frac{1}{2}x^{-\frac{1}{2}}$. *We verify this by calculating* $\lim\limits_{h\to 0}\dfrac{\sqrt{x+h} - \sqrt{x}}{h}$. *Rationalizing the numerator gives:*

$$\frac{\sqrt{x+h} - \sqrt{x}}{h} = \frac{\left(\sqrt{x+h} - \sqrt{x}\right)\left(\sqrt{x+h} + \sqrt{x}\right)}{h\left(\sqrt{x+h} + \sqrt{x}\right)}$$

$$= \frac{x+h - x}{h\left(\sqrt{x+h} + \sqrt{x}\right)} = \frac{1}{\left(\sqrt{x+h} + \sqrt{x}\right)}$$

Therefore $\lim\limits_{h\to 0}\dfrac{\sqrt{x+h} - \sqrt{x}}{h} = \lim\limits_{h\to 0}\dfrac{2}{\sqrt{x+h} + \sqrt{x}} = \dfrac{1}{\sqrt{x} + \sqrt{x}} = \dfrac{1}{2\sqrt{x}} = \dfrac{1}{2}x^{-\frac{1}{2}}$

Exercise 5

1. *Use the expression* rx^{r-1} *to determine the slope of the tangent at* $(x, f(x))$
   then give its equation at the specified point:

   (a) $f(x) = \dfrac{1}{x}$ *at* (i) $(1, 1)$,   (ii) $\left(3, \frac{1}{3}\right)$,   (iii) $\left(-2, -\frac{1}{2}\right)$

   (b) $f(x) = \dfrac{1}{\sqrt{x}}$ *at* (i) $(1, 1)$,   (ii) $\left(4, \frac{1}{2}\right)$,   (iii) $\left(9, \frac{1}{3}\right)$

   (c) $f(x) = x^{3/2}$ *at* (i) $(4, 8)$,   (ii) $\left(2, 2^{3/2}\right)$,   (iii) $\left(a, a^{3/2}\right)$

2. *Consider the function* $f(x) = x^5$.

   (a) *Determine the point on the graph where the tangent to the graph is
   horizontal.*

   (b) *There are two points on the graph where the slope of the tangent is 80.
   What are they?*

Derivative of a Function

Given a function f, we have seen that it may be possible to determine an expression
for the slope of the tangent at any point $(x, f(x))$ on its graph. We use it to define
a new function called the derivative of f.

- *The derivative of a given function f is another function, denoted by f', whose
  value $f'(x)$ is the slope of the tangent to the graph of f at $(x, f(x))$.*

- *Determining $f'(x)$ is called differentiating f with respect to x. The independent variable does not have to be x. If it is another letter, say t, then
  determining $f'(t)$ is called differentiating f with respect to t.*

- *A function that has a derivative is called a differentiable function.*

Example 6 *We showed that the slope of the tangent to the graph of $g(x) = x^2$ at (x, x^2) is $2x$. It follows that the derivative of g is the function g' with formula $g'(x) = 2x$.*

Example 7 *We showed that the slope of the tangent to the graph of $r(x) = \sqrt{x}$ at $(x, \sqrt{x})$ is $\dfrac{1}{2\sqrt{x}}$. Therefore the derivative of r is the function r' with formula $r'(x) = \dfrac{1}{2\sqrt{x}}$.*

- *In general, (by Claim 2 on page 8), if r is a real number then the slope of the tangent to the graph of $f(x) = x^r$ at (x, x^r) is rx^{r-1}. Therefore the derivative of $f(x) = x^r$ is the function f' with formula $f'(x) = rx^{r-1}$. This is called the power rule for derivatives.*

Example 8 *Let $f(x) = x\sqrt{x}$. We may write this as $f(x) = x^{3/2}$. Then by the power rule for derivatives, $f'(x) = \frac{3}{2}x^{1/2}$.*

Example 9 *Let $f(x) = \dfrac{1}{\sqrt[5]{x^7}} = \dfrac{1}{x^{7/5}} = x^{-7/5}$. Then $f'(x) = -\frac{7}{5}x^{-12/5}$*

Example 10 *Let $f(x) = mx$. Its graph is a straight line ℓ with slope m. The tangent to ℓ at any point (x, mx) is the line ℓ itself, therefore its slope is m. It follows that $f'(x) = m$. More precisely, if m is a constant then the derivative of $f(x) = mx$ is $f'(x) = m$.*

Example 11 *Let f be a constant function. Thus f has formula $f(x) = c$ where c is a constant. We have already pointed out that the slope of the tangent at any point (x, c) on the graph is 0. It follows that if $f(x) = c$, (a constant), then $f'(x) = 0$.*

Remark 12 *Consider an arbitrary function f and a point $(x, f(x))$ on its graph. The slope of the tangent at $(x, f(x))$ is the limit of the quotient $\dfrac{[f(x+h) - f(x)]}{h}$ as h approaches 0. Therefore, the derivative of f is the function f' with formula*

$$f'(x) = \lim_{h \to 0} \frac{f(x+h) - f(x)}{h}. \tag{1.4}$$

Exercise 13

1. *The graph of some function v is given below together with the tangents to the graph at the three points with x-coordinates -2, -1, and 1.5 respectively.*

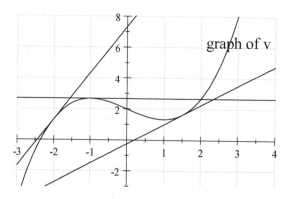

Use the tangents to estimate $v'(-2)$, $v'(-1)$ and $v'(1.5)$.

2. The graph of some function u is given below.

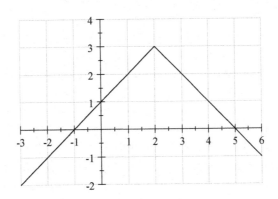

Determine $u'(a)$ when $a < 2$ and $u'(b)$ when $b > 2$. (Since one cannot draw a tangent at $(2,3)$, $u'(2)$ is undefined.)

3. Use Definition (1.4) to show that the derivative of $f(x) = mx + b$ is $f'(x) = m$ and the derivative of $g(x) = c$, (a constant), is $g'(x) = 0$.

4. Let $f(x) = \dfrac{1}{x^2}$, $x \neq 0$. Show that $\dfrac{f(x+h) - f(x)}{h} = \dfrac{-2x - h}{x^2(x+h)^2}$ then determine $f'(x)$.

5. Let $g(x) = \dfrac{1}{\sqrt{x}}$, $x > 0$. Show that $\dfrac{g(x+h) - g(x)}{h} = \dfrac{\sqrt{x} - \sqrt{x+h}}{h\left(\sqrt{x(x+h)}\right)}$. Now rationalize the numerator and determine $g'(x)$.

6. Let $u(x) = x^3 + x$. Show that $\dfrac{u(x+h) - u(x)}{h} = 3x^2 + 1 + 3xh + h^2$ then determine $u'(x)$.

7. Use the power rule for derivatives to determine $f'(x)$ given that $f(x) =$

(a)x^{23} (b) x^{-3} (c) $\dfrac{1}{x^5}$ (d) $x^{3/4}$ (e) $x^{2/3}$

(f)$\sqrt[4]{x^7}$ (g)$\dfrac{1}{x^{4/5}}$ (h) $x^2\sqrt{x}$ (i) $\dfrac{1}{x^{5/2}}$ (j) $\dfrac{1}{\sqrt[2]{x^7}}$

(k)$\sqrt[3]{x^5}$ (l)$x^{\sqrt{3}}$ (m) x^{π} (n) $x^3\sqrt{x}$ (o) $\dfrac{1}{x^3\sqrt{x}}$

Rules for Computing Derivatives
The Sum Rule

The following example illustrates what the sum rule says:

Example 14 *Consider a function like* $s(x) = x^3 + x^2$. *It is a sum of the two functions* $f(x) = x^3$ *and* $g(x) = x^2$ *whose derivatives we already know; they are*

$f'(x) = 3x^2$ and $g'(x) = 2x$. It turns out that the derivative of s is $s'(x) = 3x^2 + 2x$, (the sum of the two derivatives). This is not hard to verify:

$$\frac{s(x+h) - s(x)}{h} = \frac{(x+h)^3 + (x+h)^2 - x^3 - x^2}{h}$$

$$= \left(\frac{(x+h)^3 - x^3}{h} + \frac{(x+h)^2 - x^2}{h} \right)$$

We know that when h is close to 0, the quotients $\dfrac{(x+h)^3 - x^3}{h}$ and $\dfrac{(x+h)^2 - x^2}{h}$ are close to $3x^2$ and $2x$ respectively. It follows that their sums must be close to $3x^2 + 2x$, therefore

$$s'(x) = \lim_{h \to 0} \frac{s(x+h) - s(x)}{h} = 3x^2 + 2x.$$

- **In general, the derivative of a sum** $s(x) = f(x) + g(x)$ **is**

$$s'(x) = f'(x) + g'(x).$$

This is called the sum rule for derivatives.

To verify it, consider

$$\frac{s(x+h) - s(x)}{h} = \frac{f(x+h) - f(x)}{h} + \frac{g(x+h) - g(x)}{h}.$$

When h is close to 0, the quotients $\dfrac{f(x+h) - f(x)}{h}$ and $\dfrac{g(x+h) - g(x)}{h}$ are close to $f'(x)$ and $g'(x)$ respectively. It stands to reason that their sum must be close to $f'(x) + g'(x)$. In other words,

$$s'(x) = \lim_{h \to 0} \frac{s(x+h) - s(x)}{h} = f'(x) + g'(x).$$

Example 15 *The derivative of* $h(x) = x^4 + x^{-3}$ *is* $h'(x) = 4x^3 - 3x^{-4}$.

Example 16 *The derivative of* $u(x) = x^7 + x^5 + x^{-2} - 9$ *is* $u'(x) = 7x^6 + 5x^4 - 2x^{-3}$.

The Constant Multiple Rule for Derivatives

A constant multiple of a given function f is any function you obtain when you multiply f by a constant. For example, the following are constant multiples of $f(x) = x^2$:

$$g(x) = 4x^2, \quad u(x) = \pi^3 x^2, \quad v(x) = \frac{x^2}{8} = \frac{1}{8}x^2, \quad \text{and } w(x) = -1.3x^2.$$

Consider $v(x) = \frac{x^2}{8}$, which we have also written as $v(x) = \frac{1}{8}x^2$. We know the derivative of $f(x) = x^2$; it is $f'(x) = 2x$. It turns out that the derivative of v is obtained by simply multiplying the derivative of f by the constant $\frac{1}{8}$. That is:

$$v'(x) = \frac{1}{8} \cdot (2x) = \frac{1}{4}x = \frac{x}{4}$$

In general, if $g(x) = cf(x)$ where c is a constant, then $g'(x) = cf'(x)$

This is called the constant multiple rule for derivatives. To justify it, consider

$$\frac{g(x+h) - g(x)}{h} = \frac{cf(x+h) - cf(x)}{h} = c\left[\frac{f(x+h) - f(x)}{h}\right]$$

When h is close to 0, $\dfrac{f(x+h) - f(x)}{h}$ is close to $f'(x)$, hence we should expect $c\left[\dfrac{f(x+h) - f(x)}{h}\right]$ to be close to $cf'(x)$. Therefore $g'(x) = cf'(x)$.

With the power rule, plus these two additional rules at our disposal, it is now possible to calculate derivatives of sums and/or constant multiples of the elementary functions $f(x) = x^n$ for different values of n without using Definition 1.4 on page 10.

Example 17 *Let $f(x) = x^4 + 5x^3$. It is a sum of $g(x) = x^4$ and $h(x) = 5x^3$. Since the derivative of g is $g'(x) = 4x^3$ and the derivative of h is $h'(x) = 5(3x^2)$, it follows that the derivative of f is $f'(x) = 4x^3 + 15x^2$.*

Example 18 *Let $f(x) = \frac{2}{5x^2} - \frac{3x}{4} + 18$. We may write it as $f(x) = \frac{2}{5}x^{-2} - \frac{3}{4}x + 18$, which is a sum of $g(x) = \frac{2}{5}x^{-2}$, $h(x) = -\frac{3}{4}x$, and $u(x) = 18$. Their derivatives are $g'(x) = \frac{2}{5}(-2x^{-3}) = -\frac{4}{5}x^{-3}$, $h'(x) = -\frac{3}{4}$, and $u'(x) = 0$. Therefore*

$$f'(x) = -\frac{4}{5}x^{-3} - \frac{3}{4} + 0 = -\frac{4}{5x^3} - \frac{3}{4}$$

Exercise 19

1. *Determine the derivative of each function and simplify when possible.*

   a. $f(x) = 3x^2$ b. $v(x) = \frac{1}{2}x^2$ c. $h(x) = 6x^3 + 5x - 14$

   d. $g(x) = \dfrac{x^2}{5}$ e. $f(x) = \frac{1}{3}x^3 - 4x^2$ f. $w(x) = \frac{x}{2} + \frac{2}{\sqrt{x}}$

   g. $h(x) = 2x^{\frac{1}{2}}$ h. $u(x) = \frac{1}{4x} + \pi^2$ i. $f(x) = 3x - 4x\sqrt{x} + 2$

   j. $v(x) = \frac{2x^5}{3}$ k. $v(x) = \pi - 6x^3$ l. $u(x) = \frac{\sqrt{x}}{2} + \frac{x}{3} - \frac{x^{3/2}}{4}$

   m. $u(x) = -\frac{3}{4}x^{-3}$ n. $g(x) = \frac{2}{x} + \frac{3}{x^2} - \frac{4}{5x^3}$ o. $s(x) = 4x - \frac{1}{4x}$

   p. $h(x) = \pi^3$ q. $h(x) = \frac{2}{3}x^{3/2} - \frac{5}{3}x^{1/5}$ r. $v(x) = 4x^3 - 92x^2 + 520x$

   s. $f(x) = \frac{5x^3}{2}$ t. $g(x) = \frac{2}{3x^2} - \frac{4}{5x^3}$ u. $h(x) = \frac{3x^3}{7} + \frac{4x^5}{9} - 12$

2. *It will soon be necessary to reverse differentiation. In other words, a function f will be given, and you will be required to give a function F whose derivative is f. At this junction, you will do this by making intelligent guesses. For example, given $f(x) = 3x^2$ it is easy to come up with $F(x) = x^3$. A less obvious one is $g(x) = x^4$. Clearly, one must look for an expression involving x^5, because when we take derivatives, a power is reduced by 1. One candidate is $G(x) = \frac{1}{5}x^5$. The constant $\frac{1}{5}$ cancels the factor 5 in $5x^4$, (the derivative of x^5). Copy and complete the following table: (in each case, check your answer by taking the derivative of your guess).*

Function	Derivative	Function	Derivative
(a)	$4x^3$	(b)	x^3
(c)	x^2	(d)	$\frac{1}{2}x^5$
(e)	$-2x^{-3}$	(f)	x^{-4}
(g)	$\dfrac{5}{x^2}$ or $5x^{-2}$	(h)	$\dfrac{2}{3x^4}$
(i)	$\frac{1}{2}x^{-1/2}$	(j)	$x^{-1/2}$
(k)	$\frac{1}{3}x^{-1/3}$	(l)	$x^{-1/3}$
(m)	$4x^{-3/2}$	(n)	$x^{2/5}$
(o)	$x - 4\sqrt{x}$	(p)	$x^3 + \dfrac{1}{3x^2}$

3. *Let $f(x) = x - \frac{1}{x}$, $x \neq 0$. Determine the equation of the tangent to the graph of f at the point $(1, 0)$. Also, use the formula for the derivative to show that the slope of the tangent, at any point $\left(x, x - \frac{1}{x}\right)$ on the graph of f, cannot be smaller than 1.*

4. *Find the points, (there are two of them), on the graph of $f(x) = x^3$ where the tangents are parallel to the line $4y - 3x = 5$.*

5. *Let $g(x) = \dfrac{x}{x+1}$. Then*

$$\frac{g(x+h) - g(x)}{h} = \frac{\dfrac{x+h}{x+h+1} - \dfrac{x}{x+1}}{h} = \frac{x^2 + xh + x + h - x^2 - xh - x}{h\,(x+1+h)\,(x+1)}$$

Simplify further, then show that $g'(x) = \dfrac{1}{(x+1)^2}$.

6. *Let $v(x) = \sqrt{5x + 3}$. Then*

$$\frac{v(x+h) - v(h)}{h} = \frac{\sqrt{5x + 5h + 3} - \sqrt{5x + 3}}{h}$$

Rationalize the numerator, then determine $v'(x)$.

7. *Let $u(x) = \dfrac{1}{\sqrt{3x+1}}$.*

(a) Show that $\dfrac{u(x+h)-u(h)}{h} - \dfrac{\sqrt{3x+1}-\sqrt{3x+3h+1}}{h(\sqrt{3x+1})(\sqrt{3x+3h+1})}$, *rationalize the numerator, then determine* $u'(x)$.

(b) *Determine the equation of the tangent to the graph of* u *at* $\left(1,\frac{1}{2}\right)$.

8. *A closed box is to be constructed from a rectangular metal lamina of dimensions 12 by 18 units by cutting out rectangles of sides* x *and* $\frac{1}{2}x$ *as shown in the diagram below then fold along the dotted lines.*

(a) x *must be less than some number* b. *What is that number?*

(b) *Complete the following table:*

Value of x	1	2	2.5	3	3.5	4	5	x
Volume $V(x)$		140						

Use the table to describe how the volume of the box formed changes with x, *then use slopes of tangents to determine the largest possible volume of such a box.*

9. *Your iron works company has been contracted to design and build a* 600 m^3 *square-based open-top, rectangular steel holding tank for a company. The tank is to be made by welding thin stainless steel plates together along their edges. As the production engineer, your job is to find the dimensions of the box that will require the least area of metal. Complete the following table. The length and height of the tank are in meters, and the area* A *of metal used to construct the tank is in square meters.*

Length	2	3	6	8	10	12	14	x
Height		$\dfrac{600}{(3)^2}$						
Area $A(x)$		809						

Use the table to describe how the area of metal used changes with x, *then use slopes of tangents to determine the dimensions your shop should use.*

10. *You have to make an ice-cream cone, (more precisely, a right circular cone), with a slant length of 5 inches. (See the figures below for the definition of the slant length of a right circular cone.)*

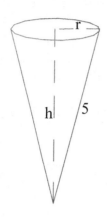

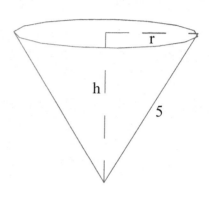

A tall and thin cone   *A wider and shorter cone*

You could make it tall and thin or wide but shorter as shown in the figures above. Let h and r, (in inches), be its height and radius respectively, and V, in cubic inches be its volume. These variables are related by the equation $V = \frac{1}{3}\pi r^2 h$. Copy and complete the following table.

h	0.5	1	1.5	2	2.5	4	4.5	h
r								
V								

Now describe how V varies as the height of the cone changes, then use an argument involving the slope of a tangent to find the largest volume of a right circular cone with a slant length of 5 inches.

11. *A paint factory has contracted you to design a 240 cubic centimeter cylindrical can, (closed at both ends). To save costs, the can should be constructed from the least amount of metal. You have to determine the height and radius of such a can. In case you do not remember: (i) The area of a circle with radius r is πr^2. (ii) If a cylinder has radius r and height h, then its volume is $\pi r^2 h$ and the area of its curved surface is $2\pi r h$. Complete the following table. The radius r and its height h are in centimeters, and the area A of metal used to construct can is in square centimeters.*

Radius	1	1.4	2	2.5	3	3.5	4.2	x
Height		$\frac{240}{\pi(1.4)^2}$			$\frac{240}{\pi(3)^2}$			
Area $A(x)$		355.2			216.5			

Use the table to describe how the area of metal used changes with r, then use slopes of tangents to determine the dimensions you should use.

Test your skills - 1, (20 minutes)

1. You are given the function $f(x) = 6 - 4x + 3x^2$.

   (a) Determine $f(x+h)$ and simplify your expression.
   (b) Find the slope of the secant line joining $(x, f(x))$ and $(x+h, f(x+h))$
   (c) Take limits in (b) to determine the slope of the tangent to the graph of f at $(x, f(x))$.
   (d) Find the equation of the tangent to the graph at $(1, 5)$.

2. Find the derivative of each function

   (a) $f(x) = 12 - x^{12}$
   (b) $g(x) = 12x + \frac{4}{x^{12}}$
   (c) $h(x) = 6x^5 - 4x^6 + 19$
   (d) $u(x) = \frac{3}{4}x^4 - 1 + \frac{5}{3x}$
   (e) $v(x) = 12x^{4/3} - \frac{8}{\sqrt{x}} + \frac{2}{x} - 3$
   (f) $w(x) = 1 - 4\pi x - 8x^{3/4}$
   (g) $6 - \left(\frac{6}{x} + \frac{x}{6}\right)$
   (h) $u(x) = 12x\sqrt{x} - \frac{5}{12x} + 9$

The General Power Rule

We start with an exercise in factoring differences of squares, cubes, etc. Most probably you have met the identity

$$X^2 - Y^2 = (X - Y)(X + Y),$$

called the formula for the difference of two squares. A formula for the difference of two cubes is

$$X^3 - Y^3 = (X - Y)\left(X^2 + XY + Y^2\right).$$

To verify it, simply expand the right hand side:

$$(X - Y)\left(X^2 + XY + Y^2\right) = x^3 + X^2Y + XY^2 - X^2Y - XY^2 - Y^3 = X^3 - Y^3$$

For a difference of two fourth powers, we choose

$$X^4 - Y^4 = (X - Y)\left(X^3 + X^2Y + XY^2 + Y^3\right).$$

There are other ways of factoring $X^4 - Y^4$, but we prefer this particular one. You may check that it is an identity by expanding the right hand side.

   Have you noticed the pattern? Use it to complete the following; (in the last identity, n is a positive integer).

$$X^5 - Y^5 = (X - Y)\,($$

$$X^6 - Y^6 = (X - Y)\,($$

$$X^7 - Y^7 = (X - Y)\,($$

$$X^n - Y^n = (X - Y)\,($$

We know how to find the derivative of any power $u(x) = x^r$ where r is any real number. We now address the derivatives of more general powers. More specifically, we address derivatives of functions like the following:

$$w(x) = \left(x - \tfrac{2}{x^2}\right)^6 \qquad h(x) = (5x - 13)^{-3} \qquad u(x) = \sqrt{x^2 + 1} = (x^2 + 1)^{1/2}$$

Write down 2 examples of your choice that fall into this category.

$$r(x) = \qquad\qquad\qquad\qquad q(x) =$$

They all have the general form $g(x) = [f(x)]^n$ where $f(x)$ is some given function and n is a given number. From the definition, the derivative of such a function g is given by

$$g'(x) = \lim_{h \to 0} \frac{[f(x+h)]^n - [f(x)]^n}{h}$$

Suppose n is a positive integer. Then we may factor $[f(x+h)]^n - [f(x)]^n$ as

$$[f(x+h) - f(x)]\left[(f(x+h))^{n-1} + (f(x+h))^{n-2} f(x) + \cdots + (f(x))^{n-1}\right].$$

Then $g'(x)$ is equal to

$$\lim_{h \to 0} \frac{[f(x+h) - f(x)]\left[(f(x+h))^{n-1} + (f(x+h))^{n-2} f(x) + \cdots + (f(x))^{n-1}\right]}{h}$$

It is convenient to write this limit as

$$\lim_{h \to 0} \left[(f(x+h))^{n-1} + (f(x+h))^{n-2} f(x) + \cdots + (f(x))^{n-1}\right] \cdot \frac{[f(x+h) - f(x)]}{h}$$

The limit of $\dfrac{[f(x+h) - f(x)]}{h}$ as h approaches 0 is $f'(x)$. It remains to figure out the limit of

$$\left[(f(x+h))^{n-1} + (f(x+h))^{n-2} f(x) + (f(x+h))^{n-3} (f(x))^2 + \cdots + (f(x))^{n-1}\right]$$

When h is close to 0, the first term $(f(x+h))^{n-1}$ in the square brackets is close to $(f(x))^{n-1}$. The second one is close to $(f(x))^{n-2} \cdot f(x)$ which also simplifies to $(f(x))^{n-1}$. The third one is also close to $(f(x))^{n-1}$. In fact, each one is close to $(f(x))^{n-1}$. Since there are n terms inside the square brackets,

$$\lim_{h \to 0} \left[(f(x+h))^{n-1} + (f(x+h))^{n-2} f(x) + \cdots + (f(x))^{n-1}\right] = n\,(f(x))^{n-1}$$

It stands to reason that when h is close to 0 then

$$\left[(f(x+h))^{n-1} + (f(x+h))^{n-2} f(x) + \cdots + (f(x))^{n-1}\right] \cdot \frac{[f(x+h) - f(x)]}{h}$$

is close to $n[f(x)]^{n-1} \cdot f'(x)$. Therefore, we have shown that if n is a positive integer, then the derivative of $[f(x)]^n$ is $n[f(x)]^{n-1} \cdot f'(x)$.

Example 20

- In the case of $g(x) = (x^2 + 4)^4$, $f(x) = x^2 + 4$, and $n = 4$ hence its derivative is $4(x^2 + 4)^3 (2x)$

- In the case of $h(x) = \left(x - \frac{2}{x^2}\right)^6$, $f(x) = x - \frac{2}{x^2}$, and $n = 6$ hence its derivative is $6\left(x - \frac{2}{x^2}\right)^5 \left(1 + \frac{4}{x^3}\right)$.

- In the case of $u(x) = (5x^4 - 2x)^3$, $f(x) = 5x^4 - 2x$, and $n = 3$ hence its derivative is $3(5x^4 - 2x)^2 (20x^3 - 2)$

- In the case of $v(x) = \left(\frac{2}{x} + \frac{x}{2}\right)^5$, $f(x) = \frac{2}{x} + \frac{x}{2}$, and $n = 5$ hence its derivative is $5\left(\frac{2}{x} + \frac{x}{2}\right)^4 \left(-\frac{2}{x^2} + \frac{1}{2}\right)$

- In the case of $w(x) = \left(4 + \frac{3}{x}\right)^9$, $f(x) = 4 + \frac{3}{x}$, and $n = 9$ hence its derivative is $9\left(4 + \frac{3}{x}\right)^8 \left(-\frac{3}{x^2}\right)$

In the proof we gave for the derivative of $[f(x)]^n$, we insisted that n should be a positive integer, but it does not have to. It is proved in Chapter 3 that

- If r is any real number then:

$$\text{The derivative of } \quad [f(x)]^r \quad \textbf{is } r\,[f(x)]^{r-1} \cdot f'(x).$$

This is called the generalized power rule for derivatives.

Example 21 The derivative of $u(x) = (5x - 13)^{-3}$ is

$$u'(x) = -3(5x - 13)^{-4}(5) = -15(5x - 13)^{-4}.$$

In this case, n is a negative integer.

Example 22 To determine the derivative of $g(x) = \dfrac{4}{2x + 5}$, first write it as an exponent $g(x) = 4(2x + 5)^{-1}$. Then, by the general power rule,

$$g'(x) = 4(-1)(2x + 5)^{-2}(2) = -8(2x + 5)^{-2}.$$

Example 23 To determine the derivative of $v(x) = \sqrt{x^2 + 1}$, first write it as $v(x) = (x^2 + 1)^{1/2}$. Then

$$v'(x) = \frac{1}{2}(x^2 + 1)^{-1/2}(2x) = x(x^2 + 1)^{-1/2} = \frac{x}{\sqrt{x^2 + 1}}$$

In this case n is a positive rational number.

Example 24 Let $w(x) = \dfrac{1}{\sqrt{2x + 3}} = (2x + 3)^{-1/2}$. Its derivative is

$$w'(x) = -\frac{1}{2}(2x + 3)^{-3/2} \cdot (2) = -(2x + 3)^{-3/2}$$

In this case n is a negative rational number.

Example 25 *Consider the function* $h(x) = \dfrac{3x}{(x-2)\,(x+1)}$. *The right hand side may be decomposed into partial fraction as*

$$\frac{3x}{(x-2)\,(x+1)} = \frac{2}{x-2} + \frac{1}{x+1}$$

If you are not familiar with decomposing rational function into partial fractions, here is one way of decomposing $h(x)$:   Because it has the two factors $(x-2)$ and $(x+1)$ in the denominator, the fraction $\dfrac{3x}{(x-2)\,(x+1)}$ is, most probably, the result of adding a fraction with denominator $(x-2)$ to another fraction with denominator $(x+1)$. Therefore it is reasonable to assume that there are numbers A and B, (to be determined), such that

$$\frac{3x}{(x-2)\,(x+1)} = \frac{A}{x-2} + \frac{B}{x+1} \tag{1.5}$$

To clear fractions, multiply both sides of (1.5) by $(x-2)\,(x+1)$. The result is

$$3x = A(x+1) + B(x-2) \tag{1.6}$$

Now substitute $x = -1$ in both sides of (1.6), (to get rid of the term involving A), then solve for B. You should get

$$-3 = B(-3), \text{ hence } B = 1$$

The next step is to substitute $x = 2$ in (1.6) (to get rid of the term involving B), and solve for A. You should get

$$6 = A(2+1) = 3A, \text{ hence } A = 2$$

Substituting the values of A and B into (1.5) gives

$$h(x) = \frac{3x}{(x-2)\,(x+1)} = \frac{2}{x-2} + \frac{1}{x+1} \tag{1.7}$$

The two fractions in the right hand side of (1.7) are called the partial fractions of $\dfrac{3x}{(x-2)\,(x+1)}$. The process of writing $\dfrac{3x}{(x-2)\,(x+1)}$ as a sum $\dfrac{2}{x-2} + \dfrac{1}{x+1}$ of simpler fractions is called splitting or decomposing $\dfrac{3x}{(x-2)\,(x+1)}$ into partial fractions.

We know how to determine the derivatives of $f(x) = \dfrac{2}{x-2} = 2\,(x-2)^{-1}$ *and* $g(x) = \dfrac{1}{x+1} = (x+1)^{-1}$. *They are* $f'(x) = 2(-1)\,(x-2)^{-2}$ *and* $g'(x) = (-1)\,(x+1)^{-2}$. *It follows that the derivative of* $h(x)$ *is*

$$h'(x) = 2(-1)\,(x-2)^{-2} + (-1)\,(x+1)^{-2} = -\frac{2}{(x-2)^2} - \frac{1}{(x+1)^2}$$

Exercise 26

1. Copy and complete the following table:

Function $[f(x)]^n$	In this case	Its derivative is
a. $(x^3 + 1)^4$	$f(x) = x^3 + 1, \; n = 4$	$12x^2 \left(x^3 + 1\right)^3$
b. $(x^2 + 3x - 4)^8$		
c. $\sqrt{x^2 + 2x + 8}$		
d. $(\sqrt{x} + 5)^{3/5}$		
e. $\sqrt{3 + \sqrt{x}}$		
f. $\dfrac{1}{(x^4 + 2x^2 + 5)^3}$		
g. $\dfrac{1}{\sqrt{x^2 + 3x + 9}}$		
h. $\sqrt{\dfrac{4x}{3} - \dfrac{2}{x}}$		
i. $\left(\sqrt{2x + 1} + x^3\right)^{-2}$		

2. Determine the derivative of each function and simplify as much as possible.

(a) $f(x) = (5x + 2)^3$

(b) $g(x) = 4\left(\frac{1}{2}x + 4\right)^2$

(c) $h(x) = \frac{3}{4}\left(\sqrt{x^2 + x}\right)$

(d) $u(t) = \dfrac{\sqrt{2t + 1}}{3}$

(e) $v(s) = 3\left(2s - 3\right)^4$

(f) $w(x) = \frac{2}{5}\left(\frac{3}{5}x + 1\right)^2$

(g) $f(y) = \sqrt{y} - \sqrt{2y + 1}$

(h) $g(x) = \dfrac{2}{3x^2} - \left(3x + \frac{1}{2}\right)^3$

(i) $h(t) = \frac{1}{2}t^2 - \dfrac{1}{\sqrt{2t^2 + 3}}$

(j) $u(x) = (2x^2 + 1)^2 + 3x^3$

(k) $w(y) = (3y - 2)^2 - (2y + 3)^3$

(l) $f(x) = \sqrt{x^3 + 2} - \sqrt{3x^2 - 1}$

3. Fill in the missing details in the following proof that if n is a positive integer then the derivative of $f(x) = x^n$ is $f'(x) = nx^{n-1}$:

We have to determine $\displaystyle\lim_{h \to 0} \dfrac{(x + h)^n - x^n}{h}$. Using the formula for $X^n - Y^n$, *(see page 17), gives*

$$\frac{(x+h)^n - x^n}{h} = \frac{[x+h-x]\left[(x+h)^{n-1} + (x+h)^{n-2}x + \cdots + x^{n-1}\right]}{h}$$

$$= \left[(x+h)^{n-1} + (x+h)^{n-2}x + \cdots + x^{n-1}\right].$$

Complete the exercise.

4. *You have to prove that if n is a positive integer then the derivative of $g(x) = \frac{1}{x^n}$ is $-nx^{-n-1} = -\frac{n}{x^{n+1}}$. To this end write g as $g(x) = \left(\frac{1}{x}\right)^n$. Since the derivative of $f(x) = \frac{1}{x}$ was shown to be $-\frac{1}{x^2}$, the general power rule implies that the derivative of g is $g'(x) = n\left(\frac{1}{x}\right)^{n-1}\left(-\frac{1}{x^2}\right)$. Show that this simplifies to $-\frac{n}{x^{n+1}}$.*

5. *Let n be a positive integer and $f(x) = x^{\frac{1}{n}}$. Prove that $f'(x) = \frac{1}{n}x^{\frac{1}{n}-1}$. More precisely, you have to show that $\lim_{h\to 0}\dfrac{(x+h)^{\frac{1}{n}} - x^{\frac{1}{n}}}{h} = \frac{1}{n}x^{\frac{1}{n}-1}$. To this end, denote $(x+h)^{\frac{1}{n}}$ by V and $x^{\frac{1}{n}}$ by U. Then $V^n = (x+h)$ and $U^n = x$. Note that*

$$V^n - U^n = (V-U)\left(V^{n-1} + V^{n-2}U + \cdots + VU^{n-2} + U^{n-1}\right)$$

Since $V^n - U^n = h$, it follows from the above identity that

$$V - U = \frac{V^n - U^n}{\left(V^{n-1} + V^{n-2}U + \cdots + VU^{n-2} + U^{n-1}\right)}$$

$$= \frac{h}{\left(V^{n-1} + V^{n-2}U + \cdots + VU^{n-2} + U^{n-1}\right)}.$$

Use this to show that

$$\frac{(x+h)^{\frac{1}{n}} - x^{\frac{1}{n}}}{h} = \frac{1}{V^{n-1} + V^{n-2}U + \cdots + VU^{n-2} + U^{n-1}}$$

then deduce that

$$\lim_{h\to 0}\frac{(x+h)^{\frac{1}{n}} - x^{\frac{1}{n}}}{h} = \frac{1}{nx^{\frac{n-1}{n}}} = \frac{1}{n}x^{\frac{1}{n}-1}$$

6. *Let m and n be integers and $f(x) = x^{\frac{m}{n}}$. Prove that $f'(x) = \frac{m}{n}x^{\frac{m}{n}-1}$. Hint: write the given function in the form $f(x) = \left(x^{\frac{1}{n}}\right)^m$. Then by the power rule and what has been proved in Exercise 5 above, the derivative of f is*

$$f'(x) = m\left(x^{\frac{1}{n}}\right)^{m-1}\left(\frac{1}{n}x^{\frac{1}{n}-1}\right)$$

Show that this simplifies to $\frac{m}{n}x^{\frac{m}{n}-1}$.

7. *Split* $f(x) = \dfrac{1}{x(x-1)}$ *into partial fractions then determine its derivative.*

8. *Split* $g(x) = \dfrac{x+7}{(x+1)(x+3)}$ *into partial fractions then determine its derivative.*

Antiderivatives

We have already encountered antiderivatives, (see problem 2 on page 14), although we did not mention the term antiderivatives. To introduce them formally, consider the table below. Take the row containing $3x^2$. We have labelled it row 1. Any function that can go in the left column of row 1 is called an antiderivative of $3x^2$. One candidate is $7 + x^3$, therefore $7 + x^3$ is an antiderivative of $3x^2$.

By the same token, any function that can go in the left column of row 2 is called an antiderivative of $20x^4$. One example is $4x^5 + 12$.

A function like $\frac{2}{3}(x+1)^{3/2}$ is a candidate for the left column of the third row, therefore it is an antiderivative of $(1+x)^{1/2}$.

	Function	Derivative
1.		$3x^2$
2.		$20x^4$
3.		$(1+x)^{1/2}$

In general, an antiderivative of a given function f is a function F whose derivative is f. For now, you have to use your knowledge of derivatives of polynomials and other powers to guess antiderivatives. For example, if you are asked to give an antiderivative of $5x^4 - \dfrac{3}{x^2} + 5$, you would pick x^5, (its derivative is $5x^4$), followed by $\dfrac{3}{x}$, (its derivative is $-\dfrac{3}{x^2}$), and $5x$, then add them to get $x^5 + \dfrac{3}{x} + 5x$. To check if it is a correct choice, simply take the derivative of $x^5 + \dfrac{3}{x} + 5x$.

Since the derivative of a constant is 0, we may add any constant to $x^5 + \dfrac{3}{x} + 5x$ and the derivative of the resulting function will still be $5x^4 - \dfrac{3}{x^2} + 5$. For example, $x^5 + \dfrac{3}{x} + 5x - 18$ also has derivative $5x^4 - \dfrac{3}{x^2} + 5$. In general, if c is any constant then $F(x) = x^5 + \dfrac{3}{x} + 5x + c$ has derivative $5x^4 - \dfrac{3}{x^2} + 5$. We call $F(x) = x^5 + \dfrac{3}{x} + 5x + c$ the **most general antiderivative** of $f(x) = 5x^4 - \dfrac{3}{x^2} + 5$.

Exercise 27

1. *Copy and complete the table by guessing the most general antiderivative of the given function:*

Function	Derivative
(a)	$6x^2 + 25$

	Function	Derivative
(b)		$4x^3 + 3x^2 - 6x + 1$
(c)		$\dfrac{1}{x^5}$
(d)		$\dfrac{4}{x^2} + \dfrac{x^4}{4}$
(e)		$3\pi x^3 - \dfrac{4}{x^3}$
(f)		$18\sqrt{x}$
(g)		$\sqrt[3]{x} + 5$
(h)		$9x^3 - 12x^2 + 7$
(i)		$\dfrac{1}{x^{5/2}}$

2. *Take the function* $f(x) = x^3 - \frac{3}{2}x$. *We know that* $F(x) = \frac{1}{4}x^4 - \frac{3}{4}x^2 + c$ *is its most general antiderivative. Say you are asked to find the particular antiderivative whose graph passes through the point* $(2, -1)$, *(the choice of* $(2, -1)$ *is arbitrary). You would have to find* c *such that* $F(2) = -1$; *i.e.* c *such that*

$$-1 = \tfrac{1}{4}(2)^4 - \tfrac{3}{4}(2)^2 + c$$

Solving gives $c = -2$. *Therefore* $F(x) = \frac{1}{4}x^4 - \frac{3}{4}x^2 - 2$ *is the required antiderivative. Follow similar steps to determine:*

(a) *The antiderivative of* $u(x) = x^{1/3} + 5$ *whose graph passes through* $(1, 7)$.

(b) *The antiderivative of* $w(x) = 9x^2 - 4x + 5$ *whose graph passes through* $(1, 0)$.

(c) *The antiderivative of* $h(x) = \dfrac{2}{x^2} - \dfrac{4}{x^3}$ *whose graph passes through* $(1, 5)$.

(d) *The antiderivative of* $h(x) = \dfrac{2}{x^2} - \dfrac{4}{x^3}$ *whose graph passes through* $(1, -8)$

(e) *The antiderivative of* $v(x) = x - \dfrac{1}{2x^2}$ *whose graph passes through* $(-2, 3)$.

(f) *The antiderivative of* $g(x) = 18\sqrt{x}$ *whose graph passes the point* (a, b).

(g) *The antiderivative of* $f(x) = 1 + 2x - 3x^2 + 4x^3$ *whose graph passes the point* $(-1, -2)$

Example 28 *To find the most general antiderivative of* $g(x) = (3x + 5)^8$, *we ask the question: "what function has derivative* $(3x + 5)^8$"? *To answer it, we note that when we take derivatives using the general power rule, the exponent of the expression we are differentiating is reduced by 1. Therefore* $(3x + 5)^8$ *must have resulted from taking the derivative of an expression involving* $(3x + 5)^9$. *But the derivative of* $(3x + 5)^9$ *is* $27(3x + 5)^8$. *Since there is no constant factor 27 in the formula for* g, *we must divide* $(3x + 5)^9$ *by 27. Therefore the most general* antiderivative of $g(x) = (3x + 5)^8$ is $G(x) = \frac{1}{27}(3x + 5)^9 + c$. Confirm by taking the derivative of $\frac{1}{27}(3x + 5)^9 + c$.

Example 29 To find the most general antiderivative of $f(x) = \dfrac{1}{(6x - 3)^2}$, start by we writing $\dfrac{1}{(6x - 3)^2}$ as $(6x - 3)^{-2}$. We again recall that when we take derivatives using the general power rule, the "exponent" is reduced by 1. This suggests that $(6x - 3)^{-2}$ must have resulted from taking the derivative of an expression involving $(6x - 3)^{-1}$. Since the derivative of $(6x - 3)^{-1}$ is $-6(x - 3)^{-2}$, and there is no constant factor -6 in the formula for f, we must divide $(6x - 3)^{-1}$ by -6. Therefore the most general antiderivative of $f(x) = \dfrac{1}{(6x - 3)^2}$ must be $F(x) = -\frac{1}{6}(6x - 3)^{-1} + c$. Confirm by taking the derivative of $-\frac{1}{6}(6x - 3)^{-1} + c$.

Example 30 To find the most general antiderivative of $h(x) = x^2\sqrt{5x^3 + 1}$, we first write the right hand side as $x^2(5x^3 + 1)^{1/2}$. Now note that the derivative of $5x^3 + 3$ is $10x^2$ and the factor x^2 appears in the formula for h. It follows, (by appealing to the general power rule), that $x^2(5x^3 + 1)^{1/2}$ must be the result of taking the derivative of an expression involving $(5x^3 + 1)^{3/2}$. Since the derivative of $(5x^3 + 1)^{3/2}$ is

$$\tfrac{3}{2}(5x^3 + 1)^{1/2}(15x^2) = \tfrac{45}{2}x^2(5x^3 + 1)^{1/2},$$

antiderivative of $h(x) = x^2\sqrt{5x^3 + 1}$ is $H(x) = \frac{2}{45}(5x^3 + 1)^{3/2} + c$. Confirm by taking the derivative of $\frac{2}{45}(5x^3 + 1)^{3/2} + c$.

Exercise 31

1. Copy and complete the table by guessing the most general antiderivative of the given function:

	Function	Derivative
(a)		$(x + 2)^{-1/2}$
(b)		$(5x - 1)^{3/2}$
(c)		$(4x - 3)^{3/4}$
(d)		$(5x + 3)^{1/2}$
(e)		$x\sqrt{x^2 + 1}$
(e)		$\dfrac{x}{\sqrt{x^2 + 1}}$

2. Find the antiderivative of $f(x) = \dfrac{1}{\sqrt{3x + 1}}$ whose graph passes through the point $(1, 5)$

3. Find the antiderivative of $f(x) = x(x^2 + 4)^{\frac{1}{2}}$ whose graph passes through the point $(0, 3)$

Another Maximization Problem

Example 32 *A farmer has 500 yards of fencing to construct three equal enclosures, shown below (not drawn to scale), for his animals.*

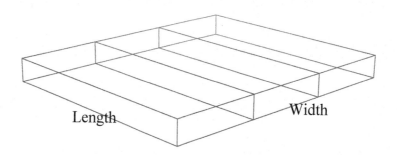

What is the largest possible area he can enclose?

The following table may shed light on how to solve this problem.

Length of enclosure	10	30	45	60	70	80	95	120
Width of enclosure	230	190	160	130	110	90	60	10
Total enclosed area	2300	5700	7200	7800	7700	7200	5700	1200

The table shows that the area increases as the length increases, until it reaches a maximum value when the length is equal to some number c between 45 and 70 yards. Increasing the length beyond c yard leads to a reduction in the enclosed area. To determine c we need an expression that gives the enclosed area in terms of the length of the enclosure. Suppose the length is x yards. Then $4x$ yards are needed to construct the two inner dividers and the two sides that are parallel to the dividers. That leaves $500 - 4x$ yards for the other two sides. It follows that the enclosure is $\frac{1}{2}(500 - 4x)$ yards wide. This implies that the total enclosed area, (in square yards), is

$$A(x) = (x)\tfrac{1}{2}(500 - 4x) = 250x - 2x^2$$

A graph of A is shown below

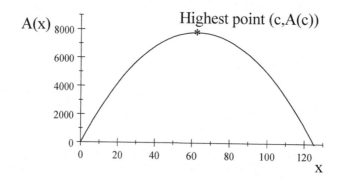

We want the value of c that gives the largest possible value of the area A. In the figure above, it is the number c corresponding to the highest point on the graph of A. Once again, c is characterized by the fact that the tangent to the graph of A at $(c, A(c))$ is horizontal, and so $A'(c) = 0$.

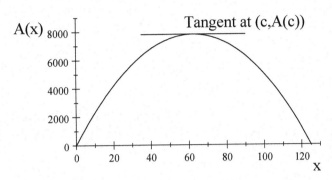

Therefore we need the solution to the equation $A'(x) = 0$. Since $A'(x) = 250 - 4x$ we must solve the equation

$$A'(x) = 250 - 4x$$

The solution is $x = \frac{250}{4}$, hence the largest possible area that may be enclosed is

$$250\left(\tfrac{250}{4}\right) - 2\left(\tfrac{250}{4}\right)^2 = 7812.5 \text{ square yards.}$$

A Minimization Problem

Example 33 *An accident patient has to be rushed by an emergency vehicle from a camp in a desert to a hospital. The camp C is 10 miles from a highway, (the distance from A to C), and the hospital H is 30 miles down the highway, (the distance from A to H), as shown in the figure below.*

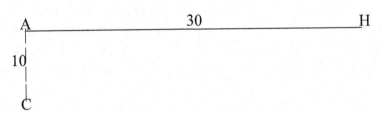

The emergency vehicle can be driven through the desert sand at 50 miles per hour, and along the highway at 90 miles per hour. What is the shortest possible time to get the patient to hospital? (In a situation like this, minutes and seconds matter.)

The driver of the emergency vehicle has a number of options. One of them is to drive directly to the hospital through the desert along the path shown below. This is a $\sqrt{1000}$ miles trip and it takes $\sqrt{1000}/50$ hours, which is close to 38 minutes.

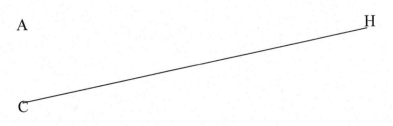

Another option is to drive 10 miles directly to the highway and then drive the 30 miles along the highway.

This takes $10/50 + 30/90$ hours, or 32 minutes. There is also an option of driving through the desert to some point B between A and H as shown below, then drive the remaining miles on the highway.

The following table gives the times for a sample of such points.

Value of x	0	2	4	6	8	10	20
Total time for trip (in hours to 3 dec. pl.).	0.533	0.515	0.504	0.500	0.501	0.505	0.558

To handle the general case, suppose he drives through the desert to a point, on the highway, that is x miles from A, then drives the remaining $30 - x$ miles on the highway. This means that he drives $\sqrt{100 + x^2}$ miles through the desert. At 50 miles per hour, it takes him $\left(\sqrt{100 + x^2}\right)/50$ hours. The remaining $30 - x$ miles on the highway take him $(30 - x)/90$ hours. Therefore, the total time, in hours, for the trip is

$$T(x) = \frac{\sqrt{100 + x^2}}{50} + \frac{30 - x}{90} = \frac{1}{50}\left(100 + x^2\right)^{1/2} + \frac{1}{3} - \frac{x}{90}$$

Clearly $0 \leq x \leq 30$. The graph of T is given below.

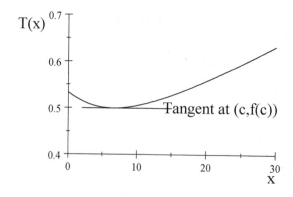

We have to find the distance c corresponding to the lowest point $(c, T(c))$ on the graph of T. Then $T(c)$ is the shortest possible time. As expected, c is characterized by the fact that the tangent to the graph of T at $(c, T(c))$ is horizontal, so that $T'(c) = 0$. Therefore, it is the solution of the equation

$$0 = T'(x) = \frac{1}{50}\left(\frac{1}{2}\right)\left(100 + x^2\right)^{-1/2}(2x) - \frac{1}{90} = \frac{x}{50\sqrt{100 + x^2}} - \frac{1}{90}$$

that is between 0 and 30. Clearing fractions and radicals gives

$$25\left(100 + x^2\right) = 81x^2$$

The positive solution of this equation is $x = \sqrt{2500/56} = 6.68$ (to 2 decimal places). Therefore the shortest possible time is $T(\sqrt{2500/56}) = 0.4996$ hours (to 4 decimal places). This is just a little less than 30 minutes.

Exercise 34

1. Suppose the farmer in Example 32 wants to construct two enclosures instead of three. What is the largest possible area he can enclose with the 500 yards of fencing?

2. Consider the patient, in Example 33 , being rushed to hospital.

    (a) Determine c when the speed through the desert is 60 miles per hour.

    (b) Determine c when the speed through the desert is v miles per hour. (Your answer should depend on v.)

    (c) When the speed through the desert is higher than some value u, the shortest time is attained by driving directly through the sand to the hospital. Use your result in part (b) to determine u.

3. In the figure below, a rectangle is drawn inside the region in the first quadrant enclosed by the two coordinate axes and the line $y = \frac{1}{2}(2 - x)$. The vertical sides of the rectangle are on the lines $x = 0$ and $x = t$.

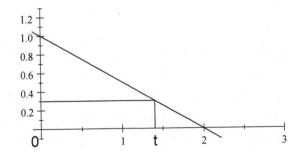

    (a) Show that the area of the rectangle is $t - \frac{1}{2}t^2$.

    (b) What is the largest possible area of such a rectangle?

4. *Consider a line segment in the **first quadrant** satisfying the following conditions: (i) it passes through $(2, 4)$, (ii) it intersects the y-axis at a point P, (iii) it intersects the x-axis at a point Q.*

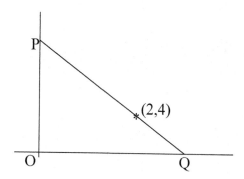

(a) *Show that when the slope of PQ is m then the coordinates of P and Q are $(0, 4 - 2m)$ and $\left(2 - \frac{4}{m}, 0\right)$ respectively. Use this information to calculate the area of triangle PQO in terms of m then complete the following table:*

Slope of PQ	-1	-1.5	-2	-2.5	-3	-3.5
Area of triangle PQO						

(b) *What is the minimum area of triangle PQO? (Defend your answer.)*

5. *You are required to design a cylindrical can, without a lid, that must contain 22 cubic centimeters of fluid and uses the least amount of material for construction. It is necessary to express the area of the material used to construct it in terms of some variable. Say you choose to express it in terms of the radius r of the can. Show that the height of the can must be $h = \frac{22}{\pi r^2}$ cm and the area of the material needed to construct it is $A = \pi r^2 + \frac{44}{r}$ square centimeters. Use this information to determine the radius r that corresponds to the smallest possible area then determine the dimensions of the can with the required property.*

6. *Consider the straight line $y = 2x + 1$.*

   (a) *Choose 4 different points on the line and calculate the distance of each of your points from $(2, -2)$.*

   (b) *A general point on the line has coordinates $(x, 2x + 1)$. Show that its distance from $(2, -2)$ is $\sqrt{5x^2 + 8x + 13}$ then use an argument involving derivatives to find the shortest distance from the line to the point $(2, -2)$.*

7. *This exercise generalizes the result of Exercise 6: Consider the straight line $y = ax + b$ and a point (u, v) in the plane. A general point on the given line has coordinates $(x, ax + b)$.*

(a) Show that if D is its the distance from (u, v) then

$$D^2 = (x - u)^2 + (ax + b - v)^2.$$

(b) Use an argument involving derivatives to show that the point on the line that is closest to (u, v) has coordinates

$$\left(\frac{u + a(v - b)}{1 + a^2}, \frac{b + a(u + av)}{1 + a^2} \right).$$

Now deduce that the shortest distance from the point (u, v) to the line $y = ax + b$ is

$$\frac{|v - au - b|}{\sqrt{1 + a^2}}$$

Test your skills - 2, (50 minutes)

1. Find the derivative of each function and simplify when possible:

(a) $f(x) = 8x - \dfrac{3}{5x^2}$ (b) $h(x) = 5(x^2 - 1)^4$ (c) $g(x) = 9\left(\sqrt{3x^2 + 4}\right)$

(d) $f(x) = \sqrt{x^2 + 2}$ (e) $h(x) = \sqrt[3]{x^2 + 2}$ (f) $u(x) = x - \dfrac{2}{3x + 2}$

(g) $v(x) = \dfrac{1}{x^2 + 1}$ (h) $w(x) = \left(3x - \dfrac{2}{3x}\right)^{2/5}$ (i) $g(x) = \dfrac{3}{\sqrt{6x + 1}}$

2. A normal to the graph of a given function f at a point $(x, f(x))$ is a line that is perpendicular to the tangent at $(x, f(x))$. Determine the equation of the normal to the graph of $f(x) = \sqrt{2x + 1}$ at the point where $x = 4$.

3. Find the most general antiderivative of $f(x) = 2 - \sqrt{3x - 1}$.

4. Consider the curve $y = \sqrt{2x + 1}$.

(a) Choose 4 different points on the curve and calculate the distance of each of your points from $(5, 0)$.

(b) A general point on the curve has coordinates $\left(x, \sqrt{2x + 1}\right)$. Show that its distance from $(5, 0)$ is $\sqrt{x^2 - 8x + 26}$ then use an argument involving derivatives to find the shortest distance from the curve to the point $(5, 0)$.

Information Provided by the Derivative

We start with a number of "new" terms. We assume that all the functions in this section are differentiable.

- We say that a function f is **increasing** on an interval $[a, b]$ if its values get bigger as the independent variable x increases. Thus f is increasing if when $a \le x_1 < x_2 \le b$ then $f(x_1) \le f(x_2)$.

Example 35 *Let* $f(x) = 3x - 1$. *It graph is given below. It is increasing on any interval* $[a, b]$. *For if* $x_1 < x_2$ *then* $3x_1 < 3x_2$, *therefore* $3x_1 - 1 < 3x_2 - 1$. *In other words, if* $x_1 < x_2$ *then* $f(x_1) \leq f(x_2)$.

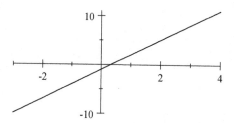

Example 36 *The function* $f(x) = x^2 - 2$, $x > 0$ *is increasing on any interval* $[a, b]$ *with* $a \geq 0$. *To see this, take any* x_1 *and* x_2 *with* $0 < x_1 < x_2$. *Then*

$$f(x_2) - f(x_1) = x_2^2 - x_1^1 = (x_2 - x_1)(x_2 + x_1)$$

Since $(x_2 + x_1) > 0$, *because it is a sum of two positive numbers, and* $(x_2 - x_1) > 0$ *because* $x_1 < x_2$, *their product* $(x_2 - x_1)(x_2 + x_1)$ *must be positive. This proves that* $f(x_1) \leq f(x_2)$.

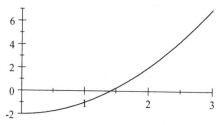

Example 37 *The function* $f(x) = x^3 + 5$ *is increasing on any interval* $[a, b]$. *To see this take any two numbers* x_1 *and* x_2 *with* $x_1 < x_2$. *Then*

$$f(x_2) - f(x_1) = x_2^3 - x_1^3 = (x_2 - x_1)(x_2^2 + x_1 x_2 + x_1^2)$$

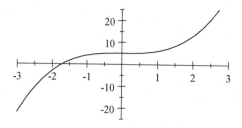

We challenge you to show that $x_2^2 + x_1 x_2 + x_1^2$ *cannot be negative. It follows that* $(x_2 - x_1)(x_2^2 + x_1 x_2 + x_1^2)$ *is non-negative which proves that* $f(x_1) \leq f(x_2)$.

The graphs in examples 35, 36 and 37 are all "forward-leaning", so they have tangents with nonnegative slopes. This is another way of saying that they have non-negative derivatives. In general, if a function f is increasing on an interval then the quotient

$$\frac{f(x + h) - f(x)}{h}$$

cannot be negative. This implies that $\lim\limits_{h\to 0}\dfrac{f(x+h)-f(x)}{h}$, which is $f'(x)$, cannot be negative. The converse is also true; that if the derivative of a function is positive on an interval then the function is increasing on the interval. This is proved, using the Mean Value Theorem, on Page 61. Therefore, to show that a given function f is increasing on an interval $[a, b]$, simply **Show that $f'(x)$ is never negative on** $[a, b]$.

A decreasing function has the opposite properties; its values decrease as the independent variable increases. More precisely:

- *A function $g(x)$ is* **decreasing** *on an interval $[a, b]$ if $g(x_2) \leq g(x_1)$ whenever $a \leq x_1 < x_2 \leq b$.*

Example 38 *Let $g(x) = 4 - x^3$. Then $g(x_2) - g(x_1) = -(x_2 - x_1)(x_2^2 + x_1 x_2 + x_1^2)$ which is negative when $x_1 < x_2$. This proves that g is decreasing on any interval $[a, b]$.*

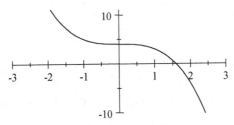

Example 39 *$f(x) = x^2 - 2$ is decreasing on any interval $[a, b]$ with $b \leq 0$. For let $x_1 < x_2 < 0$. Then $f(x_2) - f(x_1) = (x_2 - x_1)(x_2 + x_1)$ is negative because it is a product of a positive number $x_2 - x_1$ and a negative number $x_1 + x_2$. It follows that $f(x_2) \leq f(x_1)$.*

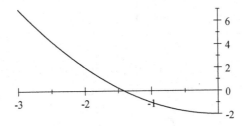

Examples 38 and 39 show that if a function is decreasing on a given interval then its graph is "back-leaning", suggesting that its derivative cannot be positive. It is also proved, (see page 61), that if the derivative of a given function is negative on an interval then the function is decreasing on the interval. Therefore, to show that a function f is decreasing on an interval $[a, b]$, simply **Show that $f'(x)$ is never positive on** $[a, b]$.

- We say that a number c is a **point of relative minimum** for a given function f if the graph of f near $(c, f(c))$ resembles a right-side up bowel.

Example 40 *Let $f(x) = x^2 + 4x + 5$. Its graph near $c = -2$ resembles a right-side up bowel, therefore $c = -2$ is a point of relative minimum for f.*

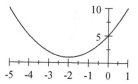

Remark 41 *The term "relative minimum for f" is used to point out that the values of f at all near-by points x are bigger than its value at c. As the graph in Example 40 shows, $f(x)$ is decreasing to the immediate left of c, and it is increasing to the immediate right of c. Therefore $f'(x)$ is negative at all near-by points x to the left of c and it is positive at the near-by points x to the right of c. At the point c itself, $f'(c) = 0$.*

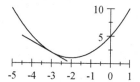 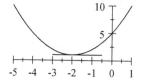 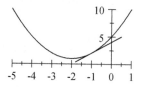

Negative slope to the left   Zero slope at the point   Positive slope to the right

- We say that a number c is a **point of relative maximum** for a given function f if the graph of f near $(c, f(c))$ resembles an upside-down bowel.

Example 42 *Let $f(x) = -2 + 2x - x^2$. Its graph near $c = 1$ resembles an upside-down bowel, therefore $c = 1$ is a point of relative maximum for f.*

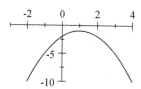

Remark 43 *The term "relative maximum" is used to point out that the values of f at all near-by points x are smaller than its value at c. As the graph in Example 42 shows, $f(x)$ is increasing to the immediate left of c, and it is decreasing to the immediate right of c. Therefore $f'(x)$ is positive at all near-by points x to the left of c and it is negative at the near-by points x to the right of c. At the point c itself, $f'(c) = 0$.*

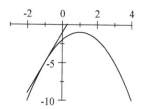 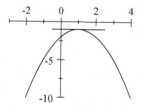 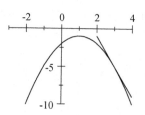

Positive slope to the left   Zero slope at the point   Negative slope to the right

We define points of relative maximum/minimum without reference to graphs as follows:

Definition 44 *Let f be a given function and c be a point in its domain.*

1. *We say that c is a point of relative maximum for f if there is an open interval (a, b) containing c such that $f(x) \leq f(c)$ for all x in (a, b).*

2. *We say that c is a point of relative minimum for f if there is an open interval (a, b) containing c such that $f(x) \geq f(c)$ for all x in (a, b).*

Every maximization (minimization) problem we have solved so far came down to determining a point of relative maximum (minimum) for some function f. We did this by determining the *numbers c such that $f'(c) = 0$*. The technical term for such a number is a *critical point of f*. More precisely;

- **A critical point** for a given function f is a number c such that $f'(c) = 0$.

In particular, a point of relative maximum for a differentiable function f is a critical point of f, and so is any point of relative minimum. But, as the examples below show, a function can have a critical point that is neither a point of relative minimum nor a point of relative maximum.

Example 45 *Let $f(x) = (x - 2)^3$. Then $f'(x) = 3(x - 2)^2$ which is zero when $x = 2$. As its graph below shows, $c = 2$ is a critical point of f that is neither a point of relative maximum nor a point of relative minimum.*

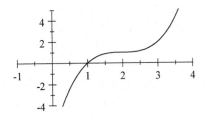

Example 46 *Let $h(x) = (x + 1)^3 (x - 1) - 1 = x^4 + 2x^3 - 2x - 2$. Then*

$$h'(x) = 4x^3 + 6x^2 - 2 = 2(x + 1)^2(2x - 1).$$

Therefore h has critical points $c_1 = -1$ and $c_2 = \frac{1}{2}$. It should be clear from the graph below that c_1 is neither a point of relative maximum nor a point of relative minimum for h.

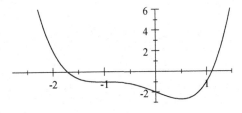

The following are steps you may follow to locate the critical points of a given function f and establish their nature without first drawing its graph:

- Step 1: *Solve the equation $f'(x) = 0$. The solutions (if any), are the critical points c of f.*

- Step 2: *For each c, determine the sign of f' at a near-by point x_1 to the left of c and the sign of f' at a near-by point x_2 to the right of c.*

  If $f'(x_1) > 0$ and $f'(x_2) < 0$ then c is a point of relative maximum.

  If $f'(x_1) < 0$ and $f'(x_2) > 0$ then c is a point of relative minimum.

  If $f'(x_1)$ and $f'(x_2)$ have the same sign then c is neither a point of relative maximum nor a point of relative minimum for f.

Example 47 *Consider $h(x) = x^4 + 2x^3 - 2x - 2$ in Example 46. We found that $h'(x) = 4x^3 + 6x^2 - 2 = 2(x+1)^2(2x-1)$. Its critical points are $c_1 = -1$ and $c_2 = \frac{1}{2}$. To establish the nature of c_1 we evaluate h' at points x_1 to its immediate left and x_2 to its immediate right. We may take $x_1 = -2$ and $x_2 = 0$. (We must choose x_2 between the consecutive critical points. A number like 1 does not qualify.) Since $h'(-2) = -4$ and $h'(0) = -2$ which are both negative, c_1 is neither a point of relative maximum nor a point of relative minimum. In the case of c_2, we may take $x_3 = 1$ to the immediate right. We already have $x_2 = 0$ to the immediate left. Since $h'(1) = 4$ which is positive, and $h'(0)$ is negative, c_2 must be a point of relative minimum.*

A visual method which is particularly useful when one is dealing with a function with many critical points, is to draw a table that shows how the sign of the derivative changes on the different intervals determined by the critical points of the given function. In the case of $h(x) = x^4 + 4x^3 - 2x - 2$ of Example 46, the intervals are $x < -1$, $-1 < x < \frac{1}{2}$ and $x > \frac{1}{2}$. They are listed in the first row of the table below. The second and third rows show the signs of the factors $2(x+1)^2$ and $(2x-2)$ of $h'(x)$ on these intervals. The fourth row shows the sign of $h'(x) = 2(x+1)^2(2x-1)$, and the last row shows the way the graph leans on each interval. For example, on the interval $x < -1$, the derivative of h is negative, therefore its graph leans backwards as indicated by the left-leaning line segment $\backslash$.

	$x < -1$	$x = -1$	$-1 < x < \frac{1}{2}$	$x = \frac{1}{2}$	$\frac{1}{2} < x$
$2(x+1)^2$	$+ + + +$	0	$+ + + + +$	$+$	$+ + + +$
$(2x-1)$	$- - - -$	$-$	$- - - - -$	0	$+ + + +$
$2(x+1)^2(2x-1)$	$- - - -$	0	$- - - - -$	0	$+ + + +$
Shape of graph	$\backslash$	$-$	$\backslash$	$-$	$/$

It is now clear that $c_1 = -1$ is neither a point of relative maximum nor a point of relative minimum, and $c_2 = \frac{1}{2}$ is a point of relative minimum.

Example 48 *Consider $f(x) = x^2 - 5x + 1$. Its derivative is $f'(x) = 2x - 5 = 0$, which is zero when $x = \frac{5}{2}$. This is its only critical point. Since $f'(x)$ is negative when $x < \frac{5}{2}$ and is positive when $x > \frac{5}{2}$, it follows that $c = \frac{5}{2}$ is a point of relative minimum for f.*

Example 49 *Let $f(x) = 3x^5 - 5x^3 + 3$. Its derivative is $f'(x) = 15x^4 - 15x^2 = 15x^2 (x - 1) (x + 1)$. Its critical points are $c_1 = -1$, $c_2 = 0$ and $c_3 = 1$. The table below shows how the sign of $f'(x)$ changes on the different intervals determined by the critical points. They are $x < -1$, $-1 < x < 0$, $0 < x < 1$ and $1 < x$.*

	$x < -1$	$x = -1$	$-1 < x < 0$	$x = 0$	$0 < x < 1$	$x = 1$	$1 < x$
$15x^2$	$+++$	$+$	$++++$	0	$++++$	$+$	$+++$
$(x - 1)$	$---$	$-$	$----$	$-$	$----$	0	$+++$
$(x + 1)$	$---$	0	$++++$	$+$	$++++$	$+$	$+++$
$h'(x)$	$+++$	0	$----$	0	$----$	0	$+++$
Shape of graph	/	—	\	—	\	—	/

It follows from the last row that $c_1 = -1$ is a point of relative maximum, $c_2 = 0$ is neither a point of relative maximum nor a point of relative minimum, and $c_3 = 1$ is a point of relative minimum. The values of f at its critical points are $f(-1) = 5$, $f(0) = 3$ and $f(1) = 1$. This information, plus the last row of the above table suggest that the graph of f has the shape shown below. It is called a sketch of the graph of f.

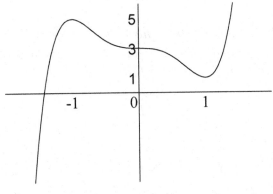

A sketch of the graph of f

38

Exercise 50

1. *In (a) to (c), the formula of a function f and its graph are given. Use the graph to estimate its critical points then determine their exact values by solving an appropriate equation. Also state the nature of each critical point.*

   (a) $f(x) = x^3 + x^2 - 8x + 5$

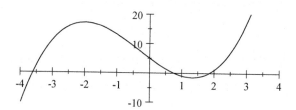

   (b) $f(x) = 4 + 18x + \frac{3}{2}x^2 - 2x^3$

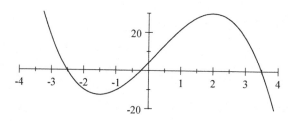

   (c) $f(x) = 2 + x^4 - 3x^3$

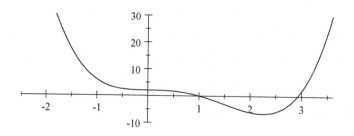

2. *In Example 49, we determined the critical points of the function $f(x) = 3x^5 - 5x^3 + 3$, established their nature, then gave a sketch of its graph. Do the same for each of the following functions:*

   a. $f(x) = x^3 - 27x + 1$ *b.* $g(x) = x^3 - x^2 - 8x + 9$

   c. $h(x) = x^3 - 3x^2 + 3x + 5$ *d.* $v(x) = (x^3 - 9x)^{5/3}$, *(5 critical points)*.

   e. $f(x) = \sqrt{x^2 - x + 5}$ *f.* $u(x) = 2 + 2x - \frac{4}{3}x^{3/2}$, *(for $x > 0$)*.

   g. $w(x) = \sqrt{8 + 2x - x^2}$,   *h.* $f(x) = x^4 - 10x^2$

3. In this exercise you have to determine the shortest distance from the point $P(0,3)$ to the parabola $y = x^2$.

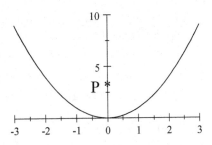

A point on the parabola has the form (x, x^2). Denote its distance from $(0,3)$ by $D(x)$. Then

$$D(x) = \sqrt{x^2 + (x^2 - 3)^2} = \sqrt{x^4 - 5x^2 + 9}$$

Therefore you have to find the smallest possible value of $\sqrt{x^4 - 5x^2 + 9}$ as x varies among the real numbers. To avoid the square root, note that it is sufficient to determine the smallest value of $(D(x))^2 = x^4 - 5x^2 + 9$. Denote it by s. The required number is $\sqrt{s}$. Determine the three critical points of $f(x) = x^4 - 5x^2 + 9$, establish their nature then describe how $D(x)$ changes with x, and find the required distance.

4. Let $a < b$ and R be the region enclosed by the graph of $f(x) = -(x-a)(x-b)$, (a parabola), and the x-axis. Let $a < x < \frac{a+b}{2}$. A rectangle is drawn inside R with one vertical side through $(x, 0)$ and the other vertical side through $(z, 0)$ as shown below.

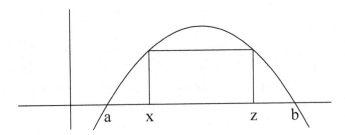

(a) Show that $z = a + b - x$ and that the area $A(x)$ of the rectangle is

$$A(x) = (x-a)(x-b)(2x - a - b)$$

(b) Take $a = 0$ and $b = 4$ and determine the critical points of the resulting function. Also establish the nature of each critical point then describe how the area changes as x changes from 0 to 2. What is its maximum value?

5. Consider the sum of a non-zero number and its reciprocal. Of course the sum depends on the number you choose. For example, the sum of 5 and its reciprocal $\frac{1}{5}$ is 5.2, whereas the sum of 10 and its reciprocal is 10.1. Find a positive number x such that the sum of x and its reciprocal is as small as possible.

40

Using a Tangent Line to Determine a "Good" Approximation

So far, the problems to which we have applied a derivative involved determining relative maximum or relative minimum values of given functions. We now apply it to a different type of problem; that of looking for a good approximation to a value of a given function. This exercise is useful in cases where an approximate value of a function is preferable to its exact value because the latter is so much more difficult to calculate. We calculate approximate values in this section by approximating a section of some suitable graph with a straight line segment. Here is an example:

Example 51 *Say we are given the function $f(x) = \sqrt{3+x}$ and we wish to esti-mate, with reasonable accuracy, its value at $x = 1.7$ (without using a calculator). We know its exact value when $x = 1$, which is pretty close to 1.7. The value is $f(1) = \sqrt{4} = 2$. To estimate $f(1.7)$, we draw a tangent to the graph of f at $(1, 2)$ as shown below.*

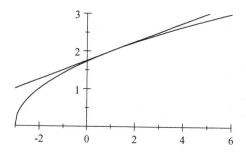

Its slope is $f'(1) = \dfrac{1}{2\sqrt{4}} = \dfrac{1}{4}$, therefore its equation is $y - 2 = \frac{1}{4}(x - 1)$ or $y = \frac{1}{4}x + \frac{7}{4}$. It is preferable to let $y = T(x)$ then write the equation as

$$T(x) = \tfrac{1}{4}x + \tfrac{7}{4}.$$

When x is "far" from 1, the point $(x, f(x))$ on the graph of f is well below the point $(x, T(x))$ on the tangent. Consequently $T(x)$ is a poor approximate value of $f(x)$. For example, in the figure below, obtained by adding the lines $x = 4$, $x = 1.7$ and $x = -1$ to the above figure, the points $(4, f(4))$ and $(-1, f(-1))$ are well below the corresponding points $(4, T(4))$ and $(-1, T(-1))$ on the tangent, therefore $T(4)$ is a poor approximate value of $f(4)$ and $T(-1)$ is a poor approximate value of $f(-1)$.

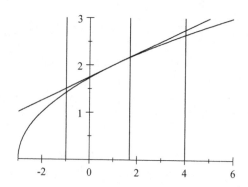

On the other hand, the point $(1.7, f(1.7))$ on the graph of f is quite close to the point $(1.7, T(1.7))$ on the tangent, therefore $T(1.7)$ should be a good approximate

value of $f(1.7)$. For that reason, we use $T(1.7)$, which is easy to evaluate, to approximate $f(1.7)$. In other words,

$$f(1.7) = \sqrt{4.7} \simeq T(1.7) = \tfrac{1}{4}(1.7) + \tfrac{7}{4} = 2.175$$

A calculator gives $\sqrt{4.7} = 2.16795$, (to 5 decimal places), which is pretty close to 2.175. The difference $f(1.7) - T(1.7)$ is called the error in the approximation.

Exercise 52

1. *Use the tangent to the graph of $f(x) = \sqrt[3]{x} = x^{1/3}$ at $(27, 3)$ to determine an approximate value of $\sqrt[3]{25.2}$. Use the value of $\sqrt[3]{25.2}$ from a calculator to estimate the error in your approximation.*

2. *Use the tangent to the graph of $g(x) = x^{3/5}$ at $(32, 8)$ to determine an approximate value of $(33.2)^{3/5}$. Use the value of $(33.2)^{3/5}$ from a calculator to estimate the error in your approximation.*

3. *Use an appropriate tangent to the graph of a suitable function to determine an approximate value of $\dfrac{1}{\sqrt{24.3}}$. Use the value of $\dfrac{1}{\sqrt{24.3}}$ from a calculator to estimate the error in your approximation.*

4. *Let $f(x) = (1 + x)^n$. Use the tangent to the graph of f at $x = 0$ to show that when x is close to 0 then $(1 + x)^n \simeq 1 + nx$. Use this to estimate (a) $\sqrt[4]{1.036}$ and (b) $(1.004)^{45}$*

Approximating Small Changes Using Differentials

We now generalize what we have done in the last section. Thus let f be a differentiable function and c be a point in the domain of f. We wish to use the points on the tangent at $(c, f(c))$ to estimate the values of f at points x close to c. The first step is to determine the equation of the tangent. Its slope is $f'(c)$, therefore its equation is given by $y - f(c) = f'(c)(x - c)$, which may be written as

$$y = f'(c)(x - c) + f(c).$$

It is convenient to write y as $T(x)$. Thus

$$T(x) = f'(c)(x - c) + f(c). \tag{1.8}$$

We now argue that since the tangent lies flat on the graph of f at $(c, f(c))$, the point $(x, f(x))$ on the graph of f is close to the point $(x, T(x))$ on the tangent, when x is close to c. Therefore $T(x)$ should be a good approximate value of $f(x)$. More precisely, if x is close to c then

$$f(x) \simeq f'(c)(x - c) + f(c) \tag{1.9}$$

It is convenient to subtract $f(c)$ from both sides to get

$$f(x) - f(c) \simeq f'(c)(x - c) \tag{1.10}$$

The number $x - c$ is viewed as a change in x or a change in the independent variable, and it is denoted by $\triangle x$ which is pronounced "delta x". Thus

$$\triangle x = x - c$$

Later on we will denote it by the more familiar symbol h we used when computing slopes of tangents. By the same token, $f(x) - f(c)$ is called the change in f, (corresponding to the change $\triangle x$). It is denoted by $\triangle f$, which is pronounced "delta f". Thus $\triangle f = f(x) - f(c)$. Now (1.10) may be written as

$$\triangle f \simeq f'(c)\triangle x$$

Some common terms:

- $\triangle f = f(x) - f(c)$ *is called the actual change in the value of* f

- $f'(c)\triangle x$ *is called the estimated change in the value of* f

- *The fraction* $\dfrac{\triangle f}{f(c)}$ *is called the* **actual fractional change** *in the value of* f *at* c.

- *The percentage* $\dfrac{\triangle f}{f(c)} \times 100$ *is called the* **actual percentage change** *in the value of* f *at* c.

- *The fraction* $\dfrac{f'(c)\triangle x}{f(c)}$ *is called the* **estimated fractional change** *in the value of* f *at* c.

- *The percentage* $\dfrac{f'(c)\triangle x}{f(c)} \times 100$ *is called the* **estimated percentage change** *in the value of* f *at* c.

Example 53 *When a sphere of radius 10 cm was heated, its radius increased to 10.03 cm. Recall that the volume of a sphere with radius r is $V(r) = \frac{4}{3}\pi r^3$. Therefore the volume of the sphere before being heated was $V(10) = \frac{4000\pi}{3}$ cu cm. The change in its radius was $\triangle r = 0.03$ cm, hence the actual change in volume is*

$$\triangle V = V(10.03) - V(10) = 12.036036\pi. \ cu \ cm$$

Since $v'(r) = 4\pi r^2$, the estimated change in volume is

$$V'(10)\triangle r = 4\pi\,(10)^2\,(0.03) = 12\pi. \ cu \ cm$$

The actual fraction change in volume is

$$\frac{12.036036\pi \times 3}{4\pi(10)^3} = 0.00903 \ cu \ cm$$

The estimated fractional change is

$$\frac{4\pi\,(10)^2\,(0.03) \times 3}{4\pi(10)^3} = 0.009 \ cu \ cm$$

The actual percentage change in volume is

$$\frac{12.036036\pi \times 3}{4\pi(10)^3} \times 100\% = 0.903\%,$$

The estimated percentage change in volume is

$$\frac{12\pi \times 3}{4\pi(10)^3} \times 100\% = 0.9\%.$$

It is convenient to convert (1.10) into an equation by introducing the error $e = f(x) - T(x)$ in the approximation. The result is

$$f(x) = T(x) + f(x) - T(x) = T(x) + e = f'(c)\,(x - c) + f(c) + e$$

which we may write as

$$f(x) - f(c) = f'(c)\,(x - c) + e \qquad (1.11)$$

Since $x - c = \triangle x$, we may write x as the sum $x = c + \triangle x$. Then (1.11) becomes

$$f(c + \triangle x) - f(c) = f'(c)\triangle x + e \qquad (1.12)$$

Because the expression $f'(c)\triangle x$ is linear in the variable $\triangle x$, the function with independent variable $\triangle x$ and formula $f'(c)\triangle x$ is called the **linear approximation of** $f(c + \triangle x) - f(c)$ **at** c. Another name for it is the **differential of** f **at** c, denoted by df_c. Thus

$$df_c\,(\triangle x) = f'(c)\triangle x$$

Note how easy it is to evaluate this function. You simply multiply the derivative of f at c with the change in x. Following the standard function notation, $f'(c)\triangle x$ is called the value of the differential df_c at $\triangle x$.

The notation df_c should not intimidate you. It is intended to emphasize that the differential depends on the point c. If you choose a different point b in the domain of f, you will get a differential of f at b which may be different from the differential at c.

We denoted $f(c + \triangle x) - f(c)$ by $\triangle f$. It follows that if the error term e in equation (1.12) is small, (which is usually the case when $\triangle x$ is small), then

$$\triangle f \simeq df_c\,(\triangle x) = f'(c)\triangle x$$

In plain words; the change in the value of f when the independent variable changes by a small amount $\triangle x$ is approximately equal to the value of the differential df_c at $\triangle x$.

Example 54 *It is required to draw a circle of radius 5 cm. Inevitably, there is an error in the measurement of its radius. What level of accuracy must we adhere to so that the error in the area of the circle does not exceed 1 square centimeter?*

Solution: *The area of a circle of radius r is $A(r) = \pi r^2$. If the error in measuring the radius is $\triangle r$ then we draw a circle of radius $(5 + \triangle r)$, instead of a circle of radius 5. The actual error in the area is*

$$\triangle A = \pi (5 + \triangle r)^2 - 25\pi = \pi \left(10\triangle r + (\triangle r)^2\right)$$

We need $\triangle r$ such that

$$\left|\pi \left(10\triangle r + (\triangle r)^2\right)\right| < 1$$

This is a quadratic inequality which is relatively hard to solve. To solve a simpler one, we determine $\triangle r$ such that the estimated error is less than 1 sq. cm. Thus we look for $\triangle r$ such that

$$|V'(5)\triangle r| = |2\pi(5)\triangle r| = |10\pi\triangle r| < 1$$

This is a linear inequality and we easily solve it to get $|\triangle r| < \frac{1}{10\pi} = 0.032$. Therefore we should use an instrument that can measure r accurately to about 0.03 cm.

Exercise 55 *When required, the volume of a sphere with radius r is $\frac{4}{3}\pi r^3$ and the volume of a box with length ℓ, width w and height h is $\ell w h$.*

1. *Give a formula for the differential of:*

   (a) $f(x) = x^3 + x$ at $c = 1$ (b) $g(x) = x\sqrt{5 - x^2}$ at $c = 2$.

2. *A spherical ornament has a radius of 1 cm. It is to be given a silver coating 0.03 cm thick. Show that the volume of paint used is $\frac{4}{3}\pi (2.03)^3 - \frac{4}{3}\pi (2)^3$ then use a differential to estimate it.*

3. *You have a cubical metallic block whose volume has to be determined. You measure its edge and find it to be 6 ± 0.2 cm. Say you compute the volume of the cube using 6 as the length of each edge. Use a differential to estimate the error in the volume you get. Also compute the estimated percentage error in the volume.*

4. *The surface area of a right circular cone with base radius r and height h is $A = \pi r\sqrt{r^2 + h^2}$. Use a differential to estimate the change in area when the radius changes from 8 cm to 8.08 cm while the height remains unchanged and equal to 12 cm.*

5. *It is required to measure the radius of a sphere then calculate its volume with an error of no more than 1% of its true value. Determine the largest estimated percentage error that can be tolerated in the measurement of r.*

6. *You are required to construct a box with a square base and a height equal to one third the length of a side of its base. Its volume must be 72 cubic feet. Use a differential to estimate how accurately the length of its base should be made so that the volume of the box is accurate to within 0.9 cubic feet.*

A Fancy Method of Approximating Some Special Sums

This is another application of formula (1.10). Consider the function $f(x) = 1 + 3x^2$. Say we want to estimate the sum

$$(0.1)\, f(0) + (0.1)\, f(0.1) + (0.1)\, f(0.2) + (0.1)\, f(0.3) + \cdots + (0.1)\, f(0.9)$$

It is the sum of the values of f at the equally spaced points 0.0, 0.1, 0.2, ..., 0.9., multiplied by 0.1. (You will meet many such sums in Chapter 4). The number 0.1 is the distance between any two of the equally spaced points 0, 0.1, 0.2, ..., 0.9, 1. Take the function $F(x) = x + x^3$. It is an antiderivative of f. Thus $F'(x) = f(x)$ for all x. By formula (1.10), when h is small then $hF'(x) \simeq F(x+h) - F(x)$. We re-write this as

$$hf(x) = hF'(x) \simeq F(x+h) - F(x)$$

In particular, if we take $h = 0.1$ and $x = 0, 0.1, 0.2, \ldots, 0.9$, we get

$$
\begin{aligned}
(0.1)\, f(0.0) &\simeq F(0.1) - F(0.0) \\
(0.1)\, f(0.1) &\simeq F(0.2) - F(0.1) \\
(0.1)\, f(0.2) &\simeq F(0.3) - F(0.2) \\
(0.1)\, f(0.3) &\simeq F(0.4) - F(0.3)
\end{aligned}
$$

$$\vdots$$

$$(0.1)\, f(0.9) \simeq F(1.0) - F(0.9)$$

When we add the terms in the ten rows, $F(0.1)$ in the first row cancels with $-F(0.1)$ in the second row; $F(0.2)$ in the second row cancels with $-F(0.2)$ in the third row; and so on. The end result is that

$$(0.1)\, f(0) + (0.1)\, f(0.1) + (0.1)\, f(0.2) + (0.1)\, f(0.3) + \cdots + (0.1)\, f(0.9) \simeq F(1) - F(0) = 2$$

We hope you are impressed.

Exercise 56 *Use a similar method to estimate each of the following special sums:*

1. $(0.2)\, f(1) + (0.2)\, f(1.2) + (0.2)\, f(1.4) + (0.2)\, f(1.6) + \cdots + (0.2)\, f(2.8)$ *when*
$$f(x) = \frac{1}{\sqrt{x}} + 4x + 3$$

2. $(0.01)\, f(0) + (0.01)\, f(0.01) + (0.01)\, f(0.02) + (0.01)\, f(0.03) + \cdots + (0.01)\, f(1.99)$ *when* $f(x) = x^3 + 5$

Test your skills - 3   (30 minutes)

1. You are given the function $f(x) = 4 - 8x + 3x^2 - \frac{1}{3}x^3$. Determine its two critical points and establish their nature.

2. Find the derivative of each function:

$$\text{(a) } g(x) = \tfrac{1}{12}(3x+5)^4 \quad \text{(b) } h(x) = \sqrt{x^2 - x + 5} \quad \text{(c) } w(x) = \sqrt{4 + \sqrt{x}}.$$

3. Guess the most general antiderivative of $f(x) = x\sqrt{3 + 4x^2}$.

4. Use the tangent to the graph of $f(x) = \sqrt{5x - 1}$ at some suitable point c to determine an approximate value of $\sqrt{8}$.

Newton's Method for Calculating Approximate Roots

The following example introduces the essential ideas of Newton's method.

Example 57 *Consider the function $f(x) = x^3 - 12x + 5$. Its graph is given below.*

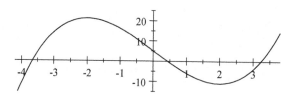

The numbers c such that $f(c) = 0$ are called the solutions, or roots, of the equation $x^3 - 12x + 5 = 0$. They are the x-intercepts of the graph of f. To the nearest whole number, they are $c_1 \simeq 3$, $c_2 \simeq 0$ and $c_3 \simeq -4$. Newton's method is a procedure for calculating a better approximate solution, given an approximate solution. For example, the procedure enables us to calculate a better approximate solution of $x^3 - 12x + 5 = 0$, given the approximate solution $c_1 = 3$. We proceed by drawing the tangent to the graph of f at $(3, f(3)) = (3, -4)$ then determine its x-intercept, which we may denote by x_1. The graph below of a magnified section of the curve and its tangent at $(3, -4)$, shows clearly that x_1 is a better approximate solution than 3.

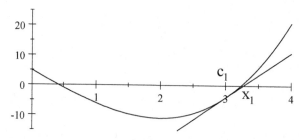

x_1 is nearer the solution than 3

To calculate the x-intercept of the tangent at $(3, -4)$, it is necessary to determine its equation. Its slope is $f'(3) = 3(3)^2 - 12 = 15$, therefore its equation is given by

$$y + 4 = 15(x - 3) \quad or \quad y = 15x - 49$$

Its x-intercept may be obtained by solving the equation

$$15x - 49 = 0$$

The result is $x_1 = \tfrac{49}{15}$. Since $f(\tfrac{49}{15}) = 0.66$, (to 2 decimal places), while $f(3) = -4$, $x_1 = \tfrac{49}{15}$ is definitely a better approximate solution than 3.

To generalize, suppose f is a given function and x_0 is an approximate solution of the equation $f(x) = 0$. (We are assuming that x_0 has been determined by some means, e.g. from a sketch of the graph of f.) Consider the tangent to the graph of f at $(x_0, f(x_0))$. Its slope is $f'(x_0)$, therefore its equation is

$$y = f(x_0) + (x - x_0)\, f'(x_0).$$

Denote its x-intercept by x_1. Then under suitable conditions, x_1 is a better approximate solution of the equation $f(x) = 0$ than x_0. You may determine x_1 by solving the equation

$$f(x_0) + (x - x_0)\, f'(x_0) = 0$$

Remove parentheses and rearrange to get $xf'(x_0) = x_0 f'(x_0) - f(x_0)$. Now divide by $f'(x_0)$, and you should get

$$x = x_0 - \frac{f(x_0)}{f'(x_0)}$$

We have therefore shown that:

- *If x_0 is an approximate solution of $f(x) = 0$ then, under suitable conditions, $x_1 = x_0 - \dfrac{f(x_0)}{f'(x_0)}$ is a better approximate solution.*

Exercise 58

1. *The graph of $f(x) = \sqrt{x} + x - 3$ below shows that the equation $\sqrt{x} + x - 3 = 0$ has a solution close to 2. Use $x_0 = 2$ as an approximate solution*

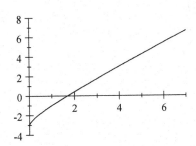

   to calculate a better approximate solution. You can actually solve this equation using the quadratic formula. Solve it and compare the result to the approximate solution.

2. *Use $x_0 = -4$ as an approximate solution to the equation $x^3 - 12x + 5 = 0$ of Example 57 to calculate a better approximate solution of the equation.*

3. *Let $f(x) = x^2 - 15$. We may view the square root of 15 as a solution to the equation $f(x) = 0$. Take $x_0 = 4$ as an approximate value of $\sqrt{15}$ and use Newton's method to calculate a better approximate value of $\sqrt{15}$.*

4. *Use a suitable function and follow the steps in question 3 above to determine an approximate value of $\sqrt[3]{29.2}$ with the help of Newton's method.*

5. *Show that if x_0 is a root of the equation $f(x) = 0$, then applying Newton's method to x_0 does not provide anything new.*

48

Repeated application of Newton's method

Consider a function f and the equation $f(x) = 0$. By Newton's method, if x_0 is an approximate solution of $f(x) = 0$ then, (under suitable conditions), a better approximate solution is

$$x_1 = x_0 - \frac{f(x_0)}{f'(x_0)}.$$

This is called the first iterate, or the first approximation given by Newton's method. It may be used to get an even better approximate solution x_2, called the second iterate, or the second approximation. As you would expect, it is the x-intercept of the tangent at $(x_1, f(x_1))$ and it is given by

$$x_2 = x_1 - \frac{f(x_1)}{f'(x_1)}$$

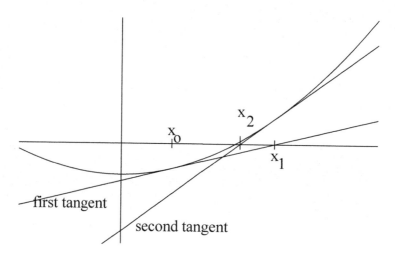

This may be used to get a third approximation

$$x_3 = x_2 - \frac{f(x_2)}{f'(x_2)}.$$

In general, if you have an nth approximation x_n, you may calculate

$$x_{n+1} = x_n - \frac{f(x_n)}{f'(x_n)}.$$

Example 59 *Consider the approximate solution $x_1 = \frac{49}{15}$ for the equation $x^3 - 12x + 5 = 0$ in Example 57. We may use it to get an even better approximate solution x_2 given by*

$$x_2 = x_1 - \frac{f(x_1)}{f'(x_1)} = \frac{49}{15} - \frac{f(\frac{49}{15})}{f'(\frac{49}{15})} = \frac{49}{15} - \frac{(\frac{49}{15})^3 - 12(\frac{49}{15}) + 5}{3(\frac{49}{15})^2 - 12} = 3.2337 \text{ to 4 dec. pl.}$$

Since $f(3.2337) = (3.2337)^3 - 12(3.2337) + 5 = 0.009804$ and $f(x_1) = f(\frac{49}{15}) = 0.66$, x_2 is definitely a better approximate solution than x_1. (Use x_2 to get x_3.)

Exercise 60

1. The graph of $f(x) = \sqrt{x} + x^2 - 3$ is given below. It shows that $x_0 = 1$ is an approximate solution of the equation $f(x) = 0$. Use it to determine the first and second approximations x_1, and x_2 respectively, given by Newton's method. Round off x_2 to 3 dec. pl.

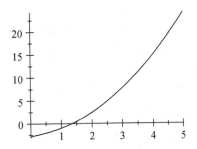

2. Sketch the graph of $f(x) = x^3 + 3x^2 + 2$ and use it to verify that the equation $x^3 + 3x^2 + 2 = 0$ has one real root. Use the sketch to estimate the root then use Newton's method to determine two better approximations.

3. There is no guarantee that Newton's method will always work. For example, consider the equation $(x + 1)^{1/3} = 0$. Of course we know its solution; it is $x = -1$. Pretend that you do not know it. The graph of $f(x) = (1 + x)^{1/3}$ is given below.

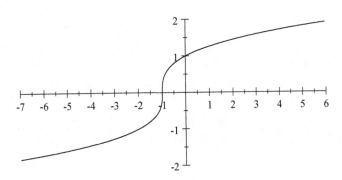

   Take $x_0 = 0$ as an approximate solution and determine x_1. Next, use x_1 to determine x_2. Also, draw the tangents at $(x_0, f(x_0))$ and $(x_1, f(x_1))$. Can you see why it fails in this case?

4. Say you want to calculate the root of $x^3 - 12x + 5 = 0$ in Example 57, which is close to 3, accurately to 5 decimal places. You compute $x_1, x_2, x_3, \ldots$ and stop when there is no change in the first five decimal places of your iterate. Recall that we found that $x_1 = \frac{49}{15} = 3.26666667$, and $x_2 = 3.23374046$. We calculate x_3 using the equation

$$x_3 = x_2 - \frac{f(x_2)}{f'(x_2)} = 3.233193843.$$

   There is a change in the fourth decimal place so we compute x_4. It turns out to be

$$x_4 = x_3 - \frac{f(x_3)}{f'(x_3)} = 3.233193694.$$

50

Since there is no change in the first 5 decimal places, we stop. We give the root as 3.23319 correct to 5 decimal places.

Use a similar procedure to determine the root of $\sqrt{x} + x^2 - 3 = 0$ that is between 1 and 2, correct to 5 decimal places.

5. *Consider the function $f(x) = x^3 + x - 3$. Since $f(1) = -1$ and $f(2) = 7$, the graph of f crosses the x-axis between $x = 1$ and $x = 2$. It follows that the equation $x^3 + x - 3 = 0$ has a root between 1 and 2. We may take $x = 1$ or $x = 2$ as an approximate root of this equation. But of the two, $x = 1$ is probably a better approximation because $f(1)$ is closer to 0 than $f(2)$. Use it to calculate the root correct to 2 decimal places.*

6. *Show that the equation $x^4 + x^3 + 4x - 1 = 0$ has a root between 0 and 1 then use Newton's method to calculate it accurately to 2 decimal places.*

Rates of Change

A rate of change is associated with a variable $f(x)$ that changes by the same amount when the independent variable x changes by one unit. Here are two examples:

Example 61 *Consider the cost of renting a truck for one day when you are charged $40.00 for the day plus 25 cents for every mile you drive. If you drive it for x miles then you are charged $f(x)$ dollars where*

$$f(x) = 40 + 0.25x$$

Clearly, $f(x)$ changes by 0.25 dollars whenever the mileage changes by one mile. In other words, when the independent variable changes from x to $x+1$ miles, $f(x)$ changes by 0.25 dollars, (from $40 + 0.25x$ to $40 + 0.25x + 0.25$ dollars). In general, when it (the mileage) changes by h miles, then $f(x)$ changes by 0.25h dollars. The number 0.25 is called the rate of change of $f(x)$ with respect to x. It happens to be equal to $f'(x)$.

Example 62 *Say you leave home, drive to a highway and start cruising at a constant speed of $\frac{7}{6}$ miles per minute, (or 70 miles per hour). Suppose you start cruising when you are 3 miles from home. Then one minute later you will be $\frac{7}{6} + 3$ miles from home, two minutes later you will be $\frac{14}{6} + 3$ miles from home, x minutes later, you will be $f(x)$ miles from home where*

$$f(x) = \frac{7x}{6} + 3.$$

In this case, $f(x)$ changes by $\frac{7}{6}$ miles whenever x changes by one minute, therefore the rate of change of $f(x)$ with respect to x is $\frac{7}{6}$, (your cruising speed). This is also the derivative of $f(x)$ with respect to x.

In general, if a variable $f(x)$ changes by the same amount when the independent variable x changes by one unit then the rate of change of $f(x)$ with respect to x is $f'(x)$.

Instantaneous Rate of Change

If $f(x)$ does not change by the same amount whenever x changes by one unit then we cannot ask for a rate of change of $f(x)$ with respect to x. Instead, we ask for an **instantaneous rate of change**. The following example illustrates the idea.

Example 63 *Say you drop a stone from the top of a tall building and simultaneously start a stop watch. Let $f(x)$ be the distance, in feet, the stone has fallen, (from the point of release), when the stop watch reads x seconds. It can be shown, (e.g. experimentally), that before it hits the ground, $f(x)$ is given by*

$$f(x) = 16x^2$$

Because the term $16x^2$ is not linear, $f(x)$ does not change by the same amount as x changes by one unit. For example,

it changes by $f(1) - f(0) = 16$ feet when x changes from 0 to 1 seconds.

it changes by $f(2) - f(1) = 48$ feet when x changes from 1 to 2 seconds.

it changes by $f(3) - f(2) = 80$ feet when x changes from 2 to 3 seconds.

*Therefore there is no number to pin down as the rate of change of $f(x)$ with respect to x. However, a rate of change at a specific time c is conceivable. Generalizing from Example 62 in which a rate of change of distance is speed, it should be the speed of the stone at time c. The technical term for it is **the instantaneous rate of change of $f(x)$ with respect to** x at time c. To visualize it, imagine attaching a speedometer to the stone. Then the instantaneous rate of change of the distance, at time c, is the speedometer reading c seconds after the stone is released. We may calculate it as follows:*

Denote it by u and consider a time $c + h$ seconds, where h is a small number. Since the stone falls smoothly, its speed also changes smoothly. This implies that when h is close to 0, then its speed at time $c + h$ does not differ appreciably from u (its speed at time c). Therefore we can assume, with reasonable accuracy, that the speed is constant between the times c and $c + h$; and equal to u feet per second. (The smaller h is, the more reasonable the approximation.) This implies that the stone falls approximately uh feet between time c and time $c + h$ seconds. The actual distance is $f(c + h) - f(c)$, hence we must have $uh \simeq f(c + h) - f(c)$, or

$$u \simeq \frac{f(c + h) - f(c)}{h}$$

Note that this is true for any small number h, and that when h is close to 0, the quotient $\dfrac{f(c + h) - f(c)}{h}$ is close to $f'(c)$. Therefore we must have

$$u = \lim_{h \to 0} \frac{f(c + h) - f(c)}{h} = f'(c) = 32c.$$

In particular, its speed at time $x = 3$ is 96 feet per second. However, because the speed is not constant, it is no longer true that when time changes by h

seconds from 3 to 3 + h seconds then the distance changes by 96h feet. The best we can now assert is that when h is close to 0 and time changes by h seconds from 3 to 3 + h seconds, then the distance changes by approximately 96h feet. This follows from the fact that

$$f(3 + h) - f(3) \simeq f'(3)h,$$

(see (1.10) on page 41). In general, when time changes by a small amount h from x to x + h seconds, then the distance changes by approximately $f'(x)h$ feet.

To generalize these observations, consider an arbitrary differentiable function $f(x)$. The **instantaneous rate of change of** $f(x)$ **with respect to** x **at a given point** c is defined to be $f'(c)$. It follows from the definition of $f'(c)$ that when h is close to 0 then

$$f(c + h) - f(c) \simeq f'(c)h.$$

In other words, when the independent variable changes by a small amount h, from c to $c + h$, the value of f changes by approximately $f'(c)h$.

Example 64 *The management of a grocery store estimate that when the price of a certain type of beef is set at x dollars per pound (x > 0), the store sells q(x) pounds of the beef daily, where*

$$q(x) = \frac{900}{x^2 + 9}$$

The derivative of q(x) is $q'(x) = -1800x(x^2 + 9)^{-2}$. This is the instantaneous rate of change of q(x) with respect to the price of the beef when the price is x dollars per pound. In particular, the instantaneous rate of change of q(x) with respect to x when the price is 3 dollars, is

$$q'(3) = -\frac{50}{3}.$$

It follows that when the price changes by a small amount h from 3 to 3+h dollars per pound, the demand of the beef (i.e. the number of pounds sold per day), changes by approximately $-\frac{50}{3}h$ pounds. For example, if the price goes up by 45 cents from 3 dollars to 3.45, (h = 0.45), dollars per pound, the demand changes by approximately

$$-\tfrac{50}{3} \times 0.45 = -7.5 \ pounds$$

(i.e. it drops by about 7.5 pounds). If the price goes down by 36 cents from 3 dollars to 2.64 dollars per pound, (h = −0.36), the demand goes up by approximately 6 pounds.

Example 65 *A rock is blasted vertically up from the ground with an initial speed of 80 feet per second. As expected, it rises, and in the process, it slows down. When it reaches the highest point of its path, it stops momentarily then reverses direction* and falls back to the ground. It can be shown that t seconds after being blasted up, its height $h(t)$, (in feet), above the ground is

$$h(t) = 80t - 16t^2, \quad 0 \le t \le 5$$

Therefore its speed at time t is $h'(t) = 80 - 32t$. This enables us to deduce the following:

- Its speed is zero when $h'(t) = 80 - 32t = 0$, therefore it reaches the highest point when $t = 2.5$.

- The maximum height it reaches must be $h(2.5) = 100$ feet, (corresponding to the time when its speed is zero).

- Since it takes 2.5 seconds to rise from the ground to the highest point, it must take 2.5 seconds to fall from the highest point to the ground. Therefore it spends 5 seconds in the air. Another way of obtaining this result is to use the formula $h(t) = 80t - 16t^2$ for the height of the rock above the ground. It is zero when $80t - 16t^2 = 0$. Solving gives $t = 0$ or $t = 5$. We know that when $t = 0$, it is just being blasted up. It follows that $t = 5$ corresponds to the instant it hits the ground on the way back, therefore it spends 5 seconds in the air.

More Motion in a Straight Line

Consider an object that moves in a straight line. Suppose its distance, at time t, from some fixed point P is $s(t)$. The instantaneous rate of change of $s(t)$ at a specific time t is called the speed of the object. A common symbol for the speed at time t is $v(t)$. Therefore

$$s'(t) = v(t)$$

If the speed itself changes with time then the instantaneous rate of change of speed at time t is called the acceleration of the object. A common symbol for acceleration at time t is $a(t)$. Therefore

$$v'(t) = a(t).$$

For a specific example, consider a projectile that is projected vertically up with a speed of 40 feet per second, from a point P which is 68 feet above the ground. It is an established fact that if gravity is the only influence on its motion then its speed decreases by 32 feet per second in every passing second. (The number 32 is called the acceleration due to gravity.) If we use the above notations then $v'(t) = a(t) = -32$. It follows that

$$v(t) = -32t + c \tag{1.13}$$

where c is some constant to be determined from the given information. Indeed, since the speed of the projectile is 40 ft. per second at time $t = 0$, it follows that

$$40 = -32(0) + c$$

54

This implies that $c = 40$. Substituting this in (1.13) gives $v(t) = -32t + 40$. But $s'(t) = v(t) = -32t + 40$. It follows that

$$s(t) = -16t^2 + 40t + b$$

where b is another constant. To determine it, we use the fact that at time $t = 0$, the projectile was 68 ft. above ground. In other words, when $t = 0$, $s = 68$. Therefore

$$68 = -16\,(0)^2 + 40\,(0) + b \qquad\qquad (1.14)$$

which gives $b = 68$. Substituting this in (1.14) gives $s(t) = -16t^2 + 40t + 68$. This is called the equation of motion for the projectile. The maximum height it reaches is obtained by determining the largest value of $s(t)$. We do the standard thing: determine its critical point(s). Since $s'(t) = -32t + 40$, the critical point is the solution of the equation

$$-32t + 40 = 0,$$

which is $t = \frac{5}{4}$, and it is easily shown to be a point of relative maximum. Therefore the maximum height it reaches is $s(\frac{5}{4})$, which is equal to 93 feet.

If we want to know the time the projectile hits the ground, we solve the equation $s(t) = 0$ for positive time t. The positive solution is $t = 3.66$ seconds to 2 decimal places.

Exercise 66

1. *A rock thrown vertically up from the surface of the moon with a speed of 20 meters per second reaches a height of $h(t) = 20t - 2t^2$ meters in t seconds.*

   (a) *Find the rock's speed at time t.*

   (b) *How long does it take the rock to reach its highest point?*

   (c) *How high does the rock go?*

   (d) *How long does it take the rock to reach half its maximum height?*

   (e) *How many minutes elapse before it returns to the surface of the moon?*

2. *Channels in the cell membrane of a living cell offer resistance to the flow of sodium ions into the cell. The flow of sodium ions is measured as a current I in micro Amps. It is related to the membrane voltage v, measured in millivolts, by the formula*

   $$I = 0.4\,(v + 40)$$

   (a) *Complete the table below*

Voltage (in millivolts).	−40	−20	0	10	15	20
Current (in micro Amps)						

(b) *Draw a graph of I on a graph paper.*

(c) *Use your graph to determine the derivative of I.*

(d) *Use an appropriate derivative formula to find the derivative of I.*

(e) *What is the physical meaning of the derivative of I?*

3. *The number of gallons of water in a tank t minutes after the tank has started to drain is*

$$A(t) = 25 \left(40 - t\right)^2. \quad 0 \le t \le 40$$

(a) *How fast is the water running out at the end of 15 minutes?*

(b) *At what time t is the water gushing out of the tank at the rate of 150 liters per minute?*

4. *The speed of blood, in centimeters per second, at a point x centimeters from the center of a given artery is given by the formula*

$$v = 1.28 - 20000x^2$$

(a) *Complete the table below*

Distance from center of artery.	0.001	0.002	0.003	0.004	0.005	0.0053
Speed of blood						

(b) *Draw a graph showing how v is related to x.*

(c) *Interpret the derivative of v for x = 0.001 cm, 0.003 cm, and 0.007 cm.*

(d) *Compute the derivative of v when v is at its maximum. Explain your results.*

5. *An experimenter on top of a 120 feet building throws a stone vertically down. Assume that he releases it with a speed of 5 feet per second and that it moves under the influence of gravity. Let s(t) be the distance of the stone from the ground, t seconds later. Show that $s(t) = -16t^2 - 5t + 120$, then calculate the time that elapses before the stone hits the ground.*

6. *An object moves along the x-axis in such a way that t seconds after starting to move, (assume that $t \ge 0$), it is a distance $x(t) = t^4 - 4t^3 + 4t^2$ feet from the origin. The object is moving forward (i.e. to the right) when its speed is positive, and it is moving backwards if the speed is negative. Give the time intervals during which it is moving, (i) forward, (ii) backwards. At what times does the object change direction?*

7. The heart pumps blood through arteries, veins, and capillaries. Blood moves faster through narrow vessels and slower through wider vessels. The formula for the speed of blood flow is $S = \dfrac{Q}{A}$ where S is speed in centimeters per second, Q is flow rate in cubic centimeters per second and A is the cross sectional area of the vessel in square centimeters. For a blood flow rate $Q = 415$ cubic millimeters per second, use the given speed formula to find the blood speed for blood moving through an artery of cross section area a) $A = 1\ mm^2$, b) $A = 2\ mm^2$, c) $A = 3\ mm^2$, and d) $A = 4\ mm^2$ and make a table to show how speed and area are related.

   (a) Draw a speed and area relation on a graph paper.

   (b) Find and interpret the derivative of S with respect to A for $A = 1\ mm^2$, $A = 2\ mm^2$, $A = 3\ mm^2$, and $A = 4\ mm^2$.

8. Alveolus are lung air sacs through which gas exchanges between air and blood takes place. The liquid lining an alveolus creates a force called surface tension T. This force causes pressure on the gases in the alveolus. The formula for the pressure is $P = \dfrac{2T}{r}$ where P is the pressure, T is surface tension and r is the radius of the alveolus

   (a) For a surface tension of $T = 3$ units, construct a table for units of pressure caused for an alveolus radius of 1, 2, 3, and 4 then draw a pressure radius relation on a graph paper.

   (b) Determine and interpret the derivative of the pressure for a radius of 0.5 and a radius of 6.

   (c) What happens to the derivative of the pressure as the radius increases?

9. The trachea (wind pipe) is tube shaped with length L cm and radius r cm. The formula for its volume is

$$V = \pi r^2 L$$

   (a) Find the volume of a trachea with radius 1.25 cm and length 12 cm.

   (b) Sketch the graph of V as a function of positive r. Why do we consider only positive values of r?

   (c) Find and give a physical interpretation of the derivative of V with respect to r.

10. A person standing on top of a 72 meter building throws a stone up in the air. It rises, above the building, reaches a maximum height then falls to the ground. You are given that its speed t seconds after being thrown up is $v(t) = 18 - 10t$ meters per second. Determine its height above the ground at time t, (hint: find the antiderivative of v that has value 72 when $t = 0$), then determine the largest height it attains.

11. *A stone is projected vertically up from the ground with an initial speed of v ft. per sec. Assume that only gravity influences its motion. Thus it accelerates at the rate of -32 ft. per sec^2.*

    (a) *Show that t seconds later, (where $0 \leq t \leq \frac{v}{16}$), its speed is $v - 32t$ ft. per sec. and its height above the ground is $vt - 16t^2$ feet.*

    (b) *Calculate the maximum height it attains.*

    (c) *For how many minutes is it in the air?*

The Mean Value Theorem and Some Applications

The Mean Value Theorem is a statement about chords and tangents drawn on **smooth** graphs. Intuitively, a function f has a **smooth** graph on an interval $[a, b]$ if, for every x in $[a, b]$, we can draw a tangent at the point $(x, f(x))$ on its graph. The left figure below is the graph of $f(x) = x^2 + x$ on the interval $[-3.5, 4.5]$, (arbitrarily chosen). Its graph is smooth since we can draw a tangent at any point $(x, f(x))$. The graph in the second figure is not smooth on the interval $[-3, 3]$ because we cannot draw a tangent at $(-1, 2)$ or $(1.5, 3.4)$.

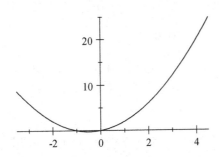
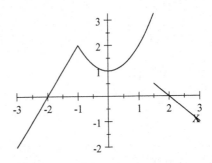

Let f have a smooth graph on an interval $[a, b]$. Let c and d be points such that $a \leq c < d \leq b$. Consider the chord PQ joining $P(c, f(c))$ and $Q(d, f(d))$ on the graph of f, (we used $f(x) = x^2 + x$, $c = -2.5$ and $d = 4$ in the figure below). Let $P'Q'$ be a variable line segment that is always parallel to PQ.

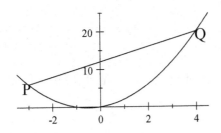
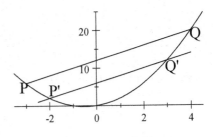

A chord PQ joining $(-3, 6)$ and $(4, 20)$ *PQ and $P'Q'$*

The mean value theorem asserts that if you slide $P'Q'$ far enough in the right direction, it becomes a tangent to the graph of f at some point $(\theta, f(\theta))$ where θ

58

is between c and d.

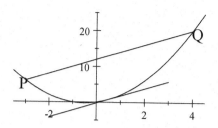

It becomes a tangent

Since the slope of PQ is $\dfrac{f(d) - f(c)}{d - c}$, the theorem asserts that there is a number θ between c and d such that

$$f'(\theta) = \frac{f(d) - f(c)}{d - c} \qquad (1.15)$$

This is conveniently written as $f(d) - f(c) = (d - c)\, f'(\theta)$.

In the case of $f(x) = x^2 + x$ and points $c = -2.5$, $d = 4$, it turns out that θ is approximately equal to 0.73. We obtained it is by solving the equation

$$2\theta + 1 = f'(\theta) = \frac{f(4) - f(-2.5)}{4 - (-2.5)} = \frac{20 - 8.75}{6.5} = \frac{11.25}{6.5}$$

The theorem is generally stated in the following form:

Theorem 67 (*The Mean Value Theorem*) *Let f have a smooth graph on an interval $[a, b]$ and c, d be points in $[a, b]$ with $c < d$. Then there is a number θ between c and d such that $f(d) - f(c) = (d - c)\, f'(\theta)$.*

Note that the theorem does not claim exactly one point θ satisfying the above conditions. It asserts that there is at least one point; leaving open the possibility of two or more. For example, let $f(x) = x^3 - 2x^2$, and choose $c = -1.5$ and $d = 3$.

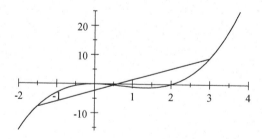

The theorem states that there is a number θ between -1.5 and 3 such that

$$f(3) - f(-1.5) = f'(\theta)\,(3 - (-1.5)) = 4.5 f'(\theta).$$

Since $f(3) - f(-1.5) = 9 - (-7.875) = 16.875$ and $f'(x) = 3x^2 - 4x$, θ satisfies the quadratic equation

$$3\left(\theta^2\right) - 4\theta = \frac{16.875}{4.5} = 3.75 \quad \text{or} \quad 3\theta^2 - 4\theta - 3.75 = 0$$

with solutions; $\theta_1 = \dfrac{4 + \sqrt{16 + 12 \times 3.75}}{6} = 1.96$ and $\theta_2 = \dfrac{4 - \sqrt{16 + 12 \times 3.75}}{6} =$ -0.64 (to 2 decimal places). Both numbers are acceptable solutions.

Exercise 68

1. The graphs of $f(x) = x^3$, $g(x) = x^3 - 3x + 4$, $h(x) = \sqrt{6 + 3x}$, and $u(x) = 4x + 1$ are given below.

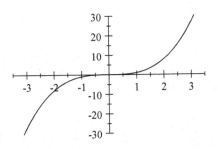

Graph of $f(x) = x^3$

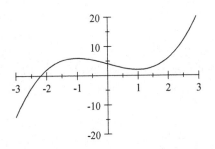

Graph of $g(x) = x^3 - 3x + 4$

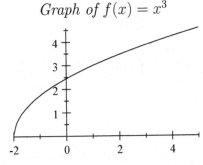

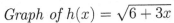

Graph of $h(x) = \sqrt{6 + 3x}$

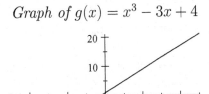

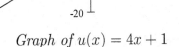

Graph of $u(x) = 4x + 1$

(a) Draw the cord joining the points $(-3, f(-3))$ and $(0, f(0))$ then find a number θ between -3 and 0 such that $f(0) - f(-3) = 3f'(\theta)$.

(b) Draw the cord joining the points $(-2, g(-2))$ and $(2, g(2))$ then find a number θ between -2 and 2 such that $g(2) - g(-2) = 4g'(\theta)$.

(c) Draw the cord joining the points $(-2, h(-2))$ and $(3, h(3))$ then find a number θ between -2 and 3 such that $h(3) - h(-2) = 5h'(\theta)$

(d) Draw the cord joining the points $(1, u(1))$ and $(4, u(4))$ then find a number θ between 1 and 4 such that $u(4) - u(1) = 3u'(\theta)$

2. Give an example of a function f with a smooth graph on an interval $[a, b]$, and with 3 or more numbers θ between a and b such that $f(b) - f(a) = (b-a)f'(\theta)$.

3. In this exercise, you have to show that $\sqrt{1 + x} \leq 1 + \frac{1}{2}x$ for all $x \geq 0$. To this end, consider the function $g(x) = 1 + \frac{1}{2}x - \sqrt{1 + x}$, $x \geq 0$. Show that $g'(x) > 0$ for all $x > 0$. Now take any $x > 0$. By the mean value theorem, there is a number θ between 0 and x such that $g(x) - g(0) = xg'(\theta)$. Explain why $xg'(\theta) > 0$ and deduce that $g(x) - g(0) > 0$ for all $x > 0$. Use this to show that $\sqrt{1 + x} \leq 1 + \frac{1}{2}x$ for all $x \geq 0$.

Generalized Mean Value Theorem

Let f and g have derivatives on an interval $[a, b]$, and $a \leq c < d \leq b$. By the Mean Value Theorem, there are points θ and α in (c, d) such that

$$f(d) - f(c) = (d - c) f'(\theta) \quad \text{and} \quad g(d) - g(c) = (d - c) g'(\alpha)$$

There is no guarantee that θ is the same as α because f and g are different functions. Assume that $g'(\alpha) \neq 0$. Then we may divide to get

$$\frac{f(d) - f(c)}{g(d) - g(c)} = \frac{f'(\theta)}{g'(\alpha)} \tag{1.16}$$

The Generalized Mean Value Theorem asserts that if $g'(x) \neq 0$ for all x in (a, b) then, indeed, there is a **single point** β in (c, d) such that

$$\frac{f(d) - f(c)}{g(d) - g(c)} = \frac{f'(\beta)}{g'(\beta)} \tag{1.17}$$

Here is a very neat proof that uses the Mean Value Theorem.

Consider the function $h(x) = f(x) - \left(\dfrac{f(d) - f(c)}{g(d) - g(c)} \right) g(x)$. Direct computations, (do them), reveal that

$$h(c) = \frac{f(c)g(d) - f(d)g(c)}{g(d) - g(c)} = h(d)$$

Therefore, by the Mean Value Theorem, there is a point β in (c, d) such that

$$0 = h(d) - h(c) = (d - c) h'(\beta) \tag{1.18}$$

Substitute $h'(\beta) = f'(\beta) - \left(\dfrac{f(d) - f(c)}{g(d) - g(c)} \right) g'(\beta)$ into (1.18) then divide by $(d - c)$ to get

$$0 = f'(\beta) - \frac{f(d) - f(c)}{g(d) - g(c)} g'(\beta)$$

Since $g'(\beta) \neq 0$, (by hypothesis), we may divide by $g'(\beta)$ and re-arrange the resulting equation to get (1.17).

Exercise 69

1. Let $f(x) = x^2 + 4x$ and $g(x) = 2x^2 + x + 2$, $x \geq 0$. Determine $\frac{f(2)-f(0)}{g(2)-g(0)}$ and $\frac{f'(x)}{g'(x)}$ then solve the equation $\frac{f(2)-f(0)}{g(2)-g(0)} = \frac{f'(x)}{g'(x)}$ for a number β in $(0, 2)$ such that $\frac{f(2)-f(0)}{g(2)-g(0)} = \frac{f'(\beta)}{g'(\beta)}$.

2. Let $f(x) = x^2 + 4x$ and $g(x) = x^3 + x - 1$, Find a number β in $(-2, 1)$ such that $\frac{f(1)-f(-2)}{g(1)-g(-2)} = \frac{f'(\beta)}{g'(\beta)}$.

Some Applications of the Mean Value Theorem

On page 33 we noted that if a function f is increasing on an interval $[a, b]$ then the tangents to its graph have positive slopes, therefore its derivatively $f'(x)$ cannot be negative. We can now show, with the help of the Mean Value Theorem, that if a function g has a positive derivative on an interval $[a, b]$ then it must be increasing on $[a, b]$.

1. *If a function f has a positive derivative on an interval $[a, b]$, then it is increasing on the interval. To see this, assume that $a \leq x < y \leq b$. By the mean value theorem, there is a number θ between x and y such that*

$$f(y) - f(x) = (y - x) f'(\theta). \tag{1.19}$$

   The right hand side of (1.19) is positive because $(y - x)$ and $f'(\theta)$ are both positive. Therefore $f(y)$ must be bigger than $f(x)$. Since x and y were arbitrary points in $[a, b]$, this proves that f is increasing on $[a, b]$.

2. *If a function f has a negative derivative on an interval $[a, b]$, then it is decreasing on the interval. To see this, assume that $a \leq x < y \leq b$. By the mean value theorem, there is a number θ between x and y such that*

$$f(y) - f(x) = (y - x) f'(\theta).$$

   This time $(y - x) f'(\theta)$ is negative, because $f'(\theta)$ is negative whereas $(y - x)$ is positive. Therefore $f(x)$ must be bigger than $f(y)$. Since x and y were arbitrary points in $[a, b]$, f is decreasing on $[a, b]$.

3. *If the derivative of a function f is zero on an interval $[a, b]$ then f is constant on $[a, b]$. For a proof, take any number x in $[a, b]$ which is bigger than a. By the mean value theorem, there is a number θ between x and a such that*

$$f(x) - f(a) = (x - a) f'(\theta) = 0. \tag{1.20}$$

   This implies that $f(x) = f(a)$. Since x was an arbitrary point in $[a, b]$, f has the constant value $f(a)$ on $[a, b]$.

Test your skills - 4,   (75 minutes)

1. Find the derivative of each function:

   (a) $f(x) = \frac{1}{4}x^4 - \frac{7}{\sqrt{x}} - 5$

   (b) $g(x) = \dfrac{(x^2 + 2)}{9} - \dfrac{2}{x + 1}$

   (c) $h(x) = \sqrt[4]{x^2 + 1}$

2. Guess the most general antiderivative of $f(x) = \dfrac{x}{\sqrt{2x^2 + 1}}$.

3. You are given that $x_0 = -2$ is an approximate solution of the equation $x^3 - 3x + 1$. Use it to get the first and second approximations x_1 and x_2 respectively, given by Newton's method. Round off your answers to 3 dec. pl.

4. A closed box is to be constructed from a rectangular metal lamina of dimensions 12 by 18 units by cutting out 4 identical squares as shown in the diagram below then fold along the dotted lines.

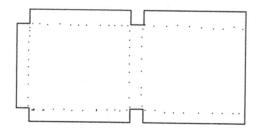

Construct a table similar to the table on page 2 then use the derivative of a suitable function to calculate the largest possible volume of such a box.

5. You may not use more than 400 sq. cm of material. What is the largest volume of a cylinder, closed at both ends, that you can construct under such a constraint?

6. A Norman window has the shape of a rectangle with height h feet surmounted by a semicircle of radius r feet. The perimeter of the window must be 20 feet. Let its area be A square feet.

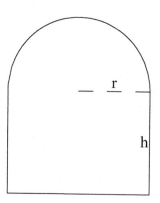

Copy and complete the following table:

r	0.5	1	1.5	2	2.5	3	4	r
h								
A								

Now describe how the area of the window varies as the radius r changes, then use an argument involving the slope of a tangent to find the largest possible area of the window.

Chapter 2   MORE GENERAL LIMITS

To be able to prove a number of results we use in this course, one must have a precise definition of a limit. So far we have considered the limit as h approaches 0 of the special quotient

$$\frac{f(x+h) - f(x)}{h}.$$

To determine it, you had to answer the question: *what number is close the values you get when you substitute small numbers h into the expression* $\frac{f(x+h) - f(x)}{h}$? The answer is the derivative of f at x, denoted by $f'(x)$. Now we are going to ask the following more general question:

Let g be a function and c a number. Suppose you choose different numbers x close, (but not equal) to c and determine the corresponding values g(x). Is there a single number l that is close to all such values?

If the answer is YES, then l is called the limit of $g(x)$ as x approaches c.

Example 70 *Consider the function g with formula* $g(x) = \dfrac{6x+6}{x^3+1}$, $x \neq -1$. *Although g(−1) is not defined, we can evaluate g(x) for all the numbers x "close" to −1. The table below gives a sample of such values.*

x	-1.5	-1.1	-1.002	-1.0001	-0.9998	-0.999	-0.9	-0.56
$\dfrac{6x+6}{x^3+1}$	1.26	1.81	1.996	1.9998	2.0004	2.002	2.21	3.202

It appears that every number x that is sufficiently close to −1 gives a value g(x) close to 2, therefore g(x) should have limit 2 as x approaches −1. We can actually defend this claim more rigorously as follows: A number x close to −1 has the form x = −1 + h where h is close to 0. (The smaller h is , the closer x = −1 + h is close to −1.) It gives a value

$$g(x) \;=\; g(-1+h) = \frac{6(-1+h)+6}{(-1+h)^3+1} = \frac{-6+6h+6}{-1+3h-3h^2+h^3+1}$$

$$=\; \frac{6h}{3h-3h^2+h^3} = \frac{6}{3-3h+h^2}$$

When h is close to 0, (which means that x = −1 + h is close to −1), then $\dfrac{6}{3-3h+h^2}$ *is close to* $\dfrac{6}{3-0+0} = 2.$

Example 71 *Let* $g(x) = \dfrac{x^2-1}{\sqrt{x}-1}$, $x \neq 1$. *Then g(x) has limit 4 as x approaches 1. (Substitute a couple of numbers x close to 1 into* $\dfrac{x^2-1}{\sqrt{x}-1}$.) *To verify it, note*

64

that a number x close to 1 has the form $x = 1 + h$ where h is close to 0. It gives a value

$$g(1+h) = \frac{(1+h)^2 - 1}{\sqrt{1+h} - 1} = \frac{2h + h^2}{\sqrt{1+h} - 1}$$

You should anticipate rationalizing the denominator to get

$$f(1+h) = \frac{h(2+h)\left(\sqrt{1+h}+1\right)}{\left(\sqrt{1+h}-1\right)\left(\sqrt{1+h}+1\right)} = \frac{h(2+h)\left(\sqrt{1+h}+1\right)}{1+h-1}$$

$$= (2+h)\left(\sqrt{1+h}+1\right).$$

Clearly, $(2+h)\left(\sqrt{1+h}+1\right)$ is close to $(2+0)\left(\sqrt{1+0}+1\right) = 4$ when h is close to 0, hence $g(x)$ has limit 4 as x approaches 1 and we may write $\lim\limits_{x\to 1}\dfrac{x^2-1}{\sqrt{x}-1} = 4$.

Exercise 72

1. Let $f(x) = \dfrac{3-x}{3-\sqrt{x+6}}$. Verify that when x is close to 3 then $f(x)$ is close to 6.

2. Set your calculator in radian mode then use it to complete the following table

h	-1	-0.05	-0.05	-0.006	0.008	0.02	0.6	0.9
$\sin h$								
$1-\cos h$								
$\dfrac{\sin h}{h}$								
$\dfrac{1-\cos h}{h}$								
$\dfrac{1-\cos h}{h^2}$								

(a) What value does the table suggest for $\lim\limits_{h\to 0}\dfrac{\sin h}{h}$?

(b) What value does the table suggest for $\lim\limits_{h\to 0}\dfrac{1-\cos h}{h}$?

(c) What value does the table suggest for $\lim\limits_{h\to 0}\dfrac{1-\cos h}{h^2}$?

3. Use a calculator to complete the following table then guess $\lim\limits_{h\to 0}\dfrac{e^h-1}{h}$.

h	-0.1	-0.02	-0.007	-0.001	0.001	0.008	0.06	0.3
$\dfrac{e^h-1}{h}$								

A Precise Definition of a Limit

Let f be a given function and c be a real number. We have so far defined the limit of f as x approaches c intuitively: it is the number that is close to all the values $f(x)$ when x is close to c. Unfortunately, this definition cannot be used to prove general statements about limits. For example, we cannot use it to prove that if f has limit l and g has limit m as x approaches a number c then $f+g$ has limit $l+m$ as x approaches c. The most we can do with it, (which is not a rigorous proof), is to take specific functions f and g which have limits l and m respectively as x approaches a specific point c and verify that the sum $f(x)+g(x)$ is close to $l+m$ when x is close to c.

In this section we derive a precise definition of a limit that can be used in proofs. We get it by replacing the vague phrase "*when x is close to c then $f(x)$ is close to L*" with a more precise statement. To this end, note that if you are asked to shade the numbers *on the number line* that you consider to be close to a given number L, you will, most probably, shade a small interval I centered at L. It may be denoted by $(L-\varepsilon, L+\varepsilon)$ where ε is some positive number. (The size of ε will depend on your choice of precision.)

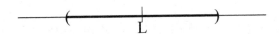

Having shaded such an interval, you would say that a number y is close to L if y is in the shaded region. In particular, the values $f(x)$ of a given function f should be close to L if, (when you plot them on the number line), they fall in the shaded interval. Therefore, to convince you that numbers x close to c give values $f(x)$ close to L, it suffices to produce an interval $(c-\delta, c+\delta)$ centered at c such that every number x in $(c-\delta, c+\delta)$ gives a value $f(x)$ in your interval $(L-\varepsilon, L+\varepsilon)$. Actually, we have to exclude the point c itself because it may not be in the domain of f. Therefore we have to ensure that every number x in $(c-\delta, c) \cup (c, c+\delta)$ gives a value $f(x)$ in $(L-\varepsilon, L+\varepsilon)$. The set $(c-\delta, c) \cup (c, c+\delta)$ is called a **punctured interval** centered at c.

If x is in the shaded region near c then $f(x)$ is in the shaded region near L.

If we are to convince every one who may show up, we must be in a position to carry out the above steps for **any** positive number ε. This suggests the following definition:

Definition 73 *A function $f(x)$ has limit L as x approaches a number c if, given **any interval** $(L-\varepsilon, L+\varepsilon)$, we can find a punctured interval $(c-\delta, c) \cup (c, c+\delta)$ such that $f(x) \in (L-\varepsilon, L+\varepsilon)$ for all x in $(c-\delta, c) \cup (c, c+\delta)$. We then write* $\lim_{x \to c} f(x) = L$.

Admittedly, this is not the definition you are likely to find in general use. The standard one is given in terms of absolute values. To make sure you are on board, we briefly review them here.

The absolute value of a given number x, (denoted by $|x|$), is the distance from the point representing x on the number line, to the point that represents 0, which we call the origin. For example, $|-5| = 5$ and $|4| = 4$. The figure below shows the distance, indicated by the dotted line, from the point representing -5 to the origin.

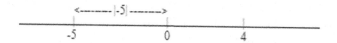

Clearly, if x is non-negative then $|x| = x$ and if x is negative then $|x|$ is obtained by simply changing the sign of x. Thus

$$|x| = \begin{cases} x & \text{if } x \geq 0 \\ -x & \text{if } x < 0 \end{cases}$$

Since a negative number is less than any non-negative number, it follows that $x < |x|$ if x is negative. Combining this with the fact that $x = |x|$ if x is non-negative, we conclude that

$$x \leq |x| \text{ for all real numbers } x$$

An alternative definition of the absolute value $|x|$ of a real number x that makes no mention of distances is the following:

$$|x| = \sqrt{(x)^2} \tag{2.1}$$

where $\sqrt{(x)^2}$ denotes the positive square root of $(x)^2$. For example,

$$\sqrt{(-5)^2} = \sqrt{25} = 5$$

confirming what we got above. It also follows from (2.1) that $|x|^2 = x^2$ for any real number x. Using (2.1), we find that if x and y are any real numbers then

$$|x + y|^2 = \left(\sqrt{(x+y)^2}\right)^2 = (x+y)^2 = x^2 + 2xy + y^2$$

$$= |x|^2 + 2xy + |y|^2 \leq |x|^2 + 2|x||y| + |y|^2 = (|x| + |y|)^2$$

In other words, if x and y are any real numbers then $|x+y|^2 \leq (|x| + |y|)^2$. When we take square roots of both sides, we get

$$|x + y| \leq |x| + |y| \tag{2.2}$$

We use this inequality repeatedly to prove a number of statements.

There is nothing special about the number 0. We can consider the distance from the point representing a number x to the point representing a number y on the number line. We denote it by $|x - y|$ or $|y - x|$ because the order in immaterial. For example, $|4 - 7|$ is the distance from the point representing 4 to the point representing 7, which is 3; and $|5 - (-6)|$ is the distance from the point representing 5 to the point representing -6, which is 11.

Turning to definition 73, we note that an interval $(l - \varepsilon, l + \varepsilon)$ centered at l consists of all the numbers y whose distance from l is less than ε. Therefore $f(x) \in (l - \varepsilon, l + \varepsilon)$ if $|f(x) - l| < \varepsilon$. Likewise the interval $(c - \delta, c + \delta)$ consists of all the numbers x whose distance from c is less than δ. When we leave out the center c of the interval, we get the numbers x whose distance from c is strictly bigger than 0 and is less than δ. Therefore, a number x is in the punctured interval $(c - \delta, c) \cup (c, c + \delta)$ if $0 < |x - c| < \delta$. Now definition 73 may be given in the following equivalent form:

Definition 74 *A function $f(x)$ has limit L as x approaches a number c if, given any positive number ε, we can find a positive number δ such that $|f(x) - L| < \varepsilon$ whenever $0 < |c - x| < \delta$.*

Example 75 *We show, using definition 74, that $f(x) = 3x + 5$ has limit 11 as x approaches 2. To this end let ε be any positive number. We have to find a number $\delta > 0$ such that $|f(x) - 11| < \varepsilon$ whenever $0 < |x - 2| < \delta$. The first step is to simplify the expression $|f(x) - 11|$:*

$$|f(x) - 11| = |3x + 5 - 11| = |3x - 6| = 3|x - 2|$$

Therefore we have to produce a number $\delta > 0$ such that $3|x - 2| < \varepsilon$ whenever $0 < |x - 2| < \delta$. Clearly $3|x - 2| < \varepsilon$ whenever $0 < |x - 2| < \frac{1}{3}\varepsilon$, therefore any $\delta \leq \frac{\varepsilon}{3}$ will do.

Example 76 *Let $g(x) = \dfrac{x - 1}{x^3 - 1}$, $x \neq 1$. Then g has limit $\frac{1}{3}$ as x approaches 1. To prove it using the precise definition of a limit, let ε be any positive number. We have to find a number $\delta > 0$ such that $\left|g(x) - \frac{1}{3}\right| < \varepsilon$ whenever $0 < |x - 2| < \delta$. A number x close to 1 may be written in the form $x = 1 + h$ where h is close to 0. This change of variable gives expressions that are easier to factor. Indeed*

$$\left|g(x) - \frac{1}{3}\right| = \left|\frac{x - 1}{x^3 - 1} - \frac{1}{3}\right| = \left|\frac{1 + h - 1}{(1 + h)^3 - 1} - \frac{1}{3}\right| = \left|\frac{h}{h^3 + 3h^2 + 3h} - \frac{1}{3}\right|$$

$$= \left|\frac{1}{h^2 + 3h + 3} - \frac{1}{3}\right| = \left|\frac{3 - h^2 - 3h - 3}{3(h^2 + 3h + 3)}\right| = \frac{|h||h + 3|}{3|h^2 + 3h + 3|}$$

We have to determine the numbers h such that

$$\frac{|h||h + 3|}{3|h^2 + 3h + 3|} < \varepsilon \tag{2.3}$$

This is a fairly difficult inequality to solve. We replace it with a simpler one as follows: Since it is the numbers x close to 1 that are relevant to the limit of g

as x approaches 1, we may restrict ourselves to numbers x in a suitable interval centered at 1. An example is $(0, 2)$. (Of course there is nothing special about this choice. Any other convenient one will also do.) Then we can put bounds on the terms $|h + 3|$ and $|h^2 + 3h + 3|$. Indeed, if $x = 1 + h$ is in $(0, 2)$ then $-1 < h < 1$, which implies that

$$|h + 3| < 4 \text{ and } |h^2 + 3h + 3| > 1.$$

(We need to replace $|h + 3| < 4$ with a bigger number because it is in the numerator of (2.3). On the other hand, $|h^2 + 3h + 3|$ is in the denominator, therefore we should replace it with a smaller number.) It now follows that if x is in $(0, 2)$ then

$$\left| g(x) - \frac{1}{3} \right| < \frac{|h| \cdot 4}{3 \cdot 1} = \frac{4|h|}{3} = \frac{4|x - 1|}{3}.$$

Recall that we need a $\delta > 0$ such that $\left| g(x) - \frac{1}{3} \right| < \varepsilon$ when $|x - 1| < \delta$. Since

$$\left| g(x) - \frac{1}{3} \right| < \frac{4|x - 1|}{3}$$

for all numbers x in $(0, 2)$, it suffices to find a $\delta > 0$ such that

$$\frac{4|x - 1|}{3} < \varepsilon$$

when $|x - 1| < \delta$. This is a fairly easy inequality to solve. Any $\delta < \frac{3\varepsilon}{4}$ will do. (In addition, the δ should be smaller than 1 because x must be in $(0, 2)$).

Example 77 *Let $g(x) = x$ and c be any real number. Then $\lim\limits_{x \to c} g(x) = c$. To prove it, let $\varepsilon > 0$ be given. We have to find a number $\delta > 0$ such that $|g(x) - c| < \varepsilon$ whenever $0 < |x - c| < \delta$. Since $|g(x) - c| = |x - c|$, the choice $\delta = \varepsilon$ will do. This proves that $\lim\limits_{x \to c} g(x) = c$.*

Example 78 *Let $g(x) = k$, where k is a constant, and c be any real number. Then $\lim\limits_{x \to c} g(x) = k$. To prove it, let $\varepsilon > 0$ be given. We have to find a number $\delta > 0$ such that $|g(x) - k| < \varepsilon$ whenever $0 < |x - c| < \delta$. But the inequality $|f(x) - k| < \varepsilon$ is satisfied by any x you choose. Therefore any positive number δ will do.*

Example 79 *Let $g(x) = \sqrt{x}$, $x > 0$, and c be any positive number. Then $\lim\limits_{x \to c} g(x) = \sqrt{c}$. For a proof, let $\varepsilon > 0$ be given. We have to find a $\delta > 0$ such that $|g(x) - \sqrt{c}| < \varepsilon$ whenever $0 < |x - c| < \delta$. Consider the expression*

$$\left| g(x) - \sqrt{c} \right| = \left| \sqrt{x} - \sqrt{c} \right|$$

To go beyond this, we have to rationalize the numerator of $|\sqrt{x} - \sqrt{c}| = \dfrac{|\sqrt{x} - \sqrt{c}|}{1}$. The result is

$$\left| \sqrt{x} - \sqrt{c} \right| = \frac{|\sqrt{x} - \sqrt{c}| \, |\sqrt{x} + \sqrt{c}|}{|\sqrt{x} + \sqrt{c}|} = \frac{|x - c|}{|\sqrt{x} + \sqrt{c}|}$$

Note that $\dfrac{|x-c|}{|\sqrt{x}+\sqrt{c}|} < \dfrac{|x-c|}{\sqrt{c}}$, *(because, in* $\dfrac{|x-c|}{\sqrt{c}}$, *we are dividing by a smaller number). Therefore*

$$|\sqrt{x}-\sqrt{c}| = \frac{|\sqrt{x}-\sqrt{c}|\,|\sqrt{x}+\sqrt{c}|}{|\sqrt{x}+\sqrt{c}|} = \frac{|x-c|}{|\sqrt{x}+\sqrt{c}|} < \frac{|x-c|}{\sqrt{c}} \qquad (2.4)$$

We are looking for a $\delta > 0$ *such that* $|\sqrt{x}-\sqrt{c}|$ *whenever* $0 < |x-c| < \delta$. *It suffices to find a* $\delta > 0$ *such that* $\dfrac{|x-c|}{\sqrt{c}} < \varepsilon$ *whenever* $0 < |x-c| < \delta$. *Clearly, any* $\delta < (\sqrt{c})\,\varepsilon$ *will do.*

Some Useful Properties of Limits

Suppose $f(x)$ has limit l and $g(x)$ has limit m as x approaches a number c. This means that when x is close to c then $f(x)$ is close to l and $g(x)$ is close to m. We should, therefore, expect

- $f(x) + g(x)$ to be close to $l + m$, suggesting that $(f+g)(x)$ has limit $l+m$ as x approaches c.

- $kf(x)$ to be close to kl for any constant k, which suggests that $kf(x)$ has limit kl as x approaches c.

- $f(x)g(x)$ to be close to lm, therefore $f(x)g(x)$ should have limit lm as x approaches c.

- $\dfrac{f(x)}{g(x)}$ to be close to $\dfrac{l}{m}$, provided $m \neq 0$, suggesting that $\dfrac{f(x)}{g(x)}$ has limit $\dfrac{l}{m}$ as x approaches c.

All these expectations can be shown to be true using the precise definition of a limit. That is:

If $f(x)$ has limit l and $g(x)$ has limit m as x approaches a number c then:

1. $(f+g)(x)$ *has limit* $l+m$ *as* x *approaches* c. *This is called the sum rule for limits and we write* $\displaystyle\lim_{x\to c}(f+g)(x) = \lim_{x\to c}f(x) + \lim_{x\to c}g(x)$.

2. *For any constant* k, $kf(x)$ *has limit* kl *as* x *approaches* c. *This is called the constant multiple rule for limits and we write* $\displaystyle\lim_{x\to c}kf(x) = k\lim_{x\to c}f(x)$.

3. $f(x)g(x)$ *has limit* lm *as* x *approaches* c. *This is called the product rule for limits and we write* $\displaystyle\lim_{x\to c}f(x)g(x) = \lim_{x\to c}f(x) \cdot \lim_{x\to c}g(x)$.

4. $\dfrac{f(x)}{g(x)}$ *has limit* $\dfrac{l}{m}$ *as* x *approaches* c, *provided* $m \neq 0$. *This is called the quotient rule for limits and we write* $\displaystyle\lim_{x\to c}\frac{f(x)}{g(x)} = \frac{\lim\limits_{x\to c}f(x)}{\lim\limits_{x\to c}g(x)}$.

Since we gave property 1 as an example of a general statement about limits which we cannot prove using the intuitive definition, we prove it here using the precise definition 74. To this end let ε be any positive number. We have to show that it is possible to find a positive number δ with the property that if $0 < |x - c| < \delta$ then

$$|f(x) + g(x) - (l + m)| < \varepsilon.$$

We are given that f has limit l as x approaches c, therefore it is possible to find a $\delta_1 > 0$ such that $|f(x) - l| < \frac{1}{2}\varepsilon$ if $0 < |x - c| < \delta_1$. By the same token, it is possible to find a $\delta_2 > 0$ such that $|g(x) - m| < \frac{1}{2}\varepsilon$ if $0 < |x - c| < \delta_2$. Take δ to be the the smaller of the two numbers δ_1 and δ_2, (we write this briefly as $\delta = \min\{\delta_1, \delta_2\}$). Then $|f(x) - l| < \frac{1}{2}\varepsilon$ and $|g(x) - l| < \frac{1}{2}\varepsilon$ if $0 < |x - c| < \delta$. Now write $f(x) + g(x) - (l + m)$ as $(f(x) - l) + (g(x) - m)$ and use (2.2) to conclude that

$$|f(x) + g(x) - (l + m)| \leq |f(x) - l| + |g(x) - l| < \tfrac{1}{2}\varepsilon + \tfrac{1}{2}\varepsilon = \varepsilon$$

whenever $|x - c| < \delta$.

You are asked to prove the other three properties in Exercise 86 on page 71, (specifically, problems 7, 10 and 11). These four properties enable us to calculate the limits of a number of functions from the knowledge of the limits of a few familiar functions.

Example 80 *Let $f(x) = kx$ where k is a constant. Let c be any number. We already know that $g(x) = x$ has limit c as x approaches c, (see Example 77). Therefore, by the constant multiple rule, $\lim_{x \to c} f(x) = \lim_{x \to c} kg(x) = kc$.*

Example 81 *Let $u(x) = x^2$ and c be any number. Again consider the function $g(x) = x$. We may write u as a product of g by itself. That is $u(x) = g(x)g(x)$. By the product rule, $\lim_{x \to c} u(x) = \lim_{x \to c} g(x) \lim_{x \to c} g(x) = c^2$. In general, if n is a positive integer and $v(x) = x^n$ then*

$$\lim_{x \to c} v(x) = c^n.$$

Example 82 *Let n be a positive integer and $w(x) = \dfrac{1}{x^n}$, $x \neq 0$. If $c \neq 0$ then by the quotient rule,*

$$\lim_{x \to c} w(x) = \frac{\lim_{x \to c} 1}{\lim_{x \to c} x^n} = \frac{1}{c^n}$$

Example 83 *Let $p(x) = a + bx + dx^2 + kx^3$ where a, b, d, and k are constants. Then a combination of the sum rule, the constant multiple rule, and the product rule reveals that $\lim_{x \to c} p(x) = a + bc + dc^2 + kc^3$.*

Example 84 *Let $g(x) = \dfrac{x^2 + 5x + 6}{x^2 - 2x - 8}$, $x \notin \{-2, 4\}$. To determine $\lim_{x \to -2} g(x)$, we factor to get*

$$g(x) = \frac{(x + 2)(x + 3)}{(x + 2)(x - 4)}$$

Since $x \neq -2$, we may divide the numerator and denominator by $(x+2)$ to get
$g(x) = \dfrac{x+3}{x-4}$. *By the quotient rule,* $\lim\limits_{x \to -2} g(x) = \dfrac{(-2+3)}{(-2-4)} = -\dfrac{1}{6}$.

Example 85 *Let* $f(x) = \dfrac{x^2 - 81}{\sqrt{x} - 3}$, $x \neq 9$. *To determine* $\lim\limits_{x \to 9} f(x)$, *we factor:*

$$f(x) = \frac{(x-9)(x+9)}{\sqrt{x}-3} = \frac{(\sqrt{x}+3)(\sqrt{x}-3)(x+9)}{(\sqrt{x}-3)}$$

Since $x \neq 9$, we may divide the numerator and denominator by $(\sqrt{x}-3)$ to get
$f(x) = (\sqrt{x}+3)(x+9)$. *We know that* $\lim\limits_{x \to 9}\sqrt{x} = 3$, *(see Example 79), therefore the product rule gives* $\lim\limits_{x \to 9} f(x) = (3+3)(9+9) = 108$

Exercise 86

1. *Use appropriate properties of limits to determine each of the following limits:*

   a) $\lim\limits_{x \to -2} (2x^3 + 3x^2 + 9)$ b) $\lim\limits_{x \to 1} \left(\dfrac{3}{\sqrt{4x+5}+6} \right)$ c) $\lim\limits_{x \to 5} \left(\dfrac{x^2 - 25}{x^2 - 6x + 5} \right)$

   d) $\lim\limits_{x \to 3} \left(\dfrac{x^2 - 9}{x^3 - 27} \right)$ e) $\lim\limits_{x \to -2} \left(\dfrac{x+2}{\sqrt{x+6}-2} \right)$ f) $\lim\limits_{x \to 2} \left(\dfrac{x^4 - 16}{x^3 - 8} \right)$

2. *Complete the following proof that $\lim\limits_{x \to 2} (x^2 + 5x + 1) = 15$:*

   Let $\varepsilon > 0$ be given. Consider
   $$|f(x) - 15| = |x^2 + 5x - 14| = |x+7|\,|x-2|$$

   We need the numbers x near 2 such that $|x+7|\,|x-2| < \varepsilon$. It is suffi-cient to restrict ourselves to numbers in an interval like $(1,3)$ centered at 2. If $x \in (1,3)$ then $|x+7| < 10$, and so

   $$|f(x) - 15| = |x+7|\,|x-2| < 10\,|x-2|. \text{ Continue with the proof}$$

3. *Use the precise definition of a limit to prove that $\lim\limits_{x \to 1} (3x^2 + 2x - 1) = 4$*

4. *Here is a method of defending the result $\lim\limits_{x \to 0} \dfrac{\sin x}{x} = 1$ suggested by the table in question 2 on page 64.*

   Consider the circle below of radius r. AOB is an angle of x radians, $(0 < x < \frac{\pi}{2})$, and BC is a tangent to the circle, hence angle OBC is a right angle. By construction, BA is perpendicular to OC, therefore angle OAB is also a right angle.

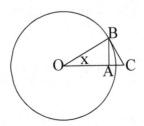

The triangle OAB, the sector OBD and the triangle OBC are given in the figures below.

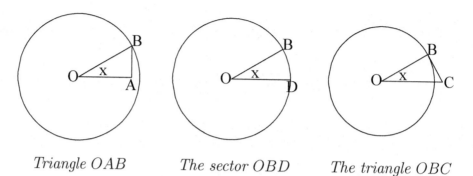

Triangle OAB The sector OBD The triangle OBC

The area of triangle OAB, is smaller than the area of sector OBD because the sector has an additional piece. Likewise, the area of sector ODB is smaller than the area of triangle OBC. Since triangle OAB has a base of length $r\cos x$ and a height $r\sin x$, its area is $\frac{1}{2}r^2 \cos x \sin x$. The sector has area $\frac{1}{2}r^2 x$, and triangle OBC has area $\frac{1}{2}r^2 \tan x$. Therefore

$$\frac{r^2 \cos x \sin x}{2} < \frac{r^2 x}{2} < \frac{r^2 \tan x}{2}$$

(a) *Simplify to* $\cos x < \dfrac{x}{\sin x} < \dfrac{1}{\cos x}$, $0 < x < \frac{\pi}{2}$. *(Give the details.)*

(b) *Deduce that the inequalities* $\cos x < \dfrac{x}{\sin x} < \dfrac{1}{\cos x}$ *also hold for* $-\frac{\pi}{2} < x < 0$. *(Hint: what is the sign of $\cos x$ and the sign of $\dfrac{\sin x}{x}$ when $-\frac{\pi}{2} < x < 0$?)*

(c) *We know that when x is close to 0 then $\cos x$ is close to 1. Use this to explain why $\lim\limits_{x \to 0}\left(\dfrac{x}{\sin x}\right)$ must be 1.*

(d) *Deduce that* $\lim\limits_{x \to 0}\dfrac{\sin x}{x} = 1$

5. *The result* $\lim\limits_{x \to 0}\dfrac{\sin x}{x} = 1$ *may be used to determine limits involving expressions of the form* $\dfrac{\sin w}{w}$ *where w is a given expression.*

Example 87 $\lim\limits_{\theta \to 0}\dfrac{\sin 4\theta}{4\theta}$ *should also be 1 because all we have to do is to let $x = 4\theta$ to transform* $\dfrac{\sin 4\theta}{4\theta}$ *into* $\dfrac{\sin x}{x}$.

Example 88 $\lim\limits_{\theta \to 0}\dfrac{\tan 3\theta}{\theta}$ *may be determined by writing*

$$\frac{\tan 3\theta}{\theta} = \frac{\sin 3\theta}{\theta \cos 3\theta} = \frac{3}{\cos 3\theta} \cdot \frac{\sin 3\theta}{3\theta}$$

When θ is close to 0, $\dfrac{3}{\cos 3\theta}$ is close to $\dfrac{3}{1} = 3$ and $\dfrac{\sin 3\theta}{3\theta}$ is close to 1. Therefore

$$\lim_{\theta \to 0} \frac{\tan 3\theta}{\theta} = 3$$

Determine the following limits:

(a) $\displaystyle\lim_{\theta \to 0} \frac{\sin \frac{1}{2}\theta}{\frac{1}{2}\theta}$ (b) $\displaystyle\lim_{\theta \to 0} \frac{\sin 4\theta}{\theta}$ (c) $\displaystyle\lim_{\theta \to 0} \frac{\sin \frac{1}{2}\theta}{4\theta}$ (d) $\displaystyle\lim_{\theta \to 0} \frac{\theta - \sin \frac{1}{3}\theta}{2\theta}$

6. Complete the following steps to determine $\displaystyle\lim_{h \to 0} \frac{1 - \cos h}{h}$ more rigorously:

$$\frac{1 - \cos h}{h} = \frac{1 - \cos h}{h} \times \frac{1 + \cos h}{1 + \cos h} = \frac{\sin^2 h}{h\,(1 + \cos h)} = \frac{\sin h}{h} \times \frac{\sin h}{1 + \cos h}$$

Therefore $\displaystyle\lim_{h \to 0} \frac{1 - \cos h}{h} =$

(a) Determine $\displaystyle\lim_{h \to 0} \frac{1 - \cos h}{h^2}$ in a similar way.

7. Complete the following proof of the constant multiple rule for limits which states that if a function f has limit l as x approaches a number c and k is a constant then kf has limit kl as x approaches c:

Start by explaining why the statement must be true if $k = 0$. Now assume that $k \neq 0$, and let ε be any positive number. You have to show that it is possible to find a positive number δ such that $|kf(x) - kl| < \varepsilon$ if $0 < |x - c| < \delta$. Use the fact that f has limit l as x approaches c to conclude that there is a positive number δ such that $|f(x) - l| < \frac{1}{k}\varepsilon$ if $0 < |x - c|$. Now explain why $|kf(x) - kl| < \varepsilon$ if $0 < |x - c| < \delta$.

8. In this exercise, you show that if $f(x)$ has limit L as x approaches a number c then the absolute values $|f(x)|$ do not differ appreciably from $|L|$. More precisely, show that if f has a limit L as x approaches c then given any positive number k, we can find a positive number δ such that $|f(x)| < |L| + k$ when $0 < |x - c| < \delta$. Follow the following steps:

Take $\varepsilon = k$. Since it is positive and f has limit L as c approaches c, there is a $\delta > 0$ with the property that if $0 < |x - c| < \delta$ then

$$|f(x) - L| < k$$

Now write $|f(x)| = |(f(x) - L) + L|$. By the inequality (2.2) on page 66, $|(f(x) - L) + L| \leq |f(x) - L| + |L|$. Use this to deduce that

$$|f(x)| < k + |L| \quad if\ 0 < |x - c|$$

9. *In this exercise, you show that if $f(x)$ has a non-zero limit L as x approaches a number c then the absolute values $|f(x)|$ are a distance away from zero when x is close to c. More precisely, show that there is a $\delta > 0$ such that $\frac{1}{2}|L| < |f(x)|$ when $0 < |x - c| < \delta$. Follow the following steps:*

> *Take $\varepsilon = \frac{1}{2}|L|$. It is non-zero, therefore, there is a $\delta > 0$ such that $|f(x) - L| < \frac{1}{2}|L|$ when $0 < |x - c| < \delta$. Write $|L| = |L - f(x) + f(x)|$. By (2.2) on page 66, $|L - f(x) + f(x)| \leq |f(x) - L| + |f(x)|$. Use this to deduce that*
> $$\tfrac{1}{2}|L| < |f(x)| \quad if\ 0 < |x - c|.$$

10. *In this exercise, you derive the product rule for limits by showing that if $f(x)$ has limit l and $g(x)$ has limit m as x approaches a number c then $f(x)g(x)$ has limit lm as x approaches c. More precisely, you show that given $\varepsilon > 0$ it is possible to find a positive number δ such that $|f(x)g(x) - lm| < \varepsilon$ when $0 < |x - c| < \delta$. Start by writing $|f(x)g(x) - lm|$ as*
$$|f(x)g(x) - lm| = |f(x)g(x) - f(x)m + f(x)m - lm|$$

(Nothing changes when you add $0 = -f(x)m + f(x)m$ to $f(x)g(x) - lm$.) Now deduce that
$$|f(x)g(x) - lm| \leq |f(x)|\,|g(x) - m| + |m|\,|f(x) - l|$$

Use the result of Exercise 8 above, with $k = 1$, to show that there is a $\delta_1 > 0$ such that
$$|f(x)g(x) - lm| < (|l| + 1)\,|g(x) - m| + |m|\,|f(x) - l|.$$

Finally show how to find a $\delta > 0$ such that $|f(x)g(x) - lm| < \varepsilon$ for all x that satisfy $0 < |x - c| < \delta$.

11. *In this exercise, you derive the quotient rule for limits by showing that if $f(x)$ has limit l and $g(x)$ has a non-zero limit m as x approaches a number c then $\dfrac{f(x)}{g(x)}$ has limit $\dfrac{l}{m}$ as x approaches c. More precisely, you show that given $\varepsilon > 0$ it is possible to find a positive number δ such that $\left|\dfrac{f(x)}{g(x)} - \dfrac{l}{m}\right| < \varepsilon$ when $0 < |x - c| < \delta$. Start by writing $\left|\dfrac{f(x)}{g(x)} - \dfrac{l}{m}\right|$ as*

$$\left|\frac{f(x)}{g(x)} - \frac{l}{m}\right| = \left|\frac{mf(x) - lg(x)}{mg(x)}\right| = \left|\frac{mf(x) - lm + lm - lg(x)}{mg(x)}\right|$$

$$\leq \frac{|f(x) - l|}{|g(x)|} + \frac{|l|\,|m - g(x)|}{|mg(x)|}$$

Use the result of Exercise 9 above to deduce that there is a $\delta_1 > 0$ such that
$$\frac{|f(x) - l|}{|g(x)|} < \frac{2\,|f(x) - l|}{m} + \frac{2\,|l|\,|g(x) - m|}{m^2}$$

Finally show how to find a $\delta > 0$ such that $\left|\dfrac{f(x)}{g(x)} - \dfrac{l}{m}\right| < \varepsilon$ for all x that satisfy $0 < |x - \dot{c}| < \delta$.

Limits From Below and Limits From Above

The following example illustrates the idea of a limit from below and a limit from above.

Example 89 *Imagine rounding off a number to the nearest whole number. This means that you replace a given number by the integer nearest to it on the number line. For example, since 5.184 is between 5 and 6, but it is closer to 5, you round it off to 5. A number in the middle of two integers will, (by agreement), be rounded off to the larger of the two integers. Thus 3.5 is rounded off to 4 and −16.5 is rounded off to −16. Let f be the function that assigns the nearest whole number to a given number x as described. Part of its graph is given below.*

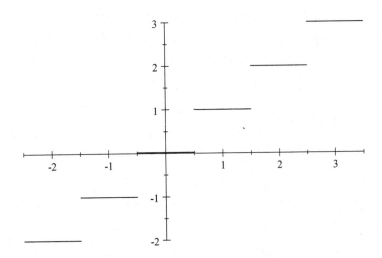

Take a number like 1.5 in the middle of two integers. Every number x that is close to 1.5 and is also to the left of 1.5 on the number line, (examples are 1.4, 1.488, and 1.4999), gives a value f(x) close to 1 (actually equal to 1). For this reason, we say that f has limit 1 as x approaches 1.5 from below, (or from the left), and write $\lim_{x \to 1.5^-} f(x) = 1$.

On the other hand, numbers x close to 1.5, but to the right of 1.5, (example, 1.51, 1.5003, and 1.500008), give values f(x) close to 2, (actually equal to 2). We say that f has limit 2 as x approaches 1.5 from above, (or from the right), and write $\lim_{x \to 1.5^+} f(x) = 2$.

The following are the general definitions of limits from below/above:

Definition 90 *A function f has limit l as x approaches a given number c from below if numbers x that are; (i) **to the left of** c, and (ii) **are close to** c, give values f(x) that are close to l. The number l is called the limit of f at c from below, or the left limit of f at c, and is denoted by $\lim_{x \to c^-} f(x)$.*

As we pointed out before giving definition 73, numbers close to l are in an interval $(l - \varepsilon, l + \varepsilon)$ for some suitable positive number ε. Likewise, numbers which are close to c and are also to the left of c are in some interval $(c - \delta, c)$ for some suitable positive number δ. This suggests the following more precise definition of a limit from the left, (or from below).

Definition 91 *A function f has limit l as x approaches a number c from below if given any positive number ε, we can find a positive number δ such that $f(x) \in (l - ε, l + ε)$ for all $x \in (c - δ, c)$. Alternatively, given any $ε > 0$ we can find a $δ > 0$ such that $|f(x) - l| < ε$ if $c - δ < x < c$.*

Definition 92 *A function f has limit m as x approaches a given number c from above if numbers x that are; (i) **to the right of** c, and (ii) **are close to** c, give values $f(x)$ that are close to m. The number m is called the limit of f at c from above, or the right limit of f at c, and is denoted by $\lim\limits_{x \to c^+} f(x)$.*

Since numbers which are close to c and are also to the right of c are in some interval $(c, c + δ)$ for some suitable positive number δ, a more precise definition of a limit from the right (or from above) is the following:

Definition 93 *A function f has limit l as x approaches a number c from above if given any positive number ε, we can find a positive number δ such that $f(x) \in (l - ε, l + ε)$ for all $x \in (c, c + δ)$. Alternatively, given any $ε > 0$ we can find a $δ > 0$ such that $|f(x) - l| < ε$ if $c < x < c + δ$.*

Example 94 *Consider the function g with formula*

$$g(x) = \begin{cases} x^2 & if \quad -2 \le x \le -1 \\ 2x + 2 & if \quad -1 < x \le 1 \\ 1 - x^2 & if \quad 1 < x \le 3 \end{cases}$$

Its graph is given below.

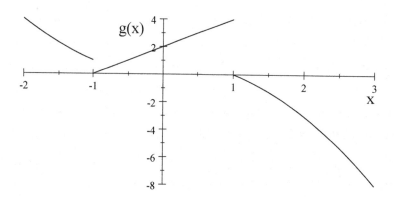

Take a number like $c = 1$. The graph and formula for g suggest that $\lim\limits_{x \to 1^+} g(x) = 0$ and $\lim\limits_{x \to 1^-} g(x) = 4$. In the case of $c = -1$, the graph and formula suggest that $\lim\limits_{x \to -1^+} g(x) = 0$ and $\lim\limits_{x \to -1^-} g(x) = 1$.

Remark 95 *If a function f has limit l as x approaches a number c then all the numbers x close to c, (whether to the left or to the right of c), give values $f(x)$ close to l. Therefore f must have limit l as x approaches c from below and from above. In other words,*

$$\lim_{x \to c^-} f(x) = l = \lim_{x \to c^+} f(x)$$

Remark 96 *It follows from Remark 95 that if the left limit of a function g at a point c is different from its right limit, (in other words, if $\lim\limits_{x \to c^-} g(x) \ne \lim\limits_{x \to c^+} g(x)$), then g has no limit at c.*

Infinite Limits

Consider a function like $f(x) = \dfrac{1}{x^2}$, $x \neq 0$. We find that numbers x close to 0 give values $f(x)$ that are large and positive. For example

$$f(0.1) = 100 \qquad f(-0.04) = 625 \qquad f(0.0005) = 4000000$$

In such a case, we say that $f(x)$ approaches ∞ as x approaches 0 and write $f(x) \to \infty$ as $x \to 0$. Alternatively, we say that f has limit ∞ as x approaches 0 and write $\lim_{x \to 0} f(x) = \infty$. But do not interpret this to mean that there is a real number ∞ with the property that $|f(x) - \infty| \to 0$ as $x \to \infty$. These statements simply mean numbers x close to 0 give values $f(x)$ that are large and positive.

Say we change the sign of $\dfrac{1}{x^2}$ and let $g(x) = -\dfrac{1}{x^2}$. We find that numbers x close to 0 give values $g(x)$ that are large and negative. For that reason, we say that $g(x)$ approaches $-\infty$ as x approaches 0 and write $g(x) \to -\infty$ as $x \to 0$. Alternatively, we say that g has limit $-\infty$ as x approaches 0 and write $\lim_{x \to 0} g(x) = -\infty$. Once again, we are not saying that there is a real number $-\infty$ with the property that $|f(x) - (-\infty)| \to 0$ as x approaches 0. We write $\lim_{x \to 0} g(x) = -\infty$ to convey the message that numbers x close to 0 give values $g(x)$ that are large and negative.

We summarize these in a definition:

Definition 97 *Let f be a given function and c be a number which need not be in the domain of f.*

1. *We say that f has limit ∞ as x approaches c, and write $\lim_{x \to 0} f(x) = \infty$, if numbers x close to c give values $f(x)$ that are large and positive.*

2. *We say that f has limit $-\infty$ as x approaches c, and write $\lim_{x \to 0} f(x) = -\infty$, if numbers x close to c give values $f(x)$ that are large and negative.*

We can have infinite one-sided limits as the next example shows.

Example 98 *Let $h(x) = \dfrac{x+1}{x-2}$, $x \neq 2$. If x **is close to 2** and is **to the right of 2** then $h(x)$ is a large positive number. For example,*

$$h(2.005) = 601, \quad h(2.001) = 3001, \quad h(2.00002) = 15001, \quad h(2.0000001) = 30000001$$

Because of this, we say that $h(x)$ approaches ∞ as x approaches 2 from above, and write

$$h(x) \to \infty \text{ as } x \to 2^+ \qquad or \qquad \lim_{x \to 2^+} h(x) = \infty.$$

*On the other hand, if x **is close to 2**, and is **to the left of 2** then $h(x)$ is a large negative number. For example,*

$$h(1.9905) = -314.79, \quad h(1.995) = -599, \quad h(1.9999998) = -15000000$$

We say that $h(x)$ approaches $-\infty$ as x approaches 2 from below, and write

$$h(x) \to -\infty \text{ as } x \to 2^- \qquad or \qquad \lim_{x \to 2^-} h(x) = -\infty.$$

Limit as x Approaches ∞ or $-\infty$

We start with limits as x approaches infinity. Only a function that is defined for all large positive numbers can have a limit as x approaches ∞. Let f be such a function. If every large positive number x gives a value $f(x)$ close to a single number l, then we say that f has limit l as x approaches ∞, and write $\lim_{x \to \infty} f(x) = l$.

Example 99 *Consider $f(x) = \dfrac{1}{x^2}$. When x is large and positive, then x^2 is a very large positive number, and so, its reciprocal is close to 0. In other words, when x is large and positive then $\dfrac{1}{x^2}$ is close to 0, therefore $\lim_{x \to \infty} \dfrac{1}{x^2} = 0$.*

Example 100 *Let g be defined by $g(x) = \dfrac{3x + 1}{x + 2}, x \neq -2$. Then $\lim_{x \to \infty} g(x) = 3$. To see this, note that when x is large and positive, the dominant term in the numerator is $3x$ and the dominant term in the denominator is x, therefore $\dfrac{3x + 1}{x + 2}$ must be close to $\dfrac{3x}{x} = 3$. Another way of arriving at the same result is to divide the numerator and denominator of $\dfrac{3x + 1}{x + 2}$ by x, (the highest power of x in the denominator). The result is $\dfrac{3x + 1}{x + 2} = \dfrac{3 + 1/x}{1 + 2/x}$. When x is large and positive, both $1/x$ and $2/x$ are small numbers close to 0, hence $\dfrac{3 + 1/x}{1 + 2/x}$ should be close to $\dfrac{3 + 0}{1 + 0} = 3$. Therefore $\lim_{x \to \infty} \dfrac{3x + 1}{x + 2} = 3$*

Limit as x Approaches $-\infty$

Only a function that is defined for all large negative numbers can have a limit as x approaches $-\infty$. Let f be such a function. If every large negative number x gives a value $f(x)$ close to a single number l, then we say that f has limit l as x approaches $-\infty$, and write $\lim_{x \to -\infty} f(x) = l$.

Example 101 *Let $f(x) = 2^x$. Then $\lim_{x \to -\infty} f(x) = 0$*

Example 102 *Let $f(x) = \dfrac{2^x - 2^{-x}}{2^x + 2^{-x}}$. When x is large and negative, the dominant term in the formula for f is 2^{-x}. Divide the numerator and denominator by 2^{-x} to get $f(x) = \dfrac{2^{2x} - 1}{2^{2x} + 1}$. Since 2^{2x} is close to 0 when x is large and negative, it follows that $\lim_{x \to -\infty} f(x) = \dfrac{0 - 1}{0 + 1} = -1$.*

Exercise 103

1. *By definition, integer(x) denotes the integer part of x. Thus integer$(5.79) = 5$ (simply throw away the decimal part), integer$(0.99) = 0$, integer$(-0.835) = 0$, integer$(-2.01) = -2$, etc. Consider the function $f(x) = $ integer(x).*

(a) Determine the following:

    i) $\lim_{x \to 2^+} f(x)$ ii) $\lim_{x \to 2^-} f(x)$ iii) $\lim_{x \to -5^+} f(x)$ iv) $\lim_{x \to -5^-} f(x)$

(b) Give a formula for $\lim_{x \to n^+} f(x)$ and $\lim_{x \to n^-} f(x)$ when n is a positive integer. Repeat when n is a negative integer.

2. Let $g(x) = x \, \text{integer}(x)$.

    (a) Determine the following:

        i) $\lim_{x \to 3^+} g(x)$ ii) $\lim_{x \to 3^-} g(x)$ iii) $\lim_{x \to -6^+} g(x)$ iv) $\lim_{x \to -6^-} g(x)$

    (b) Give a formula for $\lim_{x \to n^+} g(x)$ and $\lim_{x \to n^-} g(x)$ when n is an integer.

3. Draw the graph of $h(x) = \begin{cases} 3 - 2x & if \ x < 1 \\ 4 + x & if \ x \geq 1 \end{cases}$ then determine the following limits:

    i) $\lim_{x \to 1^+} h(x)$ ii) $\lim_{x \to 1^-} h(x)$ iii) $\lim_{x \to -1^+} h(x)$ iv) $\lim_{x \to -1^-} h(x)$

4. The expression $\lfloor x \rfloor$ denotes the largest integer that is smaller than or equal to x. For example,

$$\lfloor 6.21 \rfloor = 6, \qquad \lfloor 0.92 \rfloor = 0, \qquad \lfloor -3 \rfloor = -3, \qquad \lfloor 17 \rfloor = 17, \qquad \lfloor -0.2 \rfloor = -1.$$

    (a) Draw the graph of $f(x) = \lfloor x \rfloor$ for values of x between -5 and 4.

    (b) Determine $\lim_{x \to 2^+} f(x)$ and $\lim_{x \to 2^-} f(x)$.

    (c) Determine $\lim_{x \to -3^+} f(x)$ and $\lim_{x \to -3^+} f(x)$.

    (d) Give a formula for $\lim_{x \to n^+} f(x)$ and $\lim_{x \to n^-} f(x)$ when n is an integer.

5. Let $f(x) = \lfloor x \rfloor$, (the largest integer that is smaller than x), and $g(x) = x f(x)$.

    (a) Determine $\lim_{x \to -1^+} g(x)$ and $\lim_{x \to -1^-} g(x)$.

    (b) Determine $\lim_{x \to n^+} g(x)$ and $\lim_{x \to n^-} g(x)$ where n is an integer.

    (c) Determine $\lim_{x \to a^+} g(x)$ and $\lim_{x \to a^-} g(x)$ where a is a real number which is **not** an integer.

6. Let $f(x) = \dfrac{x^2 + 1}{x + 1}$, $x \neq -1$. Determine $\lim_{x \to -1^+} f(x)$ and $\lim_{x \to -1^-} f(x)$.

7. Let $g(x) = \dfrac{1}{x^2}$, $x \neq 0$. Determine $\lim_{x \to 0^+} g(x)$ and $\lim_{x \to 0^-} g(x)$.

8. Let $f(x) = \dfrac{3^x - 3^{-x}}{3^x + 3^{-x}}$. Determine $\lim\limits_{x \to \infty} f(x)$ and $\lim\limits_{x \to -\infty} f(x)$.

9. *Neural tissue can be electrically excited if the current across the cell membrane exceeds the threshold current. The threshold current I is related to the duration of time t that current flows across the membrane by the equation∞*

$$I(t) = \frac{a}{t} + b, \ t > 0$$

where a and b are positive constants. Sketch the graph of I then determine and give the physical interpretation of: (i) $\lim\limits_{x \to 0^+} I(t)$ *and (ii)* $\lim\limits_{x \to \infty} I(t)$.

10. ***A precise definition of an infinite limit****: Say we have to show that $f(x) = \dfrac{1}{(x-2)^2}$ has limit ∞ as x approaches 2. Then we have to convince every individual that $f(x)$ is large and positive when x is close to 2. The first question we have to address is: "what is a large positive number?" Unfortunately, the answer is: it depends on who you ask! That being the case we have to be flexible. We have to let an individual choose what he/she considers to be a large positive number. Let it be K. Then to convince him or her that numbers close to 2 give values that are large and positive, we simply have to produce a punctured interval $(c - \delta, c) \cup (c, c + \delta)$ with the property that every number x in the interval gives a value $f(x)$ that is bigger than K. To be in a position of convincing everybody that shows up, we should be able to do this for every **positive number** K. This suggests the following definition:*

> **A function $f(x)$ has limit ∞ as x approaches a number c if, given any positive number K, we can find a positive number δ such that $f(x) > K$ for all x in $(c - \delta, c) \cup (c, c + \delta)$.**

Alternatively:

> **A function $f(x)$ has limit ∞ as x approaches a number c if, given any positive number K, we can find a positive number δ such that $f(x) > K$ if $0 < |x - c| < \delta$.**

In the case of $f(x) = \dfrac{1}{(x-2)^2}$, let $K > 0$ be given. We have to find a positive number δ such that $\left| \dfrac{1}{(x-2)^2} \right| > K$ if $|x - 2| < \delta$. This inequality is satisfied by any x such that $|x - 2| < \frac{1}{\sqrt{k}}$. It follows that if we take any $\delta < \frac{1}{\sqrt{k}}$ then $\left| \dfrac{1}{(x-2)^2} \right| > K$ if $|x - 2| < \delta$.

11. *Complete the following precise definition of an infinite limit as x approaches a number from above: A function f has limit ∞ as x approaches a number c from above if given any $K > 0$ we can find a positive number δ such that ...*

Use the above precise definition to show that $f(x) = \dfrac{1}{\sqrt{x}}$ has limit ∞ as x approaches 0 from above.

12. **A precise definition of a limit as x approaches ∞ or $-\infty$:** *As we pointed out, a function $f(x)$ has limit l as x approaches ∞ if every number x that is large and positive gives a value $f(x)$ that is close to 0. Using the idea of a large positive number developed in Exercise 10 above, a precise way of saying this is the following:*

> **A function $f(x)$ has limit l as x approaches ∞ if given any positive number ε we can find a positive number K such that $|f(x) - l| < \varepsilon$ when $x > K$.**

> **A function $f(x)$ has limit m as x approaches $-\infty$ if given any positive number ε we can find a positive number N such that $|f(x) - m| < \varepsilon$ when $x < -N$.**

For an example, we show that $f(x) = \dfrac{2x}{x - 3}$ has limit 2 as x approaches ∞. To this end, let $\varepsilon > 0$. We have to find a positive number K such that $\left| \dfrac{2x}{x - 3} - 2 \right| < \varepsilon$ when $x > K$. We simplify the inequality to get

$$\left| \frac{6}{x - 3} \right| < \varepsilon$$

This is satisfied if $|x - 3| > \frac{6}{\varepsilon}$. We may assume that $x > 3$. Then $x - 3 > \frac{6}{\varepsilon}$, which simplifies to $x > 3 + \dfrac{6}{\varepsilon}$. Therefore if we take any $K > 3 + \frac{6}{\varepsilon}$ then $\left| \dfrac{2x}{x - 3} - 2 \right| < \varepsilon$ when $x > K$.

(a) Use the precise definition of a limit as x approaches ∞ to show that $f(x) = \dfrac{x}{3x + 1}$ has limit $\frac{1}{3}$ as x approaches ∞.

(a) Use the precise definition of a limit to show that $g(x) = \dfrac{4x - 1}{x + 2}$ has limit 4 as x approaches $-\infty$

Continuous Functions

Limits enable us to define continuous functions rigorously. On an intuitive level, a function f is continuous at a point c in its domain if its graph has no break at $(c, f(c))$. Thus one can plot the graph through $(c, f(c))$ without lifting the pencil off the graph paper. For example, any polynomial is continuous at every point c because polynomials have "smooth" graphs.

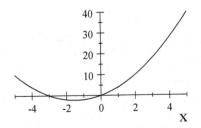 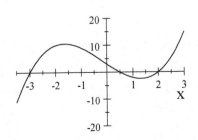

Smooth graphs

The figure below shows graphs that cannot be drawn without lifting the pencil from the graph paper at some point. The one to the left is the graph of the function f with formula

$$f(x) = \begin{cases} 2x + 1 & if \quad x < 1 \\ 3 - x & if \quad x \geq 1 \end{cases}$$

The other one is the graph of the function g defined by $g(x) = \lfloor x \rfloor x + 2$ where $\lfloor x \rfloor$ means the largest integer less than or equal to x.

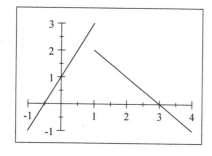 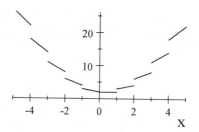

One must lift the pencil from the graph paper when plotting the graph of f through the point $c = 1$. As for g, one must lift the pencil from the graph paper at every non-zero integer point.

Clearly, a function f is continuous at a point c if numbers x close to c give values $f(x)$ close to the value $f(c)$ of the function at c, (else the graph would have a break at c). This is an imprecise way of saying that f has limit $f(c)$ as x approaches c and it suggests the following definition:

Definition 104 *A function f is continuous at a point c if c is in its domain and* $\lim_{x \to c} f(x) = f(c)$. *It is continuous on an interval I if it is continuous at every point c in I.*

Examples of functions that are continuous on every interval I are:

1. The polynomial functions $f(x) = a_n x^n + a_{n-1} x^{n-1} + \cdots + a_1 x + a_0$.

2. The exponential functions $g(x) = a^x$ where $a > 0$.

3. The trigonometric functions $h(x) = \sin x$ and $w(x) = \cos x$.

Rational functions like $f(x) = \dfrac{1}{x}$ and $g(x) = \dfrac{x + 3}{(x - 1)(x + 2)}$ generally fail to be continuous at the points where their denominators are zero. For example, f is not continuous at 0 but it is continuous everywhere else. Likewise, g is continuous at every point except 1 and -2. Their graphs are given below.

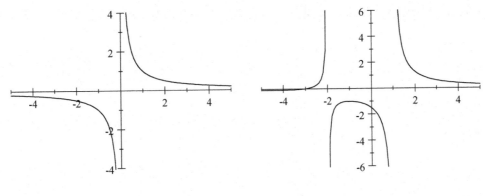

$$\text{Graph of } f(x) = \frac{1}{x} \qquad\qquad \text{Graph of } g(x) = \frac{x+3}{(x-1)\,(x+2)}$$

Exercise 105

1. *Draw the graph of the function f which has value 1 when $x = 0$ and value $\dfrac{|x|}{x}$ when $x \neq 0$. Where is it not continuous?*

2. *Let $f(x) = \text{integer}(x)$ where $\text{integer}(x)$ denotes the integer part of x. Draw the graph of f on the interval $[-4.5, 5.5]$ then give the points in $[-4.5, 5.5]$ where f is not continuous.*

Since we can plot the graph of a continuous function f without lifting the pencil off the paper, such a function has what is called the **intermediate value property**; meaning that it does not skip numbers between any two of its values. Thus, if it has value M at a number u, a value N different from M at another number w, then if Y is any number between M and N, there is a number s between u and w such that $f(s) = Y$. We used this very argument, (see Problem 5 on page 50), to locate solutions of equations. Indeed, if f is a continuous function and u, v are numbers such that $f(u) < 0$ and $f(v) > 0$ then there must be a number s between u and v such that $f(s) = 0$. More precisely, the equation $f(x) = 0$ must have a solution between u and v.

Test your skills - 5,   (50 minutes)

1. Let $f(x) = 1 + 2x^2$ and $g(x) = \sqrt{x+1}$. Determine the following limits:

   (a) $\displaystyle\lim_{x \to 5} f(x)$ (b) $\displaystyle\lim_{x \to 3} g(x)$ (c) $\displaystyle\lim_{x \to 8} (f - g)(x)$.

2. If $\displaystyle\lim_{x \to 4} f(x) = 1$ and $\displaystyle\lim_{x \to 4} g(x) = -3$, what is: (a) $\displaystyle\lim_{x \to 4}(g(x) + 5)$, (b) $\displaystyle\lim_{x \to 4} x f(x)$, and (c) $\displaystyle\lim_{x \to 4}\left(\frac{f(x)}{g(x) - 1}\right)$?

3. Given that $\displaystyle\lim_{x \to 5} f(x)$ exists and $\displaystyle\lim_{x \to 5}\left(\frac{f(x) - 5}{x - 2}\right) = 6$, determine the exact value of $\displaystyle\lim_{x \to 5} f(x)$.

4. Given that $\lim\limits_{x \to 0} \dfrac{f(x)}{x} = 6$, show that $\lim\limits_{x \to 0} f(x)$ exists and is 0

5. Let $f(x) = x^2 + 2x$ and $\varepsilon > 0$. Find $\delta > 0$ such that $|f(x) - 3| < \varepsilon$ if $0 < |x - 1| < \delta$.

6. The function f is defined on $[-5, 0]$ by

$$f(x) = \begin{cases} 2x + 10 & if \quad -5 \le x \le -3 \\ -3x - x^2 & if \quad -3 < x \le 0 \end{cases} .$$

Draw its graph and determine the following limits: (if a limit does not exist, say so and give a reason).

(i) $\lim\limits_{x \to -3+} f(x)$ (ii) $\lim\limits_{x \to -3-} f(x)$ (iii) $\lim\limits_{x \to -3} f(x)$ (iv) $\lim\limits_{x \to -4} f(x)$

7. The function f has formula $f(x) = \dfrac{x - 6}{x^2 - 16x + 60}$. Give the numbers c (if any) such that f is not continuous at c, and give reasons for your choices.

8. Determine the following limits:

(a) $\lim\limits_{x \to 0} \left(\dfrac{(1 - \cos x) \sin x}{x^3} \right)$, (b) $\lim\limits_{x \to 4} \left(\dfrac{4 - x}{\sqrt{x^2 + 9} - 5} \right)$

Chapter 3   MORE DERIVATIVES

We now address the derivatives of more functions, namely, the trigonometric functions $s(x) = \sin x$ and $c(x) = \cos x$, and the exponential function $v(x) = e^x$. We start with $\sin x$. Using trigonometric identities (proved in the Appendix), gives

$$\frac{s(x+h) - s(x)}{h} = \frac{\sin(x+h) - \sin x}{h} = \frac{\sin x \cos h + \cos x \sin h - \sin x}{h}$$

$$= \frac{(\cos h - 1)\sin x + \cos x \sin h}{h} = \frac{(\cos h - 1)}{h}\sin x + \frac{\sin h}{h}\cos x$$

You should have shown, in Exercise 4 on page 71, and Exercise 6 on page 73, that

$$\lim_{h \to 0}\frac{\sin h}{h} = 1 \quad \text{and} \quad \lim_{h \to 0}\frac{(\cos h - 1)}{h} = 0.$$

It follows from the sum and constant multiple rules for limits that

$$\lim_{h \to 0}\frac{s(x+h) - s(x)}{h} = 0 \cdot \sin x + 1 \cdot \cos x = \cos x$$

In other words, the derivative of $s(x) = \sin x$ is $s'(x) = \cos x$.

In the case of $c(x) = \cos x$,

$$\frac{c(x+h) - c(x)}{h} = \frac{\cos(x+h) - \cos x}{h} = \frac{\cos x \cos h - \sin x \sin h - \cos x}{h}$$

$$= \frac{(\cos h - 1)\cos x - \sin x \sin h}{h} = \frac{(\cos h - 1)}{h}\cos x - \frac{\sin h}{h}\sin x$$

It follows from the sum and constant multiple rules for limits that

$$\lim_{h \to 0}\frac{c(x+h) - c(x)}{h} = 0 \cdot \cos x - 1 \cdot \sin x = -\sin x$$

Therefore the derivative of $c(x) = \cos x$ is $c'(x) = -\sin x$. **Note the sign.**

For the exponential function $v(x) = e^x$,

$$\frac{v(x+h) - v(x)}{h} = \frac{e^{x+h} - e^x}{h} = \frac{e^x e^h - e^x}{h} = \frac{(e^h - 1)}{h}e^x$$

Exercise 3 on page 64 should have convinced you that $\lim_{h \to 0}\dfrac{(e^h - 1)}{h} = 1$. It follows from the constant multiple rule for limits that

$$\lim_{h \to 0}\frac{v(x+h) - v(x)}{h} = 1 \cdot e^x$$

Therefore the derivative of $v(x) = e^x$ is $v'(x) = e^x$. The following is a summary of the facts we have gathered so far

Function	x^n	$\sin x$	$\cos x$	e^x
Derivative	nx^{n-1}	$\cos x$	$-\sin x$	e^x

To this we add the sum, the constant multiple and the general power rules:

Function	$f(x) + g(x)$	$kf(x)$	$[f(x)]^n$	
Derivative	$f'(x) + g'(x)$	$kf'(x)$	$n\,[f(x)]^{n-1} \cdot f'(x)$	

Exercise 106

1. *Calculate the derivative of each function, using the rules for derivatives which we have introduced so far. When possible, simplify your answer.*

   (a) $f(x) = 3x + 4\sin x$.

   (b) $g(x) = 4\sin x + 5\cos x$.

   (c) $h(x) = \frac{\sin x}{7} = \frac{1}{7}\sin x$

   (d) $u(x) = \frac{2\cos x}{3} - \frac{4\sin x}{5}$

   (e) $v(x) = x^2 - \pi\cos x$.

   (f) $w(x) = \pi\cos x - \frac{2\sin x}{5} + \frac{4}{x}$

   (g) $s(x) = 2x^2 - 4x^3 - 3\sin x$

   (h) $q(x) = 4 - 3e^x$

   (i) $r(x) = 4x^{-1/2} + +x^2 - 6 + \dfrac{1}{x}$

   (j) $p(x) = 3x^3 - x^{3/2} + x - 19$

   (k) $f(x) = (2 + 3\sin x)^4$

   (l) $g(x) = \frac{1}{28}\,(4x + 3)^7$

   (m) $h(x) = \sqrt{4 + 3x^2}$

   (n) $v(x) = \frac{1}{9(3x+2)^3}$

   (0) $u(x) = \sqrt{4 - 3\cos x}$

   (p) $w(x) = \frac{5}{1+3e^x}$

   (q) $f(x) = \frac{2}{4x+3}$

   (r) $g(x) = \sqrt{2e^x + 3}$

2. *Find the most general antiderivative of the given function. In part (f), a is a non-zero constant.*

   a) $f(x) = 4 - 3e^x$

   b) $g(x) = 2\cos x - 3\sin x + 1$

   c) $w(x) = (x + 1)^3$

   d) $h(x) = 24\,(6x + 1)^3$

   e) $u(x) = (2x - 1)^4$

   f) $v(x) = (ax + 1)^{17}$

3. *You wish to measure the height of a building. You stand 42 feet from the base of the building, measure the angle of elevation of the top of the building and find it to be 68°. What level of accuracy should you aim for in measuring the angle so that the estimated percentage error in the height of the building does not exceed 3%?*

4. *Sketch on the same axes, the graphs of $f(x) = \cos x$, (x in radians), and $g(x) = x$. Use the sketches to determine an approximate root of the equation $\cos x = x$ then use Newton's method to calculate it accurately to 4 decimal places. (Hint: Write $\cos x = x$ as $x - \cos x = 0$.)*

The Product Rule and the Quotient Rule for Derivatives

The product rule for derivatives enables us to determine the derivative of a product $f(x)g(x)$ if we know the derivatives of $f(x)$ and $g(x)$. Contrary to what you would hope for, the derivative of $f(x)g(x)$ is NOT the product $f'(x)g'(x)$ of the individual derivatives. Here is a counterexample:

Example 107 *Let* $f(x) = x$ *and* $g(x) = 3x + 5$. *Their product is* $f(x) = f(x)g(x) = 3x^2 + 5x$ *with derivative* $f'(x) = 6x + 5$. *On the other hand,* $f'(x) = 1$, $g'(x) = 3$, *and the product of their derivatives is* $f'(x)g'(x) = (1)(3) = 3$ *which bears no resemblances to the derivative of* $f(x)g(x)$.

In fact the product rule states that if $f(x)$ and $g(x)$ have derivatives then:

The derivative of $f(x)g(x)$ **is** $f'(x)g(x) + g'(x)f(x)$.

Applying it to the product of $f(x) = x$ and $g(x) = 3x + 5$ gives the derivative of $f(x)g(x)$ to be $(1)(3x+5) + (3)(x) = 3x + 5 + 3x = 6x + 5$ as it should be.

To prove it, we have to show that

$$\lim_{h \to 0} \frac{f(x+h)g(x+h) - f(x)g(x)}{h} = f'(x)g(x) + g'(x)f(x).$$

To this end consider the rectangles in the figure below. The lengths of OA, OB, OG and OF are $f(x)$, $f(x+h)$, $g(x)$, and $g(x+h)$ respectively. It follows that $f(x+h)g(x+h)$ is the area of rectangle OBDF and $f(x)g(x)$ is the area of rectangle OAHG.

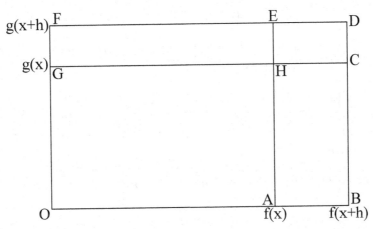

Therefore

$$f(x+h)g(x+h) - f(x)g(x) = \text{Area of ABCH} + \text{Area of EFGH} + \text{Area of CDEH}$$

This implies that

$$\frac{f(x+h)g(x+h) - f(x)g(x)}{h} = \frac{[f(x+h) - f(x)]\,g(x)}{h} + \frac{[g(x+h) - g(x)]\,f(x)}{h}$$

$$+ \frac{[f(x+h) - f(x)]\,[g(x+h) - g(x)]}{h}$$

the right hand side may be written as

$$\frac{f(x+h)-f(x)}{h}g(x) + \frac{g(x+h)-g(x)}{h}f(x) + \frac{f(x+h)-f(x)}{h}[g(x+h)-g(x)]$$

Since $\dfrac{f(x+h)-f(x)}{h}$, $\dfrac{g(x+h)-g(x)}{h}$, and $[g(x+h)-g(x)]$ have limits $f'(x)$, $g'(x)$ and 0 as respectively as h approaches 0, it follows that

$$\frac{f(x+h)-f(x)}{h}g(x) + \frac{g(x+h)-g(x)}{h}f(x) + \frac{f(x+h)-f(x)}{h}[g(x+h)-g(x)]$$

has limit $f'(x)g(x) + g'(x)f(x) + f'(x) \cdot 0 = f'(x)g(x) + g'(x)f(x)$ as h approaches 0. In other words,

$$\lim_{h \to 0} \frac{f(x+h)g(x+h) - f(x)g(x)}{h} = f'(x)g(x) + g'(x)f(x).$$

Example 108 *Let $u(x) = 3x\sin x$. We may write it as the product of $f(x) = 3x$ and $g(x) = \sin x$. Since $f'(x) = 3$ and $g'(x) = \cos x$, it follows that*

$$u'(x) = (3)(\sin x) + (\cos x)(3x) = 3\sin x + 3x\cos x.$$

Example 109 *Let $v(x) = (\sin x)^2$, (also written as $\sin^2 x$). Write it as the product $v(x) = (\sin x)(\sin x)$. Then*

$$v'(x) = (\cos x)(\sin x) + (\cos x)(\sin x) = 2\sin x\cos x$$

Example 110 *Write $f(x) = e^{2x}$ as a product $f(x) = g(x)g(x)$ where $g(x) = e^x$. Since the derivative of $g(x) = e^x$ is $g'(x) = e^x$, it follows that*

$$f'(x) = (e^x)e^x + e^x(e^x) = 2e^{2x}.$$

We can now deduce the derivative of $u(x) = e^{3x}$. Write it as $u(x) = f(x)g(x)$. By the product rule, its derivative is

$$u'(x) = (2e^{2x})e^x + e^{2x}(e^x) = 2e^{3x} + e^{3x} = 3e^{3x}.$$

What would be your guess for the derivative of $w(x) = e^{4x}$ and $u(x) = e^{nx}$ where n is a positive integer?

Example 111 *To find the derivative of $f(x) = 4e^{3x}\cos x$ we use the product rule to get*

$$f'(x) = 4(3e^{3x})\cos x + 4e^{3x}(-\sin x) = 4e^{3x}(3\cos x - \sin x)$$

Example 112 *Let f and g be given functions with $f(1) = 6$, $f'(1) = -0.9$, $g(1) = 0.8$, and $g'(1) = 7$. Let $h(x) = f(x)g(x)$. Then $h'(1) = f'(1)g(1) + g'(1)f(1) = (-0.9)(0.8) + (7)(6) = 41.28$*

Example 113 *To find the derivative of $f(x) = 3x^2 e^x \sin x$, regard the function as the product $(3x^2 e^x)\sin x$. The derivative of $3x^2 e^x$ is $6xe^x + 3x^2 e^x$. Therefore*

$$f'(x) = (6xe^x + 3x^2 e^x)\sin x + (3x^2 e^x)\cos x = 3xe^x(2\sin x + x\sin x + x\cos x)$$

The Quotient Rule for Derivatives

The quotient rule enables us to determine the derivative of a quotient $\frac{f(x)}{g(x)}$ if we know the derivatives of $f(x)$ and $g(x)$. As the next example shows, the derivative of a quotient is, in general, NOT the quotient of the corresponding derivatives.

Example 114 Let $f(x) = x$ and $g(x) = x+1$. Their derivatives are $f'(x) = 1$ and $g'(x) = 1$. The quotient of the derivatives is $\frac{f'(x)}{g'(x)} = \frac{1}{1} = 1$. But this is NOT the derivative of the quotient $f(x) = \frac{f(x)}{g(x)} = \frac{x}{x+1}$, which we found to be $f'(x) = \frac{1}{(x+1)^2}$, (see Exercise 5 on page 14).

The quotient rule states that if $f(x)$ and $g(x)$ have derivatives and $g'(x) \neq 0$ then:

$$\boxed{\text{The derivative of } \frac{f(x)}{g(x)} \text{ is } \frac{f'(x)g(x) - g'(x)f(x)}{[g(x)]^2}}$$

Applying it to the quotient $\frac{f(x)}{g(x)}$ when $f(x) = x$ and $g(x) = x+1$ gives the expected result for the derivative of $\frac{f(x)}{g(x)} = \frac{x}{x+1}$ which is

$$\frac{f'(x)g(x) - g'(x)f(x)}{[g(x)]^2} = \frac{1 \cdot (x+1) - 1 \cdot (x)}{(x+1)^2} = \frac{x+1-x}{(x+1)^2} = \frac{1}{(x+1)^2}$$

One way of proving the rule is to write $\frac{f(x)}{g(x)}$ as $f(x)[g(x)]^{-1}$ then appeal to the general power rule and the product rule. By the general power rule, the derivative of $[g(x)]^{-1}$ is $-[g(x)]^{-2}g'(x)$. Therefore, by the product rule, the derivative of $f(x)[g(x)]^{-1}$ is

$$f'(x)[g(x)]^{-1} + \left(-[g(x)]^{-2}g'(x)\right)f(x) = \frac{f'(x)}{g(x)} - \frac{g'(x)f(x)}{[g(x)]^2} = \frac{f'(x)g(x) - g'(x)f(x)}{[g(x)]^2}$$

Example 115 To find the derivative of $g(x) = e^{-x}$, we write the function as a quotient: $g(x) = \frac{1}{e^x}$. By the quotient rule,

$$g'(x) = \frac{0 \cdot e^x - 1 \cdot e^x}{(e^x)^2} = -\frac{e^x}{(e^x)^2} = -e^{-x}$$

Alternatively, write e^{-x} as $(e^x)^{-1}$ then use the general power rule.

Example 116 To find the derivatives of $t(x) = \tan x$, $w(x) = \csc x$, $f(x) = \cot x$ and $g(x) = \sec x$:

Starting with $t(x)$, write $\tan x$ as the quotient $\frac{\sin x}{\cos x}$. Then by the quotient rule,

$$t'(x) = \frac{(\cos x)(\cos x) - (\sin x)(-\sin x)}{(\cos x)^2} = \frac{\cos^2 x + \sin^2 x}{\cos^2 x} = \frac{1}{\cos^2 x} = \sec^2 x$$

Next, write w as a quotient: $w(x) = \dfrac{1}{\sin x}$. *By the quotient rule,*

$$w'(x) = \frac{0 \cdot \sin x - 1 \cdot \cos x}{(\sin x)^2} = -\frac{\cos x}{(\sin x)^2} = -\frac{\cos x}{\sin x} \cdot \frac{1}{\sin x} = -\cot x \csc x$$

In the case of $\cot x$, *write it as the quotient* $\dfrac{\cos x}{\sin x}$. *Then by the quotient rule,*

$$f'(x) = \frac{-\sin x \cdot \sin x - \cos x \cdot \cos x}{(\sin x)^2} = -\frac{1}{(\sin x)^2} = -\csc^2 x$$

Lastly, if we write $g(x) = \dfrac{1}{\cos x}$, *the quotient rule gives*

$$g'(x) = \frac{0 \cdot \cos x - 1 \cdot (-\sin x)}{(\cos x)^2} = \frac{\sin x}{(\cos x)^2} = \frac{\sin x}{\cos x} \cdot \frac{1}{\cos x} = \tan x \sec x.$$

We now have the following additional derivatives. Use them, and the earlier ones, together with the rules of derivatives we have introduced so far, to do the next set of exercises.

Function	$\tan x$	$\cot x$	$\sec x$	$\csc x$
Derivative	$\sec^2 x$	$-\csc^2 x$	$\sec x \tan x$	$-\csc x \cot x$

Exercise 117

1. Find the derivative of each function and simplify when possible.

   (a) $u(x) = 4x^2 \cos x$ (b) $v(x) = 5e^x \sin x$ (c) $w(x) = x^{-3} e^x$

   (d) $f(x) = 5x^2 - 3x \sin x$ (e) $u(x) = \frac{x}{2+x}$ (f) $h(x) = \frac{e^x}{4+e^x}$

   (g) $h(x) = 3 - 2\sin x \cos x$ (h) $v(x) = \frac{\cos x}{4 - 3\sin x}$ (i) $f(x) = \frac{x^2 - 1}{x^2 + 1}$

   j) $h(x) = 3 - x \cos x$ k) $u(x) = 7 - \tan^2 x$ l) $g(x) = \frac{\sin x - 1}{\cos x + 3}$

   m) $r(x) = 7 + x^3 \cot x$ n) $s(x) = 4x \csc x - 3x^3$ o) $v(x) = \cos^2 x - x^2$

   p) $f(x) = 2\sec^2 x - 5x^3$ q) $r(x) = 3x \tan x + \sec x$ r) $4x^3 e^x \cos x$

2. Show that if $g(x) = \frac{2}{3}x^3 e^x \sec x$ then $g'(x) = \frac{2}{3}x^2 e^x (x + 3 + x \tan x) \sec x$

3. Let $c(x) = \dfrac{e^x + e^{-x}}{2}$, $s(x) = \dfrac{e^x - e^{-x}}{2}$ and $t(x) = \dfrac{e^x - e^{-x}}{e^x + e^{-x}} = \dfrac{s(x)}{c(x)}$. Show that

   (a) $c'(x) = s(x)$ (b) $s'(x) = c(x)$ (c) $t'(x) = \dfrac{4}{(e^x + e^{-x})^2} = \dfrac{1}{c^2(x)}$

4. Determine the critical point, (there is only one), of $f(x) = \dfrac{2}{x^2 + 2x + 2}$, establish its nature, then sketch the graph.

5. The function $g(x) = \frac{6(x+1)}{x^2+3}$ has two critical points. Determine them, establish their nature, then sketch its graph.

6. Let $f(x) = \frac{x+3}{\sqrt{2x+3}}$. Show that $f'(x) = \frac{x}{(2x+3)^{3/2}}$.

7. Let $g(x) = x\sqrt{1+x^2}$. Show that $g'(x) = \frac{1+2x^2}{\sqrt{1+x^2}}$.

8. Let $f(x) = x^{-n}$. Use the quotient rule to show that $f'(x) = -nx^{-n-1}$.

9. Give the most general antiderivative of each function. (Check your answer by taking the derivative of your antiderivative.)

a) $f(x) = 4\sin x + 3\cos x - \pi$ b) $g(x) = 2 - x + \csc^2 x$

c) $h(x) = 3 - \frac{4}{5}\sec^2 x$ d) $u(x) = 5e^x - 2\sqrt{x} + 7$

e) $v(x) = 5\cot x \csc x - x^2$ f) $w(x) = 4 - \frac{2}{x^4} + e^{-x}$

g) $h(x) = e^x - e^{-x}$ h) $f(x) = 6x^2 - 5\sec x \tan x$

i) $g(x) = 2e^x + 3e^{-x}$ j) $w(x) = 5x^{-2/3} + 3\sec^2 x - 1$

The Chain Rule for Derivatives

We have already met the chain rule, (see page 19), in the form of the power rule:

$$\text{If } f(x) = [g(x)]^r \text{ then } f'(x) = r\,[g(x)]^{r-1} \times g'(x).$$

Here, $g(x)$ is any differentiable function and r is any real number. It may be easier to remember it in the form

The derivative of (**)^r is r(**   **)$^{r-1}\times$Derivative of what is in (**   **)**

You may put any differentiable expression inside (   ). Of course if you put in x, you get what you already know, that the derivative of x^n is nx^{n-1}. To get something new, you have to put in expressions like $3x + 1$, $x^2 - 1$, $4x^2$, etc. This extends to the trigonometric and the exponential functions as follows:

1. **The derivative of** $\sin($ **) is**

$$[\cos(\quad)] \times \textbf{Derivative of what is inside } (\quad).$$

(If you put x inside (   ), you get what you already know, that the derivative of $\sin x$ is $\cos x$.)

Example 118

(a) Given $f(x) = \sin 6x$, view it as $f(x) = \sin(6x)$.

Then $f'(x) = [\cos 6x] \times (6) = 6\cos 6x.$

(b) Let $g(x) = \sin(2\pi x - 1)$. Then $g'(x) = 2\pi \cos(2\pi x - 1).$

2. **The derivative of** $\cos(\quad)$ **is**

$$[-\sin(\quad)] \times \textbf{Derivative of what is inside } (\quad).$$

Example 119

(a) Given $f(x) = \cos\frac{1}{2}x$, view it as $f(x) = \cos\left(\frac{1}{2}x\right)$.

Then $\quad f'(x) = \left[-\sin\left(\frac{1}{2}x\right)\right] \times \left(\frac{1}{2}\right) = -\frac{1}{2}\sin\frac{1}{2}x$.

(b) Let $g(x) = \cos(x^3 + 2x)$. Then $g'(x) = -(3x^2 + 2)\sin(x^3 + 2x)$

3. **The derivative of** $\tan(\quad)$ **is**

$$\left[\sec^2(\quad)\right] \times \textbf{Derivative of what is inside } (\quad).$$

Example 120

(a) Given $f(x) = \tan 4x$, view it as $f(x) = \tan(4x)$.

Then $\quad f'(x) = \left[\sec^2(4x)\right] \times (4) = 4\sec^2 4x$.

(b) Let $g(x) = \tan\left(3x - \frac{1}{x}\right)$. Then $g'(x) = \left(3 + \frac{1}{x^2}\right)\sec^2\left(3x - \frac{1}{x}\right)$.

4. **The derivative of** $\csc(\quad)$ **is**

$$[-\csc(\quad)\cot(\quad)] \times \textbf{Derivative of what is inside } (\quad).$$

Example 121

(a) Given $f(x) = \csc 7x$, view it as $f(x) = \csc(7x)$.

Then $\quad f'(x) = [-\csc(7x)\cot(7x)] \times (7) = -7\csc 7x \cot 7x$.

(b) Let $g(x) = \csc(x^2)$. Then $g'(x) = -2x\csc(x^2)\cot(x^2)$

5. **The derivative of** $\sec(\quad)$ **is**

$$[\sec(\quad)\tan(\quad)] \times \textbf{Derivative of what is inside } (\quad).$$

Example 122

(a) Given $f(x) = \sec 9x$, view it as $f(x) = \sec(9x)$.

Then $\quad f'(x) = [\sec(9x)\tan(9x)] \times (9) = 9\sec 9x \tan 9x$.

(b) Let $g(x) = \sec(x + \frac{3}{x^2})$. Then $g'(x) = \left(1 - \frac{6}{x^3}\right)\sec(x + \frac{3}{x^2})\tan(x + \frac{3}{x^2})$

6. **The derivative of** $\cot(\quad)$ **is**

$$\left(-\csc^2(\quad)\right) \times \textbf{Derivative of what is inside}(\quad).$$

Example 123

(a) If $f(x) = \cot(\tfrac{1}{2}x)$ then $f'(x) = \left[-\csc^2(\tfrac{1}{2}x)\right] \times \left(\tfrac{1}{2}\right) = \tfrac{1}{2}\csc^2\tfrac{1}{2}x.$

(b) If $g(x) = \cot\left(5x - \tfrac{3}{x}\right)$ then $g'(x) = -(5 + \tfrac{3}{x^2})\csc^2\left(5x - \tfrac{3}{x}\right).$

7. **The derivative of** $e^{(\quad)}$ **is**

$$\left[e^{(\quad)}\right] \times \textbf{Derivative of what is inside}(\quad).$$

Example 124

(a) Given $f(x) = e^{4x}$, view it as $f(x) = e^{(4x)}$.

Then $f'(x) = \left[e^{(4x)}\right] \times (4) = 4e^{4x}.$

(b) If $g(x) = e^{\sin x}$ then $g'(x) = \left[e^{\sin x}\right] \times (\cos x) = e^{\sin x}\cos x$

The following is a summary of what we have described:

	Function	Derivative
1.	$(\quad)^n$	$n(\quad)^{n-1} \times$ Derivative of what is inside $(\quad)$
2.	$\sin(\quad)$	$[\cos(\quad)] \times$ Derivative of what is inside $(\quad)$
3.	$\cos(\quad)$	$[-\sin(\quad)] \times$ Derivative of what is inside $(\quad)$
4.	$e^{(\quad)}$	$\left[e^{(\quad)}\right] \times$ Derivative of what is inside $(\quad)$
5.	$\tan(\quad)$	$[\sec^2(\quad)] \times$ Derivative of what is inside $(\quad)$
6.	$\cot(\quad)$	$[-\csc^2(\quad)] \times$ Derivative of what is inside $(\quad)$
7.	$\sec(\quad)$	$[\sec(\quad)\tan(\quad)] \times$ Derivative of what is inside $(\quad)$
8.	$\csc(\quad)$	$[-\cot(\quad)\csc(\quad)] \times$ Derivative of what is inside $(\quad)$

$$(3.1)$$

Quite often, you may have to use the chain rule more than once. For an example, take $f(x) = \sqrt{\sin \pi x}$. If we write $\sqrt{\sin \pi x}$ as $(\sin \pi x)^{1/2}$ then the derivative of f is

$$\frac{1}{2}(\sin \pi x)^{-1/2} \cdot (\text{Derivative of } \sin \pi x)$$

We have to use the chain rule a second time to find the derivative of $\sin \pi x$. This turns out to be $(\cos \pi x) \cdot \pi$. Therefore

$$f'(x) = \frac{1}{2}(\sin \pi x)^{-1/2} \cdot (\cos \pi x) \cdot \pi = \frac{\pi}{2}(\sin \pi x)^{-1/2}(\cos \pi x)$$

94

Exercise 125

1. Determine the derivative of each function and simplify it when possible.

(a) $f(x) = 2\sin 3x$

(b) $w(x) = 2\sin 3x - 3\cos 2x$

(c) $v(x) = 2\sec 4x - 7$

(d) $f(x) = 2x^2 + 4\csc\frac{1}{2}x$

(e) $g(x) = e^{-x^2} - 3x^3$

(f) $h(x) = \frac{2\pi}{3}\tan \pi x$

(g) $u(x) = 4\cos 3x - 3\sec\left(\frac{1}{x}\right)$

(h) $u(x) = 2e^{\cot x} - 5$

(i) $f(x) = 4 - 7\tan^3 \pi x$

(j) $g(x) = 3x^2 \tan\sqrt{x}$

(k) $h(x) = \frac{3}{4}x^2 \sin 2x$

(l) $u(x) = \sec^2 3x - 1$

(m) $v(x) = 1 - \sqrt{3\sin x + 4}$

(n) $w(x) = \frac{4}{\sqrt{\cos 3x + 2}}$

(o) $f(x) = e^{2x}(\sin 3x - \cos 3x)$

(p) $g(x) = \cot^3 2x + 3x - 5$

(q) $h(x) = \tan 2x \csc^2\left(\frac{2x}{3}\right)$

(r) $u(x) = e^{\sqrt{1+\sec^2 2x}} - \pi x$

The Chain Rule as a Rule for Computing Derivatives of Compositions

The chain rule, in its general form, is a device for computing derivatives of compositions of functions. (You may need to review compositions of functions in the appendix.) It provides us with a method of computing the derivative of a composition $f \circ g$ of differentiable functions f and g if we know the derivative of g at a point x and the derivative of f at $g(x)$. More precisely, it states that the derivative of $f \circ g$ at x is given by the formula

$$(f \circ g)'(x) = f'(g(x)) \cdot g'(x).$$

(An outline of a proof is at the end of this section.) The results in table 3.1 on page 93 are all special cases of this general form. Here is an illustration:

Example 126 To find the derivative of $h(x) = \sec(x^2 + \pi x)$ using the formula $(f \circ g)'(x) = f'(g(x)) \cdot g'(x)$, write $h(x)$ as the composition $h(x) = f \circ g(x)$ where $g(x) = x^2 + \pi x$ and $f(x) = \sec x$. Then $f'(g(x)) = \sec(x^2 + \pi x)\tan(x^2 + \pi x)$ and $g'(x) = 2x + \pi$. Therefore

$$h'(x) = f'(g(x)) \cdot g'(x) = (2x + \pi)\left[\sec\left(x^2 + \pi x\right)\tan\left(x^2 + \pi x\right)\right]$$

For comparison to the result of using formula (7) in table 3.1, write $h(x)$ as $\sec(\quad)$. The term "$[\sec(\quad)\tan(\quad)]$" is $[\sec(x^2 + \pi x)\tan(x^2 + \pi x)]$, which corresponds to $f'(g(x))$. The term "Derivative of what is inside $(\quad)$" is $(2x + \pi)$, which corresponds to $g'(x)$.

Table 3.1 cannot be used to find the required derivative in the next example:

Example 127 To find $(f \circ g)'(2)$ given that $g(2) = 3$, $g'(2) = 8$ and $f'(3) = -\frac{3}{4}$.

By the chain rule, $(f \circ g)'(2) = f'(g(2))g'(2) = f'(3)g'(2) = -\frac{3}{4} \cdot 8 = -6$

Exercise 128

1. *In the figure below, you are given the graph of a function f and the tangents to its graph at the two points with x-coordinates -2 and 1.5. Let $g(x) = 0.1 - 0.7x$. Use the chain rule and the given tangents to estimate $(f \circ g)'(-2)$ and $(f \circ g)'(3)$.*

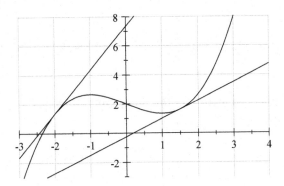

2. *Use the chain rule, in the form $(f \circ g)'(x) = f'(g(x)) \cdot g'(x)$, to find the derivative of the given function. In each case, you should write the given function as a composition of functions f and g of your choice.*

   a) $h(x) = \frac{2}{\sqrt{x^2+2}}$ b) $u(x) = \left(x^2 + \frac{1}{x}\right)^4$ c) $v(x) = \frac{2}{\cos x + 5} - 3\cos^2 x$

Another way of applying the chain rule

We illustrate with examples:

Example 129 *Let $h(x) = \tan(\pi x^3)$. We reduce it to a more familiar function by introducing a new variable $u = \pi x^3$. Then $h(x) = \tan u$. Let $f(u) = \tan u$. We can now write $h(x) = f(u) = \tan u$.*

The derivative of f, (its variable is u), is $\dfrac{df}{du} = \sec^2 u$.

The derivative of u, (its variable is x), is $\dfrac{du}{dx} = 3\pi x^2$.

The chain rule states that the derivative of h is

$$\frac{dh}{dx} = \frac{df}{du} \cdot \frac{du}{dx} = \left(\sec^2 u\right) \cdot \left(3\pi x^2\right) = 3\pi x^2 \sec^2\left(\pi x^3\right)$$

As you have probably figured out already, this is similar to writing $h(x)$ as a composition $h(x) = f \circ g(x)$ where $g(x) = \pi x^3$ and $f(x) = \tan x$. The only difference here is that we have written $g(x)$ as a new variable u; therefore we must write $f(g(x))$ as $f(u)$. In the expression $\dfrac{df}{du} \cdot \dfrac{du}{dx}$ for the derivative of h, $\dfrac{df}{du}$ is really $f'(g(x))$ and $\dfrac{du}{dx}$ is $g'(x)$.

Example 130 *Let* $h(x) = 4 \csc \left(\frac{\pi}{2x} \right) + \sec \left(\frac{\pi}{2x} \right)$. *To find its derivative, we let* $u = \frac{\pi}{2x}$. *Then* $h(x) = f(u) = 4 \csc u + \sec u$. *Clearly,* $\frac{df}{du} = -4 \csc u \cot u + \sec u \tan u$ *and* $\frac{du}{dx} = -\frac{\pi}{2x^2}$. *By the chain rule, the derivative of* h *is*

$$
\begin{aligned}
\frac{dh}{dx} &= \frac{df}{du} \cdot \frac{du}{dx} = \left(-4 \csc u \cot u + \sec u \tan u \right) \left(-\frac{\pi}{2x^2} \right) \\
&= \frac{\pi}{2x^2} \left[4 \csc \left(\frac{\pi}{2x} \right) \cot \left(\frac{\pi}{2x} \right) - \sec \left(\frac{\pi}{2x} \right) \tan \left(\frac{\pi}{2x} \right) \right]
\end{aligned}
$$

Remark 131 *The chain rule, in the form of table 3.1, is easier to use in reverse to guess antiderivatives of functions.*

Exercise 132 *Find the derivative of each function:*

i) $h(x) = 7e^{3x^2}$ *ii)* $v(x) = \frac{4}{\sqrt{\tan x}}$ *iii)* $w(x) = e^{3x} \tan 3x$ *iv)* $z(x) = (2 + x^3)^{\frac{1}{3}}$

Sketch of a Proof of the Chain Rule

By (1.11) on page 43, the derivative of a given function h at a point c is the number denoted by $h'(c)$, with the property that

$$
h(x) = h(c) + h'(c) \cdot (x - c) + e
$$

where e is an error term that is small when x is close to c. In particular, the derivative of g at c is the number $g'(c)$ with the property that

$$
g(x) = g(c) + g'(c) \cdot (x - c) + e_1 \tag{3.2}
$$

where e_1 is an error term that is small when x is close c. Likewise, the derivative of f at $g(c)$ is the number $f'(g(c))$ with the property that

$$
f(u) = f(g(c)) + f'(g(c)) \cdot (u - g(c)) + e_2 \tag{3.3}
$$

where e_2 is an error term that is small when u is close to $g(c)$. Lastly, the derivative of $f \circ g$ at c is the number $(f \circ g)'(c)$, with the property that

$$
f \circ g(x) = f(g(c)) + (f \circ g)'(c) \cdot (x - c) + e_0 \tag{3.4}
$$

where e_0 is an error term that is small when x is close c.

We are required to show that $(f \circ g)'(c) = f'(g(c)) \cdot g'(c)$. One way to do this is to expand $f \circ g(x)$ and compare the result to 3.4.

$$f \circ g(x) = f(g(x))$$

$$= f[g(c) + g'(c) \cdot (x - c) + e_1] \qquad \text{(if we use (3.2))}$$

$$= f(u) \quad \text{(if we let } u = g(c) + g'(c)(x - c) + e_1)$$

$$= f(g(c)) + f'(g(c)) \cdot (u - g(c)) + e_2 \qquad \text{(if we use (3.3))}$$

$$= f(g(c)) + f'(g(c)) \cdot [g'(c)(x - c) + e_1] + e_2$$

$$\text{(if we replace } u - g(c) \text{ with } g'(c)(x - c) + e_1)$$

$$= f(g(c)) + f'(g(c)) \cdot g'(c)(x - c) + f'(g(c)) \cdot e_1 + e_2$$

Observe that $f'(g(c))e_1 + e_2$ is an error term that is small when x is close to c. We may write it as one small number e_3. Therefore

$$f \circ g(x) = f(g(c)) + f'(g(c))g'(c)(x - c) + e_3 \qquad (3.5)$$

Now compare (3.5) to (3.4); the second term in the right hand side of (3.5) must equal the second term in the right hand side of (3.4). Conclusion:

$$(f \circ g)'(c) = f'(g(c))g'(c).$$

In general, if g is differentiable at x and f is differentiable at $g(x)$, then $f \circ g$ is differentiable at x and its derivative is

$$(f \circ g)'(x) = f'(g(x))g'(x) \qquad (3.6)$$

In the following examples, we use the chain rule in reverse to determine anti-derivatives of some trigonometric functions.

Example 133 *To find the most general antiderivative of* $f(x) = \csc \pi x \cot \pi x$ *we observe in table (3.1) that an expression involving* $\csc(\ ) \cot(\ )$ *results from taking the derivative of* $\csc(\ )$. *Therefore* $\csc \pi x \cot \pi x$ *must be the result of taking the derivative of an expression involving* $\csc \pi x$. *But the derivative of* $\csc \pi x$ *is* $-(\csc \pi x \cot \pi x)(\pi) = -\pi \csc \pi x \cot \pi x$, *and there is no constant factor* $(-\pi)$ *in the formula for* f. *This suggests that we should divide* $\csc \pi x$ *by* $-\pi$. *Therefore the most general antiderivative of* $\csc \pi x \cot \pi x$ *is* $F(x) = -\frac{1}{\pi} \csc \pi x + c$. *Confirm by taking the derivative of* $F(x)$.

Example 134 *To find the most general antiderivative* $g(x) = \dfrac{\sin x}{\cos^3 x}$, *we write the right hand side as* $(\cos x)^{-3} \sin x$. *Note that the derivative of* $\cos x$ *is* $-\sin x$ *and the factor* $\sin x$ *appears in the formula for* g. *This suggests that* $g(x)$ *must have resulted from taking the derivative of an expression involving* $(\cos x)^{-2}$. *(The exponent has to be* -2 *so that when we take a derivative, it is lowered by 1 to* -3). *Since the derivative of* $(\cos x)^{-2}$ *is* $-2(\cos x)^{-3}(\sin x) = -2(\cos x)^{-3} \sin x$, *the most general antiderivative of* $\dfrac{\sin x}{\cos^3 x}$ *must be* $G(x) = -\frac{1}{2}(\cos x)^{-2} + c$. *Confirm by taking the derivative of* $G(x)$.

Example 135 *To find the most general antiderivative of* $h(x) = \sqrt{\tan 3x}\sec^2 3x$, *first write the right hand side as* $(\tan 3x)^{1/2}\sec^2 3x$. *Now note that the derivative of* $\tan 3x$ *is* $3\sec^2 3x$, *and the factor* $\sec^2 3x$ *is present in the formula for* h. *This suggests that* $h(x)$ *is the result of taking the derivative of an expression involving* $(\tan 3x)^{3/2}$. *(The exponent has to be 3/2 so that when we take a derivative, it is lowered by 1 to 1/2). Since the derivative of* $(\tan 3x)^{3/2}$ *is* $\left(\frac{3}{2}(\tan 3x)^{1/2}\right)(\sec^2 3x)(3) = \frac{9}{2}\sqrt{\tan 3x}\sec^2 3x$, *it follows that the most general antiderivative of* $h(x) = \sqrt{\tan 3x}\sec^2 3x$ *is* $H(x) = \frac{2}{9}(\tan 3x)^{3/2} + c$. *Confirm by taking the derivative of* $H(x)$.

Exercise 136

1. Determine the derivative of each function and simplify when possible.

   a) $f(x) = (\cos x - \sin x)^{2/3}$ b) $g(x) = \tan\left(\frac{\pi}{3} - \frac{1}{x}\right)$ c) $h(x) = x\cos 2x$

   d) $u(x) = \csc 3x - \sec\sqrt{x}$ e) $v(x) = x^3 \sin^2 \pi x$ f) $w(x) = \frac{3}{4}\sec^3 2x$

   g) $f(x) = x^4 \cot\left(\frac{1}{x^3}\right)$ h) $g(x) = e^{-x^2}\sin 2x$ i) $h(x) = \sqrt[3]{1 + \cot 4x}$

   j) $u(x) = \sin\left(\frac{3x-1}{5}\right) + \frac{x}{2\pi}$ k) $v(x) = \csc^3(\pi x)$ l) $w(x) = \tan^2\left(\frac{4-\pi x}{3}\right)$

2. Let $f(x) = \frac{\sqrt{2x-x^2}}{x}$. Show that $f'(x) = \frac{-1}{x\sqrt{2x-x^2}}$.

3. Let $g(x) = \frac{\sqrt{x^2+1}}{x}$. Show that $g'(x) = \frac{-1}{x^2\sqrt{x^2+1}}$.

4. Determine the most general antiderivative of each function:

   (a) $f(x) = \frac{4}{5}\cos 2x$ (b) $g(x) = \sec 3x \tan 3x$ (c) $h(x) = e^{3x} + 5$

   (d) $u(x) = 2e^{3x}$ (e) $w(x) = 3 - 2\sin 5x$ (f) $f(x) = 2x\sqrt{x^2 + 4}$

   (g) $g(x) = \sin^3 x \cos x$ (h) $w(x) = 3e^{4x} - 4x^2$ (i) $f(x) = \csc\frac{1}{2}x \cot\frac{1}{2}x$

   (j) $g(x) = \frac{\sec^2 x}{\tan^4 x}$ (k) $h(x) = \frac{1}{2\sqrt{x+5}}$ (l) $u(x) = \frac{3\sin 2x}{\sqrt{\cos 2x+4}}$

   (m) $f(x) = \frac{4x}{\sqrt{x^2+4}}$ (n) $g(x) = \frac{4\cos x}{\sin^3 x}$ (o) $h(x) = \cos\left(\frac{5x-1}{4}\right)$

5. Let a, b, c, and d be constants. Show that $f(x) = \dfrac{a\sin x + b\cos x}{c\sin x + d\cos x}$ has derivative $f'(x) = \dfrac{ad - bc}{(c\sin x + d\cos x)^2}$.

6. In the given figure, the circle has center $C(1,0)$ and radius 1. The line segment PQ is a tangent at R to the circle and angle RCQ is $2x$ radians. Angle CRQ is a right angle because a tangent is perpendicular to a radius at

the point of tangency.

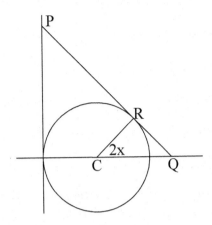

Show that the length of PQ is $\tan 2x + \cot x$, *and that as x varies, the minimum length of PQ is approximately 3.33 units.*

Higher Order Derivatives

A higher order derivative is *the derivative of a derivative*. For example, let $f(x) = 3x^2 - \frac{2}{x}$. Its derivative is $f'(x) = 6x + \frac{2}{x^2}$. The derivative of $f'(x)$ is an example of a higher order derivative of f, called the second derivative of f, and denoted by $f''(x)$. Thus

$$f''(x) = 6 - \frac{4}{x^3}.$$

One may also take the derivative of $f''(x)$. The result is called the third order derivative of f and is denoted by $f'''(x)$. Therefore $f'''(x) = \frac{12}{x^4}$. Fourth and higher order derivatives of f are calculated in a similar way.

For another example, consider $g(x) = 4x^4 - 3x^2 + 5x - 3$. Its first derivative is

$$g'(x) = 16x^3 - 6x + 5.$$

The second derivative is $g''(x) = 48x^2 - 6$. The third derivative is $g'''(x) = 96x$, the fourth is written as $g^{(4)}(x)$ and it is the constant function $g^{(4)}(x) = 96$. The fifth and higher derivatives are all zeros.

For practice, find the first and second derivative of each of the following:

(a) $f(x) = 3x^5 - 2x^3 + x - 3$ (b) $g(x) = 2\sin x - 4\cos x$ (c) $h(x) = 4e^{3x}$
(d) $u(x) = 8x\cos x - 1$ (e) $v(x) = 4 - x\sin x$ (f) $w(x) = \frac{2}{x^2}$
(g) $f(x) = \frac{4x}{7} - \frac{7}{4x^2} + 5$ (h) $g(x) = 5 - 3\sin\frac{1}{2}x$ (i) $h(x) = 2\sec \pi x$

Our first use of second order derivatives is to determine shapes of graphs. There are graphs like that of $f(x) = (x+2)^2 + 1$ and $g(x) = e^x$ shown below, shaped like

100

right-side up bowls or part of right-side up bowls.

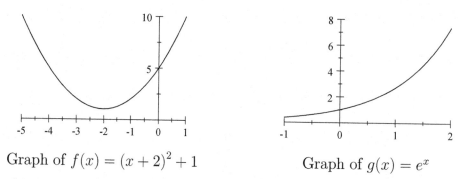

Graph of $f(x) = (x+2)^2 + 1$ Graph of $g(x) = e^x$

We call them **concave up** graphs. A more precise definition is given in Exercise 12 on page 104. Their counter-parts, like the graphs of $h(x) = -2x^2 + 5$ and $v(x) = 3 + \frac{1}{x}$, $x < 0$, are shaped like upside-down bowls or part of upside-down bowls.

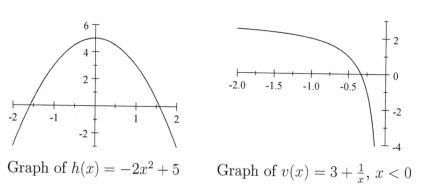

Graph of $h(x) = -2x^2 + 5$ Graph of $v(x) = 3 + \frac{1}{x}$, $x < 0$

They are called **concave down** graphs. Some graphs are concave up on some intervals and concave down on others. An example is the graph of $\sin x$, (see it on page 268). It is concave down on intervals like $[0, \pi]$, $[2\pi, 3\pi]$, etc, and concave up on $[\pi, 2\pi]$ and many others.

One key observation we can pick up from the graphs of $f(x) = (x+2)^2 + 1$ and $g(x) = e^x$ is that if the graph of a function is concave up on an interval $[a, b]$ then the derivative of the function is increasing on the interval.

Example 137 *The graph of $u(x) = \sin x$ is concave up on the interval $[-\pi, 0]$ and we have $u'(x) = \cos x$ which is increasing on $[-\pi, 0]$.*

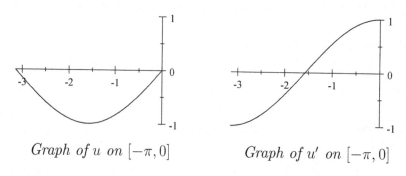

Graph of u on $[-\pi, 0]$ *Graph of u' on $[-\pi, 0]$*

Since an increasing function has a positive derivative, this suggests that if a given function has a concave up graph on an interval $[a, b]$ then its second derivative is positive on $[a, b]$, (because it is the derivative of the increasing function $f'(x)$). This is what we use here to identify concave up graphs. More precisely, *the graph of a function f is concave up on an interval $[a, b]$ if $f''(x)$, (the second derivative of f), is positive on $[a, b]$.*

The graphs of $h(x) = -2x^2 + 5$ and $v(x) = 3 + \frac{1}{x}$, $x < 0$ suggest that if a function has a concave down graph on an interval $[a, b]$ then its derivative is decreasing, therefore its second derivative should be negative on $[a, b]$. We use this observation to identify concave down graphs. More precisely, *the graph of a function f is concave down on an interval $[a, b]$ if $f''(x)$, (the second derivative of f), is negative on $[a, b]$.*

Example 138 *Consider the function $f(x) = x^3 - x^2 - 4x + 4$. Its second derivative is $f''(x) = 6x - 2$. This is positive when $x > \frac{1}{3}$ and negative when $x < \frac{1}{3}$. Therefore its graph is concave up on the interval $\left(\frac{1}{3}, \infty\right)$, and concave down on $\left(-\infty, \frac{1}{3}\right)$.*

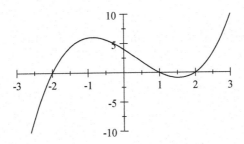

At the number $x = \frac{1}{3}$ the graph changes from being concave down, (to the left of $\frac{1}{3}$), to being concave up. Such a number is called a **point of inflection** for the function.

Example 139 *Let $g(x) = -2x^3$. Then $g''(x) = -12x$ which is positive on $(-\infty, 0)$ and negative on $(0, \infty)$. The graph is concave up on $(-\infty, 0)$ and concave down on $(0, \infty)$.*

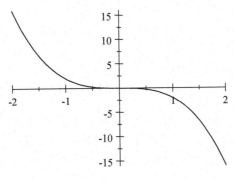

Its point of inflection is $x = 0$.

The Second Derivative Test for Maxima/Minima

If a number c is a point of relative maximum for a function f then the graph of f must be concave down on some interval $[a, b]$ containing c, (see the point $c = -1$

in the figure below). We observed that in such a case, the second derivative of f should be negative on the interval. In particular, $f''(c)$ should be negative.

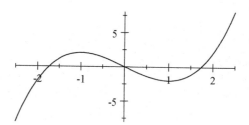

The exact opposite happens when c is a point of relative minimum for f. In that case the graph of f must be concave up on some interval $[a, b]$ containing c, therefore $f''(x)$ should be positive on $[a, b]$. In particular, $f''(c)$ should be positive.

We already noted that a point of relative maximum or relative minimum for a function f is a critical point of f. Therefore the above observations lend support to the following **second derivative test** for the nature of a critical point:

- If c is a critical point of a function f and $f''(c)$ is negative, then c is a point of relative maximum.

- If c is a critical point of a function f and $f''(c)$ is positive, then c is a point of relative minimum.

For functions whose second derivatives are easy to determine, the second derivative test may be quicker to apply than the slope method we used in Example 49 on page 37.

Example 140 *Consider the function $f(x) = 4 + 18x + \frac{3}{2}x^2 - 2x^3$ we met earlier, (page 38). Its critical points are the numbers x such that*

$$f'(x) = 18 + 3x - 6x^2 = -3(2x + 3)(x - 2) = 0$$

They are $x = 2$ and $x = -\frac{3}{2}$. To establish their nature, we determine the second derivative of f then evaluate it at 2 and $-\frac{3}{2}$. It turns out to be $f''(x) = 3 - 12x$ and

$$f''(2) = 3 - 24 = -21$$

which is negative. Therefore $c = 2$ is a point of relative maximum for f. On the other hand,

$$f''(-\tfrac{3}{2}) = 3 - 12(-\tfrac{3}{2}) = 21$$

which is positive. Therefore $c = -\frac{3}{2}$ is a point of relative minimum for f.

Remark 141 *Unfortunately, if $f''(c) = 0$ at a critical point c, then the second derivative test shades no light on the nature of the critical point. As the example below show, it could be a point of relative minimum, a point of relative maximum, or neither.*

Example 142 Let $f(x) = 1 + x^4$, $h(x) = 3 - (x + 1)^4$, and $g(x) = 2(x - 1)^3 - 1$. The second derivative of f ,(it happens to be $12x^2$), is zero at its critical point $c = 0$, which is a point of relative minimum. The second derivative of h ,(it happens to be $-12(x + 1)^2$), is zero at its critical point $c = -1$, which is a point of relative maximum. The second derivative of g, (it happens to be $12(x - 1)$), is zero at its critical point $c = 1$ which is neither a point of relative maximum nor a point of relative minimum.

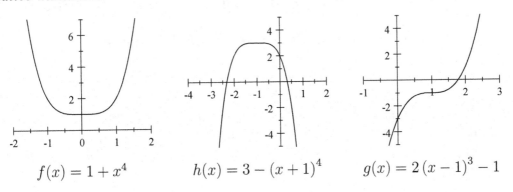

$$f(x) = 1 + x^4 \qquad h(x) = 3 - (x + 1)^4 \qquad g(x) = 2(x - 1)^3 - 1$$

> **If you get a critical point c such that $f''(c) = 0$, use the slope method we introduced in Example 37 to establish its nature.**

Exercise 143

1. Determine the interval or intervals where the graph of $w(x) = x^4 - x^2 + 3$ is concave up or concave down. What are the points of inflection of w (if any)?

2. Determine the interval or intervals where the graph of $h(x) = x^5 - x^3 + 5$ is concave up or concave down. What are the points of inflection of h (if any)?

3. In each case determine the critical point(s) of the given function then use the second derivative test to establish the nature of every critical point.

   (a) $u(x) = x^2 + 3x + 1$ (b) $g(x) = x^4 - 4x^2 - 3$ (c) $f(x) = xe^x$

4. For the function $f(x) = \sqrt{x^2 + 4x + 7}$, which of the two tests; the slope method of Example 49 on page 37 and then the second derivative test, is quicker to establish the nature of its critical point?

5. Use an appropriate method to establish the nature of each critical point of the given function, then sketch its graph.

   (a) $h(x) = x^2(6 - x)$ (b) $w(x) = x^4 - 2x^2 + 3$ (c) $g(x) = x^6 - 3x^4$

   (e) $f(x) = \frac{x^2 - 1}{x - 4}$ (f) $v(x) = x + 4/x$ (g) $u(x) = \frac{x}{9 + x^2}$

6. Show that if $f(x) = \frac{x^2}{\sqrt{x^2 - 4}}$ then $f'(x) = \frac{x^3 - 8x}{(x^2 - 4)^{3/2}}$. Now determine its vertical asymptotes and its critical points, (there are three of them). Establish the nature of each critical point then sketch the graph.

7. *You are given f with formula $f(x) = x^4 - 32x + 4$.*

    (a) *Determine its critical point, (it has only one), and establish its nature.*

    (b) *Sketch the graph of f and use it to estimate the roots of the equation $x^4 - 32x + 4 = 0$*

    (c) *Use Newton's method to determine the smallest root accurately to 3 decimal places. (Only the smallest root is required more accurately.)*

8. *Let n be a positive integer. Determine the nth derivative of:*

    (a) $f(x) = \sin x$ (b) $g(x) = e^{ax}$ (c) $u(x) = xe^x$ (d) $h(x) = \frac{1}{x}$

9. *In the given figure, the circle has center O and radius 2 units. BD and OC intersect at R, $ABDE$ is a rectangle, BCD is an isosceles triangle and so is EFA, with $CB = CD = FA = FE$. Each of the angles BOC and DOC is x radians.*

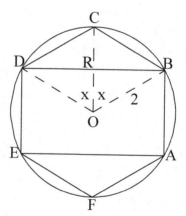

    (a) *Show that RC has length $2 - 2\cos x$ and BD has length $4\sin x$.*

    (b) *Prove that the area of the polygon $ABCDEF$ is $8\sin x + 8\sin x \cos x$.*

    (c) *What is the largest possible area of such a polygon?*

10. *According to the product rule, the derivative of $f(x)g(x)$ is $f'(x)g(x)+g'(x)f(x)$. Show that the second derivative of $f(x)g(x)$ is*

$$f''(x)g(x) + 2f'(x)g'(x) + g''(x)f(x)$$

    (a) *What is the third derivative of $f(x)g(x)$?*

    (b) *What is the nth derivative of $f(x)g(x)$, where n is a positive integer?*

11. *Sketch on the same axes, the graphs of $f(x) = \sin x$, (x in radians), and $g(x) = x - 1$. Use the sketches to determine an approximate root of the equation $\sin x = x - 1$ then use Newton's method to calculate it accurately to 4 decimal places.*

12. **A more precise definition of a concave up graph:**

(a) *The graph of a function f is concave up on an interval if it satisfies the following condition: Given any two points $P(a, f(a))$ and $Q(b, f(b))$ on the graph, no point R on the chord PQ is below the graph of f. (See the figure below.)*

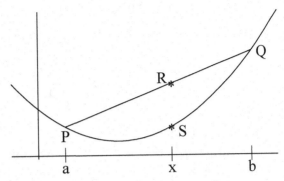

The equation of the chord PQ is $y(x) = \left[\frac{f(b)-f(a)}{b-a}\right](x-a) + f(a)$. A number x between a and b may be written as

$$x = (1-t)\,a + tb$$

where $0 \le t \le 1$. If $R(x, y(x))$ is a point on the chord PQ and S is the point on the graph of f with the same x coordinate, show that R has coordinates $((1-t)\,a+bt, (1-t)f(a)+tf(b))$ and S has coordinates $((1-t)\,a+bt, f((1-t)a+tb))$. Use the fact that S must be below R in order for the graph to be concave up to deduce that

$$f((1-t)a + tb) \le (1-t)f(a) + tf(b)$$

*for all $0 \le t \le 1$. Thus **the graph of a function f is concave up if for any points a and b in its domain and any number t between 0 and 1**, $f((1-t)a + tb) \le (1-t)f(a) + tf(b)$.*

(b) *In this part of the question, you are required to show that if the second derivative of a function f is positive on an interval $[a, b]$ then its graph is concave up; i.e. $f((1-t)a + tb) \le (1-t)f(a) + tf(b)$ for all points a and b in the interval and for all $0 \le t \le 1$. To get started, use the mean value theorem to show that there is a number θ between b and $(1-t)a+tb$ such that*

$$f(b) - f((1-t)a + tb) = (b-a)(1-t)f'(\theta) \qquad (3.7)$$

Multiply both sides of (3.7) by t and solve to get

$$tf(b) = tf((1-t)a + tb) + t(1-t)(b-a)f'(\theta) \qquad (3.8)$$

Also show that there is a number α between a and $(1-t)\,a + tb$ such that

$$(1-t)f(a) = -t(1-t)(b-a)f'(\alpha) + (1-t)f((1-t)\,a + tb) \quad (3.9)$$

Add (3.8) to (3.9) and deduce that

$$(1-t)f(a) + tf(b) = f((1-t)a + tb) + t(1-t)(b-a)(f'(\theta) - f'(\alpha))$$

Use the Mean value Theorem to show that $(f'(\theta) - f'(\alpha)) \ge 0$ then deduce that $(1-t)f(a) + tf(b) \ge f((1-t)a + tb)$

L'Hopital's Rule.

If f and g are functions such that $f(a) = g(a) = 0$ then when we try to compute $\lim_{x \to a} \dfrac{f(x)}{g(x)}$ by substituting $x = a$ into the numerator and denominator, we may end up with the undefined expression $\frac{0}{0}$. In such a case, L'Hopital's rule may save the day. It states that if $f(a) = g(a) = 0$ and $\lim_{x \to a} \dfrac{f'(x)}{g'(x)}$ exists then

$$\lim_{x \to a} \frac{f(x)}{g(x)} = \lim_{x \to a} \frac{f'(x)}{g'(x)}.$$

To prove it, take any number x close to a. By the Generalized Mean Value Theorem, there is a number y between a and x such that

$$\frac{f(x) - f(a)}{g(x) - g(a)} = \frac{f'(y)}{g'(y)}$$

Since $f(a) = g(a) = 0$, it follows that

$$\lim_{x \to a} \frac{f(x)}{g(x)} = \lim_{x \to a} \frac{f(x) - f(a)}{g(x) - g(a)} = \lim_{y \to a} \frac{f'(y)}{g'(y)}$$

The variable y does not change the value of the limit. Therefore we may write

$$\lim_{x \to a} \frac{f(x)}{g(x)} = \lim_{x \to a} \frac{f'(x)}{g'(x)}$$

Example 144 *Consider* $h(x) = \dfrac{x^3 - 1}{x^{14} - 1}$. *Let* $f(x) = x^3 - 1$ *and* $g(x) = x^{14} - 1$. *Then* $f(1) = g(1) = 0$ *and*

$$\lim_{x \to 1} \frac{f'(x)}{g'(x)} = \lim_{x \to 1} \left(\frac{3x^2}{14x^{13}} \right) = \frac{3}{14}.$$

By L'Hopital's rule, $\lim\limits_{x \to 1} h(x) = \lim\limits_{x \to 1} \dfrac{x^3 - 1}{x^{14} - 1} = \lim\limits_{x \to 1} \dfrac{f(x)}{g(x)} = \lim\limits_{x \to 1} \dfrac{f'(x)}{g'(x)} = \frac{3}{14}$

Example 145 *Consider* $h(x) = \frac{\sin 2x^2}{x^2}$. *Let* $f(x) = \sin 2x^2$ *and* $g(x) = x^2$. *Then* $f(0) = g(0) = 0$ *and*

$$\lim_{x \to 0} \frac{f'(x)}{g'(x)} = \lim_{x \to 0} \frac{(\cos 2x^2) \cdot 4x}{2x} = \lim_{x \to 0} 2 \cos 2x^2 = 2.$$

By L'Hopital's rule, $\lim\limits_{x \to 0} h(x) = \lim\limits_{x \to 0} \frac{\sin 2x^2}{x^2} = \lim\limits_{x \to 0} \frac{f'(x)}{g'(x)} = 2$

The rule extends to cases in which derivatives also vanish. For example, suppose f and g are such that $f(a) = f'(a) = 0$ and $g(a) = g'(a) = 0$. If $\lim_{x \to a} \dfrac{f''(x)}{g''(x)}$ exists then

$$\lim_{x \to a} \frac{f(x)}{g(x)} = \lim_{x \to a} \frac{f''(x)}{g''(x)}.$$

In general, if $f(a) = f'(a) = \cdots = f^{(n-1)} = 0;$ $g(a) = g'(a) = \cdots = g^{(n-1)} = 0,$ and $\lim_{x \to a} \dfrac{f^{(n)}(x)}{g^{(n)}(x)}$ exists, then

$$\lim_{x \to a} \frac{f(x)}{g(x)} = \lim_{x \to a} \frac{f^{(n)}(x)}{g^{(n)}(x)}.$$

Example 146 *Consider* $\lim_{x \to 0} \dfrac{1 - \cos 2x}{3x^2}$. *Let* $f(x) = 1 - \cos 2x$ *and* $g(x) = 3x^2$. *Then* $f(0) = f'(0) = 0$ *and* $g(0) = g'(0) = 0$. *However,*

$$\lim_{x \to 0} \frac{f''(x)}{g''(x)} = \lim_{x \to 0} \frac{4 \cos 2x}{6} = \frac{2}{3}.$$

By L'Hopital's rule,

$$\lim_{x \to 0} \frac{1 - \cos 2x}{3x^2} = \lim_{x \to 0} \frac{f(x)}{g(x)} = \lim_{x \to 0} \frac{f''(x)}{g''(x)} = \frac{2}{3}$$

Exercise 147 *Calculate the following limits:*

$(a) \lim_{x \to 0} \left(\dfrac{\sin 3x}{\tan 5x} \right)$ $(b) \lim_{x \to 2} \left(\dfrac{x^2 - 4}{x^3 - 8} \right)$ $(c) \lim_{x \to 0} \dfrac{\tan(3x^2)}{4x^2}$

$(d) \lim_{x \to 0} \dfrac{x - \sin x}{2x^2}$ $(e) \lim_{x \to 0} \left(\dfrac{x - \sin x}{x - x \cos x} \right)$ $(f) \lim_{x \to -1} \left(\dfrac{x^4 - 2x^2 + 1}{x^3 + 4x^2 + 5x + 2} \right)$

Another version of L'Hopital's rule

Suppose f and g are functions such that $f(x) \to \infty$ (or $-\infty$) and $g(x) \to \infty$ (or $-\infty$) as x approaches a, (a could be finite or infinite.)

If $\lim_{x \to a} \dfrac{f'(x)}{g'(x)}$ exists then $\lim_{x \to a} \dfrac{f(x)}{g(x)}$ also exists and it is equal to $\lim_{x \to a} \dfrac{f'(x)}{g'(x)}$.

The proof happens to be considerably harder, so we skip it. Here are examples where it is used.

Example 148 *Let* $f(x) = 3x + 5$ *and* $g(x) = 4 - 5x$. *Then* $f(x) \to \infty$ *and* $g(x) \to -\infty$ *as* $x \to \infty$. *Since* $\lim_{x \to \infty} \dfrac{f'(x)}{g'(x)} = \lim_{x \to \infty} -\dfrac{3}{5}$, *it follows that* $\lim_{x \to \infty} \dfrac{3x + 5}{4 - 5x} = -\dfrac{3}{5}$.

Example 149 *Consider* $\lim_{x \to \infty} \dfrac{3x}{e^x}$. *Let* $f(x) = 3x$ *and* $g(x) = e^x$. *Then* $f(x) \to \infty$ *and* $g(x) \to \infty$ *as* $x \to \infty$. *But* $\lim_{x \to \infty} \dfrac{f'(x)}{g'(x)} = \lim_{x \to \infty} \dfrac{3}{e^x} = 0$, *therefore* $\lim_{x \to \infty} \dfrac{3x}{e^x} = 0$.

Example 150 *Let* $f(x) = x^2 + 5x - 1$ *and* $g(x) = 2x^2 - 3x + 5$. *Then* $f(x), g(x),$ $f'(x),$ *and* $g'(x)$ *all approach* ∞ *as* $x \to \infty$. *Since* $\lim_{x \to \infty} \dfrac{f''(x)}{g''(x)} = \lim_{x \to \infty} \dfrac{2}{4} = \dfrac{1}{2}$, *it follows that* $\lim_{x \to \infty} \dfrac{x^2 + 5x - 1}{2x^2 - 3x + 5} = \dfrac{1}{2}$.

Exercise 151 *Determine the required limits. In part (c), n is a positive integer.*

$$(a)\ \lim_{x\to\infty}\left(\tfrac{107x}{x^2+1}\right) \qquad (b)\ \lim_{x\to\infty}\tfrac{x^2}{e^x} \qquad (c)\ \lim_{x\to\infty}\tfrac{x^n}{e^x}$$

Remark 152 *Problems 8 and 10 on page 124 provide applications of L'Hopital's rule to two other forms of indeterminate limits.*

Taylor Polynomials

On page 40 we approximated a section of a graph with straight line segment and went on to use the line to compute useful approximate values. A straight line is the graph of a first degree polynomial, therefore we essentially approximated a given function with a first degree polynomial. Clearly, a tangent line cannot approximate a really "curved" graph on a large interval. For example, the tangent at $\left(\tfrac{\pi}{3},\tfrac{\sqrt{3}}{2}\right)$ to the graph of $f(x)=\sin x$, (shown below), does not "accurately" approximate the graph on a large interval like $[0,2]$.

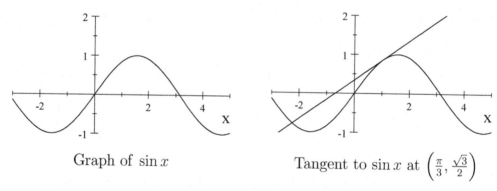

Graph of $\sin x$ Tangent to $\sin x$ at $\left(\tfrac{\pi}{3},\tfrac{\sqrt{3}}{2}\right)$

When we need better approximations on larger intervals, we have to consider approximating with parabolas or higher degree polynomials. In this section, we consider such polynomials called Taylor polynomials. To illustrate the process, we consider the graph of $\sin x$ near the point $x=\tfrac{\pi}{3}$.

The tangent at $\left(\tfrac{\pi}{3},\tfrac{\sqrt{3}}{2}\right)$ to the graph of $f(x)=\sin x$ is the line $p_1(x)$ that passes through $\left(\tfrac{\pi}{3},\tfrac{\sqrt{3}}{2}\right)$ and has the same first order derivative at $x=\tfrac{\pi}{3}$ as f. Since $f'(\tfrac{\pi}{3})=\cos\tfrac{\pi}{3}=\tfrac{1}{2}$, it has slope $\tfrac{1}{2}$, therefore its equation is $y=\tfrac{\sqrt{3}}{2}+\tfrac{1}{2}\left(x-\tfrac{\pi}{3}\right)$. The first degree polynomial

$$p_1(x)=\tfrac{\sqrt{3}}{2}+\tfrac{1}{2}\left(x-\tfrac{\pi}{3}\right)$$

is called the **first order Taylor polynomial** for f at $x=\tfrac{\pi}{3}$.

The **second order Taylor polynomial** for f at $x=\tfrac{\pi}{3}$ is the parabola that passes through $\left(\tfrac{\pi}{3},\tfrac{\sqrt{3}}{2}\right)$ and has the same first and second order derivatives as f at $x=\tfrac{\pi}{3}$. In other words, it is the polynomial $p_2(x)$ of degree 2 that satisfies the following conditions: $p_2(\tfrac{\pi}{3})=f(\tfrac{\pi}{3})$, $p_2'(\tfrac{\pi}{3})=f'(\tfrac{\pi}{3})$ and $p_2''(\tfrac{\pi}{3})=f''(\tfrac{\pi}{3})$. To guarantee that its graph passes through $\left(\tfrac{\pi}{3},f(\tfrac{\pi}{3})\right)$, it should have the form

$$p_2(x)=f(\tfrac{\pi}{3})+a\left(x-\tfrac{\pi}{3}\right)+b\left(x-\tfrac{\pi}{3}\right)^2,$$

where a and b are constants. Since $p_2'(x) = a + 2b\left(x - \frac{\pi}{3}\right)$, and $p_2''(x) = 2b$, the condition $f'\left(\frac{\pi}{3}\right) = p_2'\left(\frac{\pi}{3}\right)$ implies that $a = f'\left(\frac{\pi}{3}\right)$, and the condition $f''\left(\frac{\pi}{3}\right) = p_2''\left(\frac{\pi}{3}\right)$ implies that $b = \frac{1}{2}f''\left(\frac{\pi}{3}\right)$. Therefore

$$p_2(x) = \frac{\sqrt{3}}{2} + f'\left(\tfrac{\pi}{3}\right)\left(x - \tfrac{\pi}{3}\right) + \tfrac{1}{2}f''\left(\tfrac{\pi}{3}\right)\left(x - \tfrac{\pi}{3}\right)^2$$

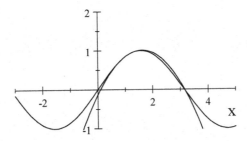

The graphs of $p_2(x)$ and $f(x) = \sin x$ on the interval $[-3, 5]$

There is no doubt that p_2 approximates f more accurately than p_1 on a bigger interval. Let us skip the third degree Taylor polynomial and go straight to the fourth degree. We denote it by $p_4(x)$. It is the fourth degree polynomial whose graph passes through $\left(\frac{\pi}{3}, f(\frac{\pi}{3})\right)$ and, in addition, has the same first, second, third and fourth order derivatives as f at $\frac{\pi}{3}$. In other words

$$p_4(\tfrac{\pi}{3}) = f(\tfrac{\pi}{3}),\ p_4'(\tfrac{\pi}{3}) = f'(\tfrac{\pi}{3}),\ p_4''(\tfrac{\pi}{3}) = f''(\tfrac{\pi}{3}),\ p_4'''(\tfrac{\pi}{3}) = f'''(\tfrac{\pi}{3})\ \text{and}\ p_4^{(4)}(\tfrac{\pi}{3}) = f^{(4)}(\tfrac{\pi}{3})$$

It turns out that $p_4(x)$ is the polynomial

$$f\left(\tfrac{\pi}{3}\right) + f'\left(\tfrac{\pi}{3}\right)\left(x - \tfrac{\pi}{3}\right) + \tfrac{1}{2}f''\left(\tfrac{\pi}{3}\right)\left(x - \tfrac{\pi}{3}\right)^2 + \tfrac{1}{3!}f'''\left(\tfrac{\pi}{3}\right)\left(x - \tfrac{\pi}{3}\right)^3 + \tfrac{1}{4!}f^{(4)}\left(\tfrac{\pi}{3}\right)\left(x - \tfrac{\pi}{3}\right)^4$$

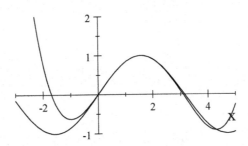

The graphs of $p_4(x)$ and $f(x) = \sin x$ on the interval $[-3, 5]$.

In general, the **nth order Taylor polynomial** for f at $x = \frac{\pi}{3}$ is the polynomial p_n whose graph passes through $\left(\frac{\pi}{3}, f(\frac{\pi}{3})\right)$, and has the same first, second, $\ldots$, nth derivatives as f at $\frac{\pi}{3}$. It has formula

$$p_n(x) = f\left(\tfrac{\pi}{3}\right) + f'\left(\tfrac{\pi}{3}\right)\left(x - \tfrac{\pi}{3}\right) + \tfrac{1}{2!}f''\left(\tfrac{\pi}{3}\right)\left(x - \tfrac{\pi}{3}\right)^2 + \cdots + \tfrac{1}{n!}f^{(n)}\left(\tfrac{\pi}{3}\right)\left(x - \tfrac{\pi}{3}\right)^n.$$

To generalize, let f be a function that has derivatives of all orders on some interval containing a point c. The **nth order Taylor polynomial for f at c**, denoted by p_n, is the polynomial that has value $f(c)$ at c, and has the same first, second, $\ldots$, nth derivatives as f at c. Its formula is

$$p_n(x) = f(c) + f'(c)(x - c) + \tfrac{1}{2!}f''(c)(x - c)^2 + \cdots + \tfrac{1}{n!}f^{(n)}(c)(x - c)^n.$$

110

Example 153 *Let* $f(x) = \sqrt{2x+3} = (2x+3)^{1/2}$ *and* $c = 3$. *Then*

$$f'(x) = (2x+3)^{-1/2}, \qquad f''(x) = -(2x+3)^{-3/2},$$

$$f'''(x) = 3(2x+3)^{-5/2}, \qquad f^{(4)}(x) = -15(2x+3)^{-7/2}, \text{ etc.}$$

It follows that $f(3) = 3$, $f'(3) = \frac{1}{3}$, $f''(3) = -\frac{1}{27}$, $f'''(3) = \frac{1}{81}$ *and* $f^{(4)}(3) = -\frac{5}{729}$. *Now we can determine some of its Taylor polynomials at c. The first four are*

$$p_1(x) = 3 + \tfrac{1}{3}(x-3), \qquad p_2(x) = 3 + \tfrac{1}{3}(x-3) - \tfrac{1}{54}(x-3)^2$$

$$p_3(x) = 3 + \tfrac{1}{3}(x-3) - \tfrac{1}{54}(x-3)^2 + \tfrac{1}{486}(x-3)^3$$

$$p_4(x) = 3 + \tfrac{1}{3}(x-3) - \tfrac{1}{54}(x-3)^2 + \tfrac{1}{486}(x-3)^3 - \tfrac{5}{17496}(x-3)^4$$

The graphs of f *and the polynomials* p_1, p_2, p_3, *and* p_4 *are shown in the figure below.*

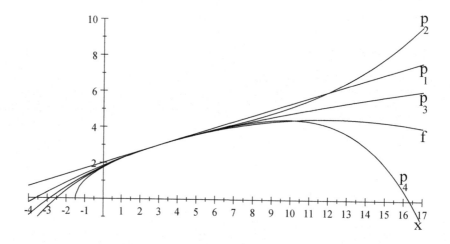

Say we are asked to use p_3 *to find an approximate value of* $\sqrt{8}$. *We look for* x *such that* $2x+3 = 8$. *It turns out to be* $x = 2.5$. *We then argue that since 2.5 is not very far from 3, it should be the case that* $f(2.5) \simeq p_3(2.5)$. *Therefore*

$$\sqrt{8} = f(2.5) \simeq p_3(2.5) = 3 - \tfrac{1}{3}(0.5) - \tfrac{1}{54}(0.5)^2 - \tfrac{1}{486}(0.5)^3$$

This is close to 2.8284. To 5 decimal places, a calculator gives 2.82843 for $\sqrt{8}$.

Remark 154 *Taylor polynomials may be used to approximate functions whose values are hard to evaluate. Evaluating a polynomial is a relatively easy operation because it involves only multiplications and additions. It is shown later in the book that if:*

(a) $p_n(x)$ *is the nth Taylor polynomial of a function* f *at a point* c
(b) x *is a number in the domain of* f
(c) M *is the largest value of* $\left|f^{(n+1)}(x)\right|$ *on the interval with end-points* c *and* x

then

$$|f(x) - p_n(x)| \le \frac{M\left|(x-c)^{n+1}\right|}{(n+1)!}.$$

In particular, if f is a polynomial of degree n then $M = 0$, therefore $p_n(x) = f(x)$ for all x.

Exercise 155

1. *Determine the required polynomial:*

   (a) *The 3rd order Taylor polynomial for $f(x) = \sqrt{1+3x}$ at $c = 1$.*

   (b) *The 3rd order Taylor polynomial for $g(x) = \cos x$ at $c = \frac{\pi}{2}$.*

   (c) *The 5th order Taylor polynomial for $h(x) = \sin x$ at $c = 0$.*

   (d) *The 5rd order Taylor polynomial for $h(x) = \tan x$ at $c = 0$.*

   (e) *The nth order Taylor polynomial for $f(x) = e^x$ at $c = 0$.*

2. *Determine the 3rd order Taylor polynomial for $f(x) = (3+x)^{2/3}$ at $c = 5$ and use it to give an approximate value of $(8.3)^{2/3}$.*

3. *Determine the 3rd order Taylor polynomial for $f(x) = (3x+1)^{1/2}$ at $c = 1$ and use it to give an approximate value of $\sqrt{3.4}$.*

4. *Find constants a_0, a_1, a_2, a_3 and a_4 such that*

$$1 + x^4 = a_0 + a_1(x-1) + a_2(x-1)^2 + a_3(x-1)^3 + a_4(x-1)^4$$

5. *Let a and b be constants, k be a given positive integer and f be defined by $f(x) = (a + bx)^k$.*

   (a) *Determine all the Taylor polynomials of f at 0.*

   (b) *Since $f(x)$ is a polynomial of degree k, $f(x) = p_k(x)$ for all x, (by Remark 154). Use this to deduce that for all real numbers a and b;*

$$(a+b)^k = a^k + ka^{k-1}b + \tfrac{1}{2!}k(k-1)a^{k-2}b^2 + \cdots + kab^{k-1} + b^k$$

$$(3.10)$$

   This is called the Binomial Theorem, and the coefficients

$$1, \quad k, \quad \frac{k(k-1)}{2!}, \quad \frac{k(k-1)(k-2)}{3!}, \quad \ldots \quad k, \quad 1$$

   of the terms a^k, $a^{k-1}b$, $a^{k-2}b^2$, $a^{k-3}b^3$, $\ldots$, ab^{k-1}, b^k are called binomial coefficients. They are denoted respectively by

$$\binom{k}{0}, \quad \binom{k}{1}, \quad \binom{k}{2}, \quad \binom{k}{3}, \quad \ldots \quad \binom{k}{k-1}, \quad \binom{k}{k}$$

   OR

$$C_0^k, \quad C_1^k, \quad C_2^k, \quad C_3^k, \quad \ldots \quad C_{k-1}^k, \quad C_k^k$$

6. *Let a and b be positive and n be a positive integer. Use (3.10) to show that* $(a+b)^n > a^n + nb$.

7. *Let r be a real number and* $f(x) = (1+x)^r$. *Show that the Taylor polynomials for f at c = 0 are*

$$p_1(x) = 1 + rx, \qquad p_2(x) = 1 + rx + \frac{r(r-1)}{2!}x^2,$$

$$p_3(x) = 1 + rx + \frac{r(r-1)}{2!}x^2 + \frac{r(r-1)(r-2)}{3!}x^3,$$

$$\vdots$$

$$p_n(x) = 1 + rx + \frac{r(r-1)}{2!}x^2 + \frac{r(r-1)(r-2)}{3!}x^3 + \cdots + \frac{r(r-1)\cdots(r+1-n)}{n!}x^n$$

Let $r = \frac{1}{2}$.

(a) *Use the third Taylor polynomial for* $f(x) = \sqrt{1+x}$ *at c = 0 to find an approximate value of* $f(\frac{3}{4}) = \frac{\sqrt{7}}{2}$.

(b) *The Taylor polynomials for* $\sqrt{1-x}$ *may be obtained from the corresponding Taylor polynomials for* $\sqrt{1+x}$ *by simply replacing x wherever it is with* $(-x)$. *Determine the fourth Taylor polynomial for* $f(x) = \sqrt{1-x}$ *at c = 0 and use it to find an approximate value of* $\frac{\sqrt{8}}{3}$.

Implicit Differentiation

Before we differentiate functions implicitly, we need to introduce other notations for the derivatives of a given function f, and introduce the concept of a function defined implicitly. The first step is to denote $f(x)$ by y, (or any other convenient letter, but y is the the most popular). Then:

The first order derivative of f may be denoted by $\dfrac{dy}{dx}$.

The second order derivative of f may be denoted by $\dfrac{d^2y}{dx^2}$.

In general, the nth order derivative of f may be denoted by $\dfrac{d^n y}{dx^n}$.

According to the chain rule, the derivative of $[f(x)]^2$ is $2[f(x)]f'(x)$ or simply $2f(x)f'(x)$. If we write $f(x)$ as y then $[f(x)]^2$ may be written as y^2 and $2f(x)f'(x)$ as $2y\dfrac{dy}{dx}$. Therefore the derivative of y^2 is $2y\dfrac{dy}{dx}$. In general:

The derivative of y^n *is* $ny^{n-1}\dfrac{dy}{dx}$.

The derivative of $\cos y$ *is* $-\sin y\dfrac{dy}{dx}$.

The derivative of e^y *is* $e^y\dfrac{dy}{dx}$.

The derivative of $\sec y$ *is* $\sec y\tan y\dfrac{dy}{dx}$.

The derivative of $\cot y$ *is* $-\csc^2 y\dfrac{dy}{dx}$.

The derivative of $\sin y$ *is* $\cos y\dfrac{dy}{dx}$. We hope the pattern is clear

Turning to functions defined implicitly, take a function like $f(x) = \sqrt[3]{5x^2 + 7}$. We say it is defined directly because, given any number x in its domain, we evaluate the value $f(x)$ of f at x by simply substituting x into the right hand side. But this same function may be defined by the equation

$$(f(x))^3 - 5x^2 = 7 \tag{3.11}$$

This time, when a number, e.g. -2, is given, we do not get the value $f(-2)$ of f at -2 directly from substituting -2 into (3.11). We have to solve yet another equation, (which is $(f(-2))^3 - 20 = 7$) to get it. We say that equation (3.11) defines f indirectly or *implicitly*.

The following are more examples of functions defined implicitly. For convenience, we have written $f(x)$ as y. Thus, in each case, y is defined implicitly.

a. $3xy = 4$ b. $\sin xy = \frac{1}{2}x$ c. $x^2 + y^2 = 9$

d. $\sqrt{xy} - 4y^2 = 12$ e. $\sqrt{x+y} - 4x^2 = y$ f. $\frac{x+3}{y} = 4x^2 + y^2$.

It may be possible to write a function defined implicitly in a direct form. For example, if y is defined implicitly by $3xy = 4$, we may solve for y to get $y = \frac{4}{3x}$. However, there are functions for which the implicit form may be hard to convert into a direct form. Try y defined implicitly by

$$\frac{x+3}{y} = 4x^2 + y^2$$

The following is a procedure, called implicit differentiation or differentiating implicitly, for determining the derivative of a function y that is defined implicitly:

Take the derivative of *each term* in the equation defining y, using the rules for derivatives. The result should be an equation with at least one term involving the derivative of y. Solve the equation for the derivative.

Example 156 *Let y be defined implicitly by $y^3 - 5x^2 = 7$. To find $\frac{dy}{dx}$ we take the derivatives of the terms y^3, $-5x^2$, and 7, one at a time, to get*

$$3y^2\frac{dy}{dx} - 10x = 0$$

We now solve for $\frac{dy}{dx}$ and the result is $\frac{dy}{dx} = \frac{10x}{3y^2}$.

Example 157 *Let y be defined implicitly by $\frac{x+3}{y} = 4x^2 + y^2$. To find $\frac{dy}{dx}$, we have to find the derivative of each term in the equation. Before we do so, it is a good idea to first clear the fractions. Therefore, multiply both sides by y to get*

$$x + 3 = 4x^2y + y^3.$$

(This saves us a trip to the quotient rule.) Now take the derivatives of the terms x, 3, $4x^2y$, and y^3 one at a time. Note that $4x^2y$ is a product of $(4x^2)$ and y, therefore

determining its derivative requires the use of the product rule. Its derivative is $8xy + 4x^2\frac{dy}{dx}$. The result of taking the derivative of each term is

$$1 + 0 = 8xy + 4x^2\frac{dy}{dx} + 3y^2\frac{dy}{dx}.$$

Re-arrange this to get $1 - 8xy = (4x^2 + 3y^2)\frac{dy}{dx}$, then divide by $4x^2 + 3y^2$ to get the derivative:

$$\frac{dy}{dx} = \frac{1-8xy}{4x^2+3y^2}$$

Example 158 *Let y be defined implicitly by $\sin xy = \frac{1}{2}x - y^2$. To find $\frac{dy}{dx}$, take the derivative of the terms $\sin xy$, $\frac{1}{2}x$, and y^2 one at a time. Note that the term $\sin xy$ has the form $\sin(\ )$, therefore its derivative is*

$$\cos xy \times \text{Derivative of } (xy) \text{ which equals } (\cos xy)\left(y + x\frac{dy}{dx}\right)$$

The result of taking the derivative of each term is

$$(\cos xy)\left(y + x\frac{dy}{dx}\right) = \frac{1}{2} - 2y\frac{dy}{dx}$$

Re-arranging gives $(x\cos xy + 2y)\frac{dy}{dx} = \left(\frac{1}{2} - y\cos xy\right)$. We then solve for $\frac{dy}{dx}$ and the result is

$$\frac{dy}{dx} = \frac{1-2y\cos xy}{2x\cos xy + 4y}$$

To get higher order derivatives, take derivatives more than once using the fact that the derivative of $\frac{dy}{dx}$ is denoted by $\frac{d^2y}{dx^2}$, the derivative of $\frac{d^2y}{dx^2}$ is denoted by $\frac{d^3y}{dx^3}$, ...

Example 159 *Let y be defined implicitly by $x^2y - 3y^3 = -2x$, and assume that we are required to find $\frac{dy}{dx}$ and $\frac{d^2y}{dx^2}$ at the point $(1,1)$ on the graph of y. Differentiating each term gives $2xy + x^2\frac{dy}{dx} - 9y^2\frac{dy}{dx} = -2$ which we may re-arrange as*

$$\left(x^2 - 9y^2\right)\frac{dy}{dx} = -2 - 2xy. \tag{3.12}$$

As before, we solve for $\frac{dy}{dx}$. The result is

$$\frac{dy}{dx} = \frac{-2-2xy}{x^2-9y^2} = \frac{2+2xy}{9y^2-x^2} \tag{3.13}$$

The value of $\frac{dy}{dx}$ at $(1,1)$ is obtained by substituting $x = 1$ and $y = 1$ into (3.13), and it is $\frac{4}{8} = \frac{1}{2}$. To get the second derivative, you may take the derivatives of both sides of (3.13), but if you wish to avoid using the quotient rule, take the derivative of both sides of (3.12) to get

$$\left(2x - 18y\frac{dy}{dx}\right)\frac{dy}{dx} + \left(x^2 - 9y^2\right)\frac{d^2y}{dx^2} = -\left(2y + 2x\frac{dy}{dx}\right).$$

We have just shown that at $(1,1)$, $\frac{dy}{dx} = \frac{1}{2}$, therefore the value of $\frac{d^2y}{dx^2}$ at $(1,1)$ is given by

$$[2 - 18(\tfrac{1}{2})](\tfrac{1}{2}) + (-8)\frac{d^2y}{dx^2} = -[2 + 2(\tfrac{1}{2})]$$

Solve to get $\frac{d^2y}{dx^2} = -\dfrac{1}{16}$.

Example 160 *Let y be defined implicitly by $x^2 + xy + y^2 = 7$. The point $(1, -3)$ is on the graph of y, (because when we substitute $x = 1$ and $y = 3$ in the left hand side of the equation, we get 7). Suppose we have to find the equation of the tangent at $(1, 3)$. We first differentiate implicitly and the result is*

$$2x + y + x\frac{dy}{dx} + 2y\frac{dy}{dx} = 0 \qquad (3.14)$$

We then substitute $x = 1$ and $y = -3$ in (3.14), and solve for the derivative, to get $\frac{dy}{dx} = -\frac{1}{5}$. Therefore the slope of the tangent at $(1, -3)$ is $-\frac{1}{5}$. Its equation must be given by

$$(y - (-3)) = -\tfrac{1}{5}(x - 1),$$

which simplifies to $y = -\frac{1}{5}x - \frac{14}{5}$.

Exercise 161

1. Use implicit differentiation to find $\frac{dy}{dx}$ given that y is defined implicitly by:

   (a) $xy + x^2y^2 = 5.$ (b) $x^3 - xy + y^3 = 1.$ (c) $x^{1/2} + y^{1/2} = 1.$

   (d) $x^2 = \frac{x-y^2}{x+y}$ (e) $x^2 + xy - y^2 = 1.$ (f) $x \sin y + 2y = 0$

2. Show that if $e^y - x = 0$ then $\frac{dy}{dx} = \frac{1}{e^y}$ and deduce that $\frac{dy}{dx} = \frac{1}{x}$.

3. Show that if $\tan y - x = 0$ then $\frac{dy}{dx} = \frac{1}{\sec^2 y}$ and deduce that $\frac{dy}{dx} = \frac{1}{1+x^2}$.

4. In each question, verify that the given point is on the curve defined implicitly by the given equation then find the equation of the tangent at that point.

   a) $xy + x^2y^2 = 6$, $(1, 2)$ b) $x^2y^2 = 4y$, $(2, 1)$

   c) $y^2 + 2xy - 4 = 0$, $(0, 2)$ d) $y^2 - 3x^2y = \cos x$, $(0, 1)$

   e) $x^3y^3 = 9y$, $(1, 3)$ f) $\cos y + y + x = 2$, $(1, 0)$

   g) $x^2 + y^2 - xy = 7$, $(2, -1)$ h) $xe^y - 3y = 1$, $(1, 0)$

5. The function y is defined implicitly by $x^2 + 2x + y^2 - 3y = 0$. Determine the points on its graph where it has horizontal tangents. Also determine the points where it has vertical tangents, (i.e. the points where the tangent has infinite slope).

6. Find the second derivative $\frac{d^2y}{dx^2}$ (or y'') at the given point, if $y = f(x)$ is defined implicitly by the given equation: (a) $x^3 + y^3 = 16$ at $(2, 2)$

   (b) $xy + y^2 = 1$ at $(0, -1)$ (c) $e^{xy} + 2y - 3x = \sin y$ at $\left(\frac{1}{3}, 0\right)$

7. Find the second derivative $\frac{d^2y}{dx^2}$ (or y'') given that y is defined implicitly by the equation:

   (a) $x^2 + y^2 = 4$ (b) $y^2x^2 + 3x - 4y = 5$ (c) $e^{xy} + 2y - 3x = \sin y$

8. *Show that the tangent to the curve $y^2 = x^3 - 6x + 4$ at the point $A(-1, 3)$ intersects the curve at another point B and give the coordinates of B.*

9. *Find the two points where the curve $x^2 + xy + y^2 = 7$ crosses the x-axis, and show that the tangents to the curve at these points are parallel. What are their equations?*

10. *A normal to a given curve at a point (a, b) on the curve is a line that is perpendicular to the tangent at (a, b). Find the equation of the normal to the parabola $y^2 - 4x = 0$ at the point $(1, 2)$.*

Related Rates

There are instances in which two or more changing quantities are related by an equation. Here are two examples:

Example 162 *Imagine a peaceful pond. Suppose you drop a stone in its middle. A circular ripple will spread outward.*

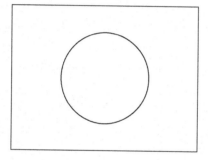 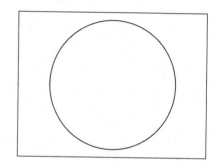

Ripple at some instant   *Ripple a little later*

At any time t before the ripple hits the edges of the pond, its radius $r(t)$ and the area $A(t)$ it encloses are related by the equation

$$A(t) = \pi \left[r(t) \right]^2 \quad \text{or simply} \quad A = \pi r^2$$

Example 163 *A ladder 5 meters long leans against a vertical wall. Say it starts sliding down.*

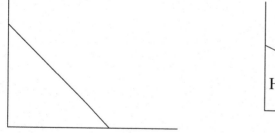 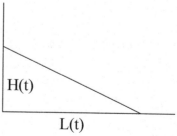

Ladder sliding down the wall

At a time t before the top of the ladder hits the bottom of the wall, let $H(t)$ be the distance from the bottom of the wall to the top of the ladder and $L(t)$ be the distance from the foot of the ladder to the bottom of the wall. Clearly, as time changes, $H(t)$ and $L(t)$ also change but they are related by the equation

$$[H(t)]^2 + [L(t)]^2 = 25 \quad \text{or simply} \quad H^2 + L^2 = 25$$

In general, suppose we have two quantities f and g which change with time and are related by some equation. If we know the rate of change of one of them, then, given some more relevant information, we may be able to determine the rate of change of the other. The standard procedure is to differentiate implicitly, (with respect to time), both sides of the equation connecting f and g.

Example 164 *Consider the two related variables A and r in Example 162. The equation relating them is*

$$A = \pi r^2. \tag{3.15}$$

Say we know that the radius r is increasing at the constant rate of 0.5 meters per sec. (that is $\frac{dr}{dt} = 0.5$). Suppose we are required to determine the rate at which the area is changing when $r = 4$ meters. We would differentiate both sides of (3.15) implicitly with respect to t. The derivative of A is $\frac{dA}{dt}$ and the derivative of πr^2 is $\pi(2r)\frac{dr}{dt}$, hence

$$\frac{dA}{dt} = 2\pi r \frac{dr}{dt}$$

Substitute $r = 4$ and $\frac{dr}{dt} = 0.5$ to get

$$\frac{dA}{dt} = 2\pi(4)(0.5) = 4\pi$$

Therefore when $r = 4$, the area is <u>increasing</u> *(because the derivative is positive) at the rate of 4π square meters per second.*

Example 165 *Assume that the foot of the ladder in Example 163 is sliding away from the bottom of the wall at the constant rate of 0.1 meter per second, (i.e. $\frac{dL}{dt} = 0.1$. Say we want to determine the rate at which its top is sliding when it is 3 meters from the floor. We would differentiate both sides of the equation $H^2 + L^2 = 25$ implicitly with respect to t to get*

$$(2H)\frac{dH}{dt} + (2L)\frac{dL}{dt} = 0$$

It follows that when $H = 3$ then

$$2(3)\frac{dH}{dt} + (2L)(0.1) = 0.$$

The geometry of the problem provides additional information: By Pythagorath's theorem, when $H = 3$, $L = 4$. Therefore, when $H = 3$ then

$$\frac{dH}{dt} = -\frac{0.4}{3}$$

In other words, the top of the ladder is <u>sliding down</u> *at the rate of $\frac{0.4}{3}$ meters per second.*

Exercise 166

1. An observer is 6 kilometers away from the launch site of a rocket. A few seconds after it is launched vertically up, he observes that the angle of elevation of the rocket, from where he is watching, is 30° and it is increasing at the rate of r radians per second. Determine the speed of the rocket at that instant, in terms of r.

2. Crude oil spills out of a damaged tanker at the rate of 50 liters per minute. It spreads in a circle with a thickness of 0.8 centimeters. Given that one cubic meter equals 1000 liters, determine the rate at which the radius of the spill is increasing when it reaches (a) 100 meters, (b) 900 meters. Why does the rate decrease as the radius increases?

3. A boy is 1.6 meters tall and he walks in a straight line past a street light at a speed of 45 meters per minute. You may assume that he passes directly under it. The light is 5 meters above the horizontal ground.

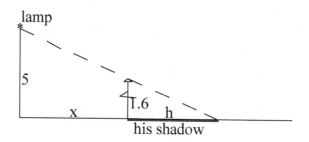

Show that when he is x meters from the point directly under the light, then the length h of his shadow is related to x by the equation

$$\frac{h}{1.6} = \frac{x+h}{5}$$

At what rate is his shadow lengthening? (In other words, what is $\frac{dh}{dt}$?)

4. A boy standing 2 meters from a street lamp throws a ball vertically up. The lamp is 12 meters above the ground.

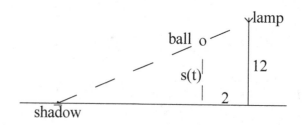

The distance s of the ball from the ground, t seconds after being released is $s(t) = 15t - 5t^2$ meters. Determine the speed of the ball's shadow along the ground, 1 second after the ball is thrown.

5. *Water is pumped into a hemispherical tank of radius 6 meters at the rate of 30 liters per minute. What is the rate at which the water level is rising when it reaches 4 meters? Assume that one cubic meter of water is equal to 1000 liters. You may use the fact, proved in Exercise 8, on page 185 that the volume of water in a hemispherical tank of radius R meters, when it is h meters deep, is $\pi \left(Rh^2 - \frac{h^3}{3} \right)$ cubic meters*

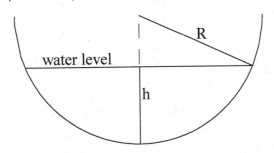

6. *The figure below shows a boy, 1.5 meters tall, walking in a straight line towards a bright lamp on the floor at a speed of 0.4 meters per second. Behind him, there is a vertical wall which is 7 meters from the lamp. Part of the boy's shadow is cast onto the wall. Let the height of the shadow on the wall be h meters when the distance of the boy from the wall is x meters.*

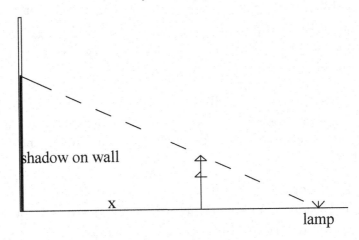

Show that $h(7-x) = 10.5$ then calculate the rate at which h is changing when he is 2 meters from the lamp.

Test your skills - 6, (90 minutes)

1. Find the derivative of each function

   (a) $r(x) = 4x^4 + 3x^2 + 5x - 1$

   (b) $b(x) = 1 + 4e^x + 2\sqrt{x} - 4x$

   (c) $t(x) = 2x^{-2} + 6x + 8$

   (d) $w(x) = 3\sin x - 2\cos x + \frac{3}{4} - \frac{4}{x}$

   (e) $h(x) = 5x^2 e^{3x}$

   (f) $f(x) = \frac{\sin x}{\cos x - 2\sin x}$

   (g) $u(x) = 4 + 3e^{4x}\tan 2x$

   (h) $g(x) = 3x\sin x - 4x^2 e^x - 6$

   (i) $p(x) = 2e^x \left(x - \frac{1}{x} \right)^4$

   (j) $q(x) = 5\csc 4x + \frac{3}{4}\cot 2x$

   (k) $w(x) = x^2 e^{\sin 3x}$

   (l) $s(x) = 4x^2 \cos\left(\frac{2}{x} \right)$

2. Find the most general antiderivative of each given function:

    a) $f(x) = 4\sec^2 3x + 1$ b) $g(x) = \cos 2\left(\frac{4-x}{3}\right) - 9x$

    c) $h(x) = x^3\sqrt{1+x^4}$ d) $u(x) = (1 - 3x)^8$

    e) $w(x) = (1 + \tan x)^2 \sec^2 x$ f) $v(x) = \left(\sqrt{\sin x}\right)\cos x$

3. Find the *first* and the *second* order derivative of each function:

    (a) $f(x) = \frac{4}{3}x^3 + \frac{3}{x^3}$ (b) $v(x) = -3x\sin x$ (c) $w(x) = 3e^{2x}\sin 3x$

4. Imagine a rectangle inside a circle of radius 1. What is the largest possible area of such a rectangle?

5. The function $y = f(x)$ is defined implicitly by the equation $x^2 - 2xy + y^3 = 2$.

    (a) Determine $\frac{dy}{dx}$ in terms of x and y.

    (b) Determine the slope and equation of the tangent to the graph of y at the point $(1, -1)$.

6. The function $g(x) = x^3 + 6x^2 + 9x + 1$ is given.

    (a) Find $g'(x)$ then determine its critical points.

    (b) Establish the nature of each critical point then sketch the graph of g.

    (c) The sketch should show that $x = 0$ is an approximate solution of the equation $x^3 + 6x^2 + 9x + 1 = 0$. Use Newton's method to obtain a better approximate solution than $x = 0$.

7. A stone was thrown vertically up in the air and it was found that t seconds later, it was at a point $s(t)$ feet from the ground where $s(t) = 128t - 16t^2$, $0 \le t \le 8$. Determine the speed of the stone at time t and the largest height it attains.

Derivatives of Inverse Functions

Implicit differentiation enables us to determine the derivatives of inverse functions. In this section, we determine the derivatives of arcsin x, arccos x, arctan x, and ln x.

To find the derivative of arcsin x

Let $f(x) = \sin x$, $-\frac{1}{2}\pi \le x \le \frac{1}{2}\pi$. Its inverse is $f^{-1}(x) = \arcsin x$, also written as $\sin^{-1}(x)$, (which you should not mistake for $1/\sin x$). Its graph is shown below.

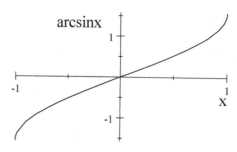

Graph of arcsin x

To simplify notation, write $y = \arcsin x$, (which one may read as "y is the angle whose sine is x"). Then

$$\sin y = x \qquad (3.16)$$

Differentiate both sides of (3.16) implicitly and solve for the derivative of y. You should get

$$\frac{dy}{dx} = \frac{1}{\cos y}.$$

But we need a derivative expressed in terms of x. To get it, we turn to the identity

$$\cos^2 y + \sin^2 y = 1$$

It implies that $\cos y = \pm\sqrt{1 - \sin^2 y} = \pm\sqrt{1 - x^2}$. Since the slope of the tangent at any point on the graph of y is positive, (see its graph), we must take the positive sign. Therefore if $y = \arcsin x$ then

$$\frac{dy}{dx} = \frac{1}{\sqrt{1 - x^2}}$$

To find the derivative of $\arccos x$

Let $g(x) = \cos x$, $0 \le x \le \pi$. Its inverse is $g^{-1}(x) = \arccos x$, with graph shown below.

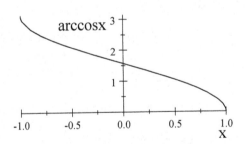

Graph of $\arccos x$

Once again, write $y = \arccos x$, (to be read as "y is the angle whose cosine is x"). Then

$$\cos y = x$$

When we differentiate implicitly and re-arrange the resulting equation we get

$$\frac{dy}{dx} = -\frac{1}{\sin y} = \pm\sqrt{\frac{1}{1 - \cos^2 x}} = \pm\frac{1}{\sqrt{1 - x^2}}$$

Since the slope of the tangent at any point on the graph of $\arccos x$ is negative, we must take the negative sign. Therefore if $y = \arccos x$ then

$$\frac{dy}{dx} = -\frac{1}{\sqrt{1 - x^2}}$$

The derivative of arctan x

Let $h(x) = \tan x$, $-\frac{1}{2}\pi \le x \le \frac{1}{2}\pi$. Its inverse is $h^{-1}(x) = \arctan x$ with graph shown below.

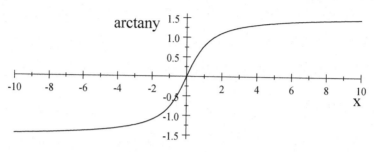

Graph of $\arctan x$

Write $y = \arctan x$, to be read "y is the angle whose tangent is x". It follows that $\tan y = x$. Take derivatives implicitly to get

$$\left(\sec^2 y\right) \frac{dy}{dx} = \left(1 + \tan^2 y\right) \frac{dy}{dx} = 1$$

We have used the identity $\sec^2 y = 1 + \tan^2 y$. Since $1 + \tan^2 y = 1 + x^2$, it follows that

$$\frac{dy}{dx} = \frac{1}{1 + x^2}$$

To find the derivative of $\ln x$

Let $w(x) = e^x$, where x is any real number. Its inverse is called the natural logarithm function and is denoted by $\ln x$. Its graph is given below.

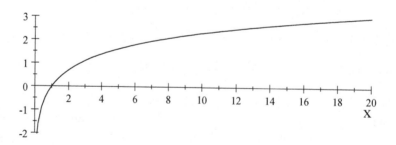

Graph of $\ln x$

As before write $y = \ln x$. Then $e^y = x$. Differentiate implicitly and solve for the derivative. Since $e^x = x$, the result should be

$$\frac{dy}{dx} = \frac{1}{x} \tag{3.17}$$

Note that the logarithm function $\ln x$ is defined for positive values of x. For $x < 0$, we have to consider another function, namely $g(x) = \ln(-x)$. By the chain rule, its derivative is

$$g'(x) = \frac{1}{(-x)} \cdot (-1) = \frac{1}{x}$$

Thus the derivative of $\ln(-x)$ is also $\frac{1}{x}$. The two results may be stated as follows:

$$\text{If } h(x) = \ln|x| \text{ then } h'(x) = \frac{1}{x}$$

We summarize all these results in a table:

Function	$\arcsin x$	$\arccos x$	$\arctan x$	$\ln x$
Derivative	$\dfrac{1}{\sqrt{1-x^2}}$	$-\dfrac{1}{\sqrt{1-x^2}}$	$\dfrac{1}{1+x^2}$	$\dfrac{1}{x}$

The chain rule extends to these functions as follows:

Function	Derivative
arcsin ( )	$\dfrac{1}{\sqrt{1-(\quad)^2}} \times$ Derivative of what is in ( )
arccos ( )	$-\dfrac{1}{\sqrt{1-(\quad)^2}} \times$ Derivative of what is in ( )
arctan ( )	$\dfrac{1}{1+(\quad)^2} \times$ Derivative of what is in ( )
ln | |	$\dfrac{1}{\boxed{}} \times$ Derivative of what is in | |

Exercise 167

1. *Calculate the derivative of each function and simplify your answer as much as possible. Where they appear, a and b are constants.*

   a) $f(x) = x^2 \ln x$

   b) $g(x) = \arcsin bx$

   c) $u(x) = (\ln x)\, e^x$

   d) $h(x) = 4 \arccos \frac{x}{3}$

   e) $w(x) = \frac{3}{4} \arcsin x$

   f) $v(x) = 2 \arctan bx$

   g) $f(x) = \frac{2 \arcsin(6x)}{3}$

   h) $g(x) = \ln(9x) - 16$

   i) $f(x) = \frac{5 \arctan x}{3}$

   j) $v(x) = \arctan \left(\frac{1}{x}\right)$

   k) $f(x) = \arcsin \left(\frac{1}{x}\right)$

   l) $u(x) = \arctan \left(x^2\right)$

   m) $g(x) = \sqrt{\ln x}$

   n) $h(x) = \ln(\sin x)$

   o) $v(x) = \ln \left(1 + 4x^3\right)$

   p) $w(x) = \ln \left(1 + bx\right)$

   q) $u(x) = \frac{4}{5} \arcsin(\frac{x}{4})$

   r) $z(x) = \ln(bx)$

   s) $h(x) = 5 \arctan(8x)$

   t) $f(x) = \arctan \left(\frac{ax}{b}\right)$

   u) $w(x) = 2 \arcsin \left(\sqrt{x}\right)$

   v) $f(x) = \arcsin \left(ax + b\right)$

   w) $v(x) = \ln \frac{x(x+5)}{x+3}$

   x) $z(x) = \arcsin \left(\frac{ax}{b}\right)$

2. *A lake polluted by coliform bacteria is treated with bacteria agents. Environmentalists estimate that t days after the treatment the number N of viable bacteria per milliliter will be given by*

$$N(t) = 10t - \ln \left(\tfrac{t}{10}\right) - 30.$$

(a) Find the number of viable bacteria per milliliter for $t = 3$, $t = 4.5$, and $t = 7.5$ days.

(b) Find and interpret the derivative of N for $t = 5$ days and $t = 9$ days.

3. Show that:

(a) If $f(x) = \arcsin(\cos x)$ then $f'(x) = -1$

(b) If $g(x) = \cos(\arcsin x)$ then $g'(x) = -\dfrac{x}{\sqrt{1-x^2}}$

(c) If $w(x) = \ln(\sec x + \tan x)$ then $w'(x) = \sec x$.

(d) If $u(x) = \ln(\sec x)$ then $u'(x) = \tan x$

(e) If $h(x) = \arccos\left(\dfrac{1}{x}\right)$ then $h'(x) = \dfrac{1}{x\sqrt{x^2-1}}$

(f) If $f(x) = \arctan\left(\dfrac{b}{x}\right)$ where b is a constant then $f'(x) = -\dfrac{b}{b^2+x^2}$

4. Given $h(x) = \cot(\arcsin x)$, determine $h'(x)$ and simplify your answer.

5. Given $g(x) = \ln(\csc x)$, determine $g'(x)$ and simplify your answer.

6. To find the derivative of a complicated function like $f(x) = \ln\dfrac{(x^2+1)(3x-1)}{(x^4+5)}$, first expand the logarithm expression:

$$\ln\frac{(x^2+1)(3x-1)}{(x^4+5)} = \ln(x^2+1) + \ln(3x-1) - \ln(x^4+5) \qquad (3.18)$$

The derivatives of the expressions in the right hand side of (3.18) are much easier to determine. We obtain

$$f'(x) = \frac{2x}{x^2+1} + \frac{3}{3x-1} - \frac{4x^3}{x^4+5}$$

Determine the derivative of each function below in a similar way:

(a) $\ln\left[x^3(1+\sin x)^4\right]$ (b) $\ln\dfrac{x^2}{(x+2)(3x-5)}$

(c) $\ln\dfrac{(x^2+3)^3}{\sqrt{3x-5}}$ (d) $\ln\sqrt{\dfrac{x+4}{x^2+4}}$

7. Let p be a positive number. Use L'Hopital's rule to show that $\lim\limits_{x\to\infty}\dfrac{\ln x}{x^p} = 0$.

8. Assume that you are required to determine $\lim\limits_{x\to\frac{\pi}{4}}(\tan x)^{1/(x-\frac{\pi}{4})}$. A direct substitution of $\frac{\pi}{4}$ does not work because the exponent becomes infinite when $x = \frac{\pi}{4}$. Let $f(x) = (\tan x)^{1/(x-\frac{\pi}{4})}$. Show that

$$\ln(f(x)) = \frac{\ln(\tan x)}{x - \frac{\pi}{4}}.$$

Now use L'Hopital's rule to verify that $\lim\limits_{x\to\frac{\pi}{4}}\ln(f(x)) = 2$, then deduce the value of $\lim\limits_{x\to\frac{\pi}{4}}(\tan x)^{1/(x-\frac{\pi}{4})}$.

9. Let n be a positive integer and $f(x) = x^n - \ln x$, $x \geq 1$. By applying the Mean Value Theorem to f on the interval $[1, x]$, show that $f(x) - f(1) > 0$ for all $x > 1$. Use this to deduce that $x^n > 1 + \ln x$ for all $x > 1$.

10. Suppose you are required to determine $\lim_{x \to 0^+} x \ln x$. In this form, it does not fall into one of the two forms to which we applied L'Hopital's rule on page 106 or 108. However, if we write $x \ln x$ as $\left(\dfrac{\ln x}{\frac{1}{x}} \right)$, it falls into the second form. Use L'Hopital's rule to determine $\lim_{x \to 0^+} \left(\dfrac{\ln x}{\frac{1}{x}} \right)$.

Now that we have the derivative of $\ln x$, we can determine derivatives of other functions involving exponents.

Example 168 *We know that the derivative of the exponential function $u(x) = e^x$, to base e is $u'(x) = e^x$. However, the derivatives of exponential functions to other bases are a little different. For example, consider $f(x) = 2^x$. Its derivative is not quite 2^x. To determine it, write $y = 2^x$ then take logarithms of both sides to base e. The result is*

$$\ln y = \ln (2^x) = x \ln 2$$

Now that x is no longer an exponent, we can easily take derivatives implicitly, with respect to x, (remember that $\ln 2$ is a constant), to get

$$\frac{1}{y} \frac{dy}{dx} = \ln 2.$$

Since $y = 2^x$, solving for $\dfrac{dy}{dx}$ gives

$$f'(x) = \frac{dy}{dx} = (\ln 2) y = (\ln 2) 2^x.$$

You should be able to show, in a similar way, that if b is any positive number then the derivative of $v(x) = b^x$ is $v'(x) = (\ln b) b^x$. This fact is worth recording:

The derivative of b^x is $(\ln b) b^x$

Example 169 *To determine the derivative of $g(x) = (3x)^{x^2}$, we write $g(x)$ as y to get $y = (3x)^{x^2}$ then take logarithms of both sides to base e. We end up expressing $\ln y$ as a product of two familiar functions:*

$$\ln y = x^2 \ln 3x$$

Taking derivatives of both sides, with respect to x, gives

$$\frac{1}{y} \frac{dy}{dx} = 2x \ln 3x + x^2 \cdot \frac{1}{3x} \cdot 3 = 2x \ln 3x + x = x (2 \ln 3x + 1).$$

It follows that $g'(x) = \dfrac{dy}{dx} = x (2 \ln 3x + 1) y = x (2 \ln 3x + 1) (3x)^{x^2}$.

Example 170 *Let b be a positive number that is not equal to 1. To determine the derivative of $h(x) = \log_b x$, we use the change of base formula to write h in terms of the natural logarithm as*

$$h(x) = \frac{\ln x}{\ln b}$$

It follows that $h'(x) = \dfrac{1}{\ln b} \cdot \dfrac{1}{x} = \dfrac{1}{(\ln b)\, x}.$

Exercise 171

1. Here is another method of determining the derivative of $f(x) = b^x$: Since e^x is the inverse of $\ln x$, $b = e^{\ln b}$. Therefore

$$f(x) = \left(e^{\ln b}\right)^x = e^{x \ln b}$$

Now use the chain rule to determine $f'(x)$.

2. Determine the derivative of each function:

   a) $u(x) = 5^{2x}$
   b) $v(x) = \log_{10}\left(3 + 2x^2\right)$
   c) $w(x) = (1+x)^{1/x}$

   d) $f(x) = x^{2x}$
   e) $g(x) = \ln \frac{3x(x+2)}{(x^2+1)(x^2+5)}$
   f) $z(x) = \left(1 + \frac{1}{x}\right)^x$

   g) $g(x) = \left(\sqrt{x}\right)^{x^2}$
   h) $h(x) = (x^x)^x$
   i) $u(x) = (x)^{x^x}$

3. Use the chain rule to find the derivative of

   a) $f(x) = \arctan \sqrt{x}$
   b) $g(x) = \arcsin\left(x^2\right)$
   c) $h(x) = \arccos\left(\frac{1}{x}\right)$

4. Let a be a constant. Show that the derivative of $w(x) = \ln\left(x + \sqrt{a^2 + x^2}\right)$ is $w'(x) = \frac{1}{\sqrt{x^2+a^2}}$

5. Let a be a constant and $v(x) = \ln\left(\frac{a + \sqrt{a^2 + x^2}}{x}\right)$. Show that $v'(x) = -\frac{a}{x\sqrt{a^2+x^2}}$.

6. The pH of a substance is defined by the formula

   $$\text{pH} = -\log\left[H^+\right]$$

   where $\left[H^+\right]$ is the concentration of hydrogen ions in the substance, measured in moles per liter, (which is abbreviated to moles/L). A certain chemical reaction causes the concentration of hydrogen ions in a substance to increase at the rate of 0.004 (moles/L)/sec. Find and interpret the rate of change in pH when $\left[H^+\right] = 0.15$ moles/L.

7. When oxyhemoglobin is reduced to hemoglobin, the hemoglobins are able to bind more H^+ ions in the blood thus regulating the acid level of blood. A formula to calculate a patient's blood pH level is

   $$\text{pH} = 6.1 + \log\left(\frac{x}{y}\right),$$

where x is the base concentration and y is the acid concentration. During hyperventilation, oxyhemoglobin is not reduced to hemoglobin quickly. The alveolar carbon dioxide pressure decreases, blood base level decrease and blood acid level increases. This condition is known as respiratory alkalosis where the arterial blood pH level exceeds 7.5

(a) Find the blood base to acid concentration ratio (x/y) for alkalosis.

(b) During alkalosis if $\dfrac{dx}{dt} = 0.5$ (mEq/L)/sec and $-\dfrac{dy}{dt} = 0.3$ (mEq/L)/sec when $x = 28$ mEq/L and $y = 2.6$ mEq/L, what is the rate of change in pH?

(c) Interpret and explain scenario (b) above.

(d) How many moles per liter of hydrogen ions must be present for alkalosis to be diagnosed?

8. Acid rain is toxic to vegetation and aquatic life. In northeastern U.S. forest soils are naturally acidic and their surface waters are mildly alkaline (basic). Acid rain increases the amount of positive hydrogen ions in the soil. If rainfall over a lake causes the pH to decrease at the rate of $-\frac{d\text{pH}}{dt} = 0.8$/hr, find the rate of change in hydrogen ions in the lake.

9. Give the most general antiderivative of the given function:

a) $f(x) = \frac{3}{x}$ b) $g(x) = \frac{2}{\sqrt{1-x^2}}$ c) $h(x) = \frac{3}{2(\sqrt{1-x^2})}$

d) $h(x) = \frac{5}{4(x^2+1)}$ e) $u(x) = \frac{5}{4x^2+1}$ f) $v(x) = \frac{4}{5(x+2)}$

g) $w(x) = \frac{2}{x^2} - \frac{4}{x}$ h) $f(x) = \frac{x}{\sqrt{1-x^2}}$ i) $g(x) = x + \frac{1}{\sqrt{1-x^2}}$

j) $h(x) = x - \frac{1}{x^2+1}$ k) $u(x) = \frac{5}{x} - \frac{1}{x^2+1}$ l) $v(x) = \frac{6}{x} - \frac{1}{\sqrt{1-x^2}}$

The Hyperbolic Functions

The hyperbolic sine, cosine and tangent are defined in this section. You are asked to determine some of their properties in the exercises.

The hyperbolic sine function

The hyperbolic sine function is denoted by $\sinh x$, (pronounced "sine hyperbolic x", or "shine x" or "sinch x"). Its domain is the set of all real numbers and its formula is $\sinh x = \frac{1}{2}\left(e^x - e^{-x}\right)$. A section of its graph is given below.

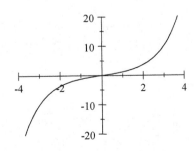

Unlike the trigonometric sine function, $\sinh x$ is not periodic and its values can exceed 1 or go below -1. It follows from the formula that:

1. $\sinh 0 = 0$.

2. $\sinh x \to \infty$ as $x \to \infty$ and $\sinh x \to -\infty$ as $x \to -\infty$.

3. $\sinh x$ has no critical point because its derivative $\frac{1}{2}\left(e^x + e^{-x}\right)$ is positive for all real numbers x.

The hyperbolic cosine function

The hyperbolic cosine function is denoted by $\cosh x$, (pronounced "cosine hyperbolic x", or "kosh x"). Its domain is the set of all real numbers and its formula is $\cosh x = \frac{1}{2}\left(e^x + e^{-x}\right)$. A section of its graph is given below.

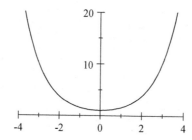

Like the hyperbolic sine function, it is not periodic and its values can exceed 1. Its formula implies that:

1. $\cosh 0 = 1$.

2. $\cosh x \to \infty$ as $x \to \infty$ and $\cosh x \to \infty$ as $x \to -\infty$.

3. It has a critical point $x = 0$, which is a point of relative minimum.

The hyperbolic tangent function

The hyperbolic tangent function is denoted by $\tanh x$, (pronounced "tan hyperbolic x"). Its domain is the set of all real numbers and its formula is

$$\tanh x = \frac{\sinh x}{\cosh x} = \frac{e^x - e^{-x}}{e^x + e^{-x}}.$$

A section of its graph is given below

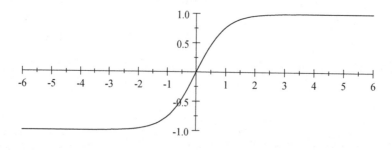

It has no critical points and all its values are between -1 and 1. This is because $e^x + e^{-x}$ is bigger than $|e^x - e^{-x}|$ for all real numbers x.

Three more hyperbolic functions

The hyperbolic secant, cosecant and cotangent are defined the same way the corresponding trigonometric functions are defined. Thus

$$\coth x = \frac{1}{\tanh x}, \quad \operatorname{sech} x = \frac{1}{\cosh x}, \quad \text{and} \quad \operatorname{csch} x = \frac{1}{\sinh x}.$$

These functions satisfy identities that should remind you of some trigonometric identities. Examples:

1. Since $\cosh x + \sinh x = e^x$ and $\cosh x - \sinh x = e^{-x}$, it follows that

$$1 = e^x e^{-x} = (\cosh x + \sinh x)(\cosh x - \sinh x) = \cosh^2 x - \sinh^2 x$$

   The corresponding trigonometric identity is $1 = \cos^2 x + \sin^2 x$.

2. Since $\cosh^2 x - \sinh^2 x = 1$, dividing both sides of the identity by $\cosh^2 x$ gives $1 - \tanh^2 x = \operatorname{sech}^2 x$. The corresponding trigonometric identity is $\sec^2 x = 1 + \tan^2 x$.

3. Since $\cosh^2 x - \sinh^2 x = 1$, dividing both sides of the identity by $\sinh^2 x$ gives $\coth^2 x - 1 = \operatorname{csch}^2 x$. The corresponding trigonometric identity is $\csc^2 x = 1 + \cot^2 x$.

4. $\sinh(x+y) = \sinh x \cosh y + \cosh x \sinh y$. To prove this, note that $\sinh x \cosh y + \cosh x \sinh y$ may be expanded as

$$\tfrac{1}{2}\left(e^x - e^{-x}\right)\tfrac{1}{2}\left(e^y + e^{-y}\right) + \tfrac{1}{2}\left(e^x + e^{-x}\right)\tfrac{1}{2}\left(e^y - e^{-y}\right)$$

   This expands into

$$\tfrac{1}{4}\left(e^{x+y} + e^{x-y} - e^{y-x} - e^{-x-y}\right) + \tfrac{1}{4}\left(e^{x+y} - e^{x-y} + e^{y-x} - e^{-x-y}\right),$$

   which simplifies to $\tfrac{1}{2}\left(e^{x+y} - e^{-x-y}\right) = \sinh(x+y)$

   The corresponding trigonometric identity is $\sin(x+y) = \sin x \cos y + \cos x \sin y$

Exercise 172

1. *Show that the derivative of* $\sinh x$ *is* $\cosh x$.

2. *Show that the derivative of* $f(x) = \cosh x$ *is* $f'(x) = \sinh x$, *(NOT* $-\cosh x$).

3. *Use the quotient rule to show that the derivative of* $g(x) = \tanh x$ *is* $g'(x) = \frac{\cosh^2 x - \sinh^2 x}{\cosh^2 x} = \operatorname{sech}^2 x$ *and that of* $h(x) = \coth x$ *is* $h'(x) = -\operatorname{csch}^2 x$.

4. *Show that the derivative of* $u(x) = \operatorname{sech} x$ *is* $u'(x) = -\operatorname{sech} x \tanh x$ *and the derivative of* $v(x) = \operatorname{csch} x$ *is* $v'(x) = -\operatorname{csch} x \coth x$

5. *Prove the following identities:*

(a) $\sinh(-x) = -\sinh x$

(b) $\cosh(-x) = \cosh x$

(c) $\cosh(x+y) = \cosh x \cosh y + \sinh x \sinh y$

(d) $\cosh 2x = 2\sinh^2 x + 1$

(e) $\sinh(x-y) = \sinh x \cosh y - \cosh x \sinh y$

(f) $\cosh 2x = 2\cosh^2 x - 1$

(g) $\cosh(x-y) = \cosh x \cosh y - \sinh x \sinh y$

(h) $\sinh 2x = 2\sinh x \cosh x$

We record (d), (f) and (h) for later use in the following form:

$$i)\ \sinh^2 x = \tfrac{1}{2}(\cosh 2x - 1) \qquad ii)\ \cosh^2 x = \tfrac{1}{2}(\cosh 2x + 1)$$
$$iii)\ \sinh x \cosh x = \tfrac{1}{2}\sinh 2x. \tag{3.19}$$

6. *Show that:* (a) $\cosh(\ln x) = \tfrac{1}{2}\left(x + \tfrac{1}{x}\right)$, (b) $\sinh(\ln x) = \tfrac{1}{2}\left(x - \tfrac{1}{x}\right)$, (c) $\tanh(\ln x) = \frac{x^2-1}{x^2+1}$.

7. *Use the chain rule to find the derivative of each function. Where they appear, a and b are constants.*

(a) $f(x) = 4\sinh 3x$ (b) $g(x) = a\cosh bx$ (c) $h(x) = \sqrt{3 + \cosh 2x}$

(d) $v(x) = 5\cosh(x^2)$ (e) $z(x) = \ln(\sinh ax)$ (f) $w(x) = 4\tanh\sqrt{x}$

8. *Determine the most general antiderivative of each function:*

a) $f(x) = 3\sinh 4x + 4\cosh 3x$ b) $g(x) = \text{sech}^2 2x - 3$

c) $h(x) = \sin 2x + \sinh 3x + 7$ d) $u(x) = 5e^{2x} + \text{csch}^2\tfrac{1}{2}x$

Inverses of the hyperbolic functions

Consider $f(x) = \sinh x$. Its graph reveals that it is a one-to-one function, therefore it has an inverse, which we denote by $\sinh^{-1} x$. To obtain a formula for $\sinh^{-1} x$, we write the equation

$$y = \sinh x \tag{3.20}$$

and solve for x in terms of y. Clearly, (3.20) implies that

$$y = \frac{e^x - e^{-x}}{2} = \frac{e^{2x}-1}{2e^x}$$

This may be re-arranged as $(e^x)^2 - 2ye^x - 1 = 0$, which is a quadratic equation in e^x. By the quadratic formula

$$e^x = \frac{2y \pm \sqrt{4y^2+4}}{2} = y \pm \sqrt{y^2+1}$$

Since e^x cannot be negative, and $\sqrt{y^2+1} > y$, we must choose the positive sign. Therefore $e^x = y + \sqrt{y^2+1}$. This implies that $x = \ln\left(y + \sqrt{y^2+1}\right)$. Therefore

$$\sinh^{-1} y = \ln\left(y + \sqrt{y^2+1}\right) \tag{3.21}$$

Now consider $g(x) = \cosh x$, for $x \geq 0$, (we have to restrict the values of x to get a one-to-one function). Its inverse is denoted by $\cosh^{-1} x$. To determine its formula, we write the equation

$$y = \cosh x$$

then solve for x in terms of y. Go through the same routine as above. You should obtain

$$e^x = y \pm \sqrt{y^2 - 1}$$

Since $e^x \to \infty$ as $y \to \infty$ while $y - \sqrt{y^2 - 1} \to 0$ as $y \to \infty$, we must choose the positive sign. Therefore $e^x = y + \sqrt{y^2 - 1}$. This implies that $x = \ln\left(y + \sqrt{y^2 - 1}\right)$. Therefore

$$\cosh^{-1} y = \ln\left(y + \sqrt{y^2 - 1}\right), \quad y \geq 1 \tag{3.22}$$

Lastly, note that the graph of $h(x) = \tanh x$ indicates that h is one-to-one, therefore it has an inverse, denoted by $\tanh^{-1} x$. Write $y = \tanh x = \frac{e^x - e^{-x}}{e^x + e^{-x}}$ and re-arrange to get $e^{2x} = \left(\frac{1+y}{1-y}\right)$. Deduce that

$$\tanh^{-1} y = \tfrac{1}{2} \ln\left(\frac{1+y}{1-y}\right), \quad -1 < y < 1$$

Exercise 173

1. *Verify that if* $u(x) = \cosh^{-1} x$, $v(x) = \sinh^{-1} x$, *and* $w(x) = \tanh^{-1} x$ *then*

$$u'(x) = \frac{1}{\sqrt{x^2 - 1}}, \quad v'(x) = \frac{1}{\sqrt{x^2 + 1}}, \quad w'(x) = \frac{1}{1 - x^2}$$

2. *We showed above that if* $y = \cosh x$ *and* $y \geq 0$ *then* $e^x = y + \sqrt{y^2 - 1}$. *This implies that*

$$e^{-x} = \frac{1}{e^x} = \frac{1}{y + \sqrt{y^2 - 1}} = \frac{y - \sqrt{y^2 - 1}}{\left(y + \sqrt{y^2 - 1}\right)\left(y - \sqrt{y^2 - 1}\right)} = y - \sqrt{y^2 - 1}$$

Since $\sinh x = \tfrac{1}{2}(e^x - e^{-x})$, *it follows that* $\sinh x = \sqrt{y^2 - 1}$. *Show in a similar way that if* $y = \sinh x$ *then* $\cosh x = \sqrt{y^2 + 1}$.

These results will be used in integration by substitution. We mark them for later reference:

$$\left. \begin{array}{l} \textit{If } \cosh x = y \textit{ then } \sinh x = \sqrt{y^2 - 1}. \\ \textit{If } \sinh x = y \textit{ then } \cosh x = \sqrt{y^2 + 1}. \end{array} \right\} \tag{3.23}$$

3. *The function* $f(x) = \sec x$ *is one-to-one on* $\left[0, \tfrac{1}{2}\pi\right]$ *therefore we may define an inverse function* $g(x) = \operatorname{arcsec} x$.

Show that its derivative is $g'(x) = \frac{1}{x\sqrt{x^2 - 1}}, x > 1$

In exercise 3 on page 124 you showed that the derivative of $\arccos\left(\frac{1}{x}\right)$ is also $\frac{1}{x\sqrt{x^2-1}}$. Is $\operatorname{arcsec} x$ related to $\arccos\left(\frac{1}{x}\right)$ or this is pure chance? Defend your answer.

4. *The function $f(x) = \csc x$ is one-to-one on $\left[0, \frac{1}{2}\pi\right]$ therefore we may define an inverse function $h(x) = \operatorname{arccsc} x$. Determine its derivative.*

Test your skills - 7,　(90 minutes)

1. Find the derivative of each function

   (a) $h(x) = \frac{1}{3}x^{1/3} - \frac{1}{4}x^{-1/4}$　　　　(b) $f(x) = 5x^2 e^{3x}$

   (c) $v(x) = -5\cos\left(\arcsin 3x\right)$　　　　(d) $g(x) = 1 - \frac{2}{3}\tanh\frac{3}{2}x$

   (e) $u(x) = e^{2x}\left(2 + e^x\right)^{-1}$　　　　(f) $w(x) = 10^{2x}\cosh 2x$

   (g) $f(x) = 4\arctan 3x - \pi x$　　　　(h) $g(x) = -\log_5\left(3x - 4\right)$

   (i) $v(x) = 5e^{2x}\sin x \cosh 3x$　　　　(j) $w(x) = \frac{5}{3}\sinh^2 6x - 2x^2$

2. Show that the derivative of $f(x) = \frac{\sin x - 2\cos x}{\cos x + 2\sin x}$ is $f'(x) = \frac{5}{(\cos x + 2\sin x)^2}$

3. Find the first and the second order derivatives of $f(x) = a\sec 3x$ where a is a constant.

4. Show that if $f(x) = \frac{\sqrt{x^2-4}}{4x}$ then $f'(x) = \frac{1}{x^2\sqrt{x^2-4}}$.

5. Determine the critical point of $g(x) = xe^x$, (there is only one), verify that it is a point of relative minimum, then sketch the graph of g.

6. Use L'Hopital's rule, or any other valid method, to calculate the following limits:

   a) $\lim\limits_{x\to 0}\frac{x+2\sin x}{x\cos x}$　　b) $\lim\limits_{x\to 0}\left(1+x\right)^{1/x}$　　c) $\lim\limits_{x\to\pi/4}\left(\frac{1-\tan x}{\sin x - \cos x}\right)$

7. A function y is defined implicitly by $x^2 + xy^2 - y = 5$. Use implicit differentiation to determine $\frac{dy}{dx}$. Now find the equation of the tangent to the graph of y at the point $(2, 1)$.

8. Determine the 3rd order Taylor polynomial for $f(x) = \sqrt{1 + 2x}$ at $c = 4$ and use it to calculate an approximate value of $\sqrt{7.8}$

9. Let $f(x) = \ln\left(x + \sqrt{a^2 + x^2}\right)$ where a is a constant. Show that $f'(x) = \frac{1}{\sqrt{a^2+x^2}}$.

10. Find the most general antiderivative of $f(x) = \frac{\sin x}{\cos^3 x}$.

Chapter 4   INTEGRATION

What we have done so far is called differential calculus. The focus has been on determining slopes of tangents and using them to solve a range of problems. To determine the slope of a tangent at a point $(x, f(x))$ on the graph a given function f, we found that we could not use the "*Rise*" over "*Run*" because we had only one point on the tangent, and two are needed to determine a *rise* and a *run*. We went around this by approximating the required slope with the slopes of secant lines then deduced the exact value from the approximations. (The slope of a secant line can be calculated because we know two points on the line.)

Integration address the following problem:
Given a function whose graph is above the x-axis on an interval $[a, b]$, what is the area of the region R between the graph and the x-axis?

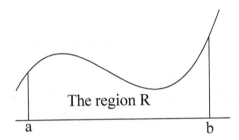

The region R

a   b

If the region is enclosed by straight line segments, school geometry suffices to solve the problem. Here are some examples:

Example 174 *Let f be the constant function $f(x) = 5$, $-2 \le x \le 4$. Its graph on the interval $[-2, 4]$ is a horizontal straight line segment from $(-2, 5)$ to $(4, 5)$. The region enclosed by the x-axis and the graph of f*

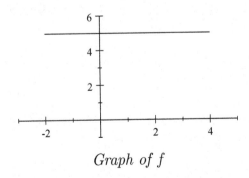

Graph of f

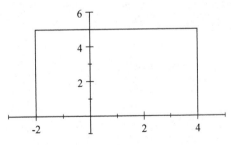

Region enclosed by the x-axis and graph

is a rectangle. Its area is 30 square units, (obtained by multiplying its length, which is 6 units, by its width, which is 5 units).

Example 175 *Let $f(x) = 4 - x$, $-1 \le x \le 3$. Its graph on the interval $[-1, 3]$ is a straight line segment from $(-1, 5)$ to $(3, 1)$, shown in the left diagram below. The region enclosed by the x-axis and its graph is shown in the right diagram. It*

134

consists of a 4 × 1 rectangle and a right triangle sitting on top of the rectangle.

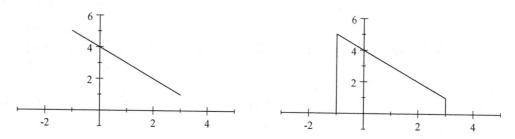

Its area can be calculated using school geometry. It is $4 \times 1 + \frac{1}{2}(4)(4) = 12$ square units

Example 176 *Consider the triangle enclosed by the x-axis and the graphs of $f(x) = \frac{1}{3}x + 1$ and $g(x) = 2x - 4$. Its base AB and height CD have length 5 and 2 units respectively, hence its area is $\frac{1}{2}(5)(2) = 5$ square units.*

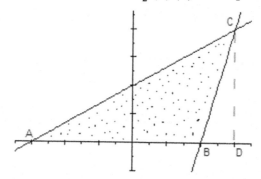

Unfortunately, if the region is enclosed by more complicated curves than line segments, school geometry does not work. This is when we turn to Calculus. Just as we did with slopes of tangents, we **calculate approximate values** of the required area and use the approximations to deduce its exact value. The following example illustrates the process:

Example 177 *A city owns a piece of land, (shown in the figure below), between a straight road, which we may imagine to be part of the x-axis, and a river shaped like the graph of the parabola $f(x) = x^2 + 3$, $0 \le x \le 2$ where x and $f(x)$ are in*

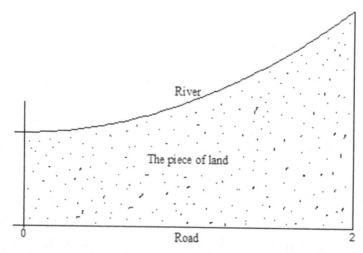

miles. Imagine being assigned the task of calculating its area in order to assess its value. The complicating factor is that the boundary of the river is not a straight line, therefore high school geometry does not work. As suggested above, it is reasonable to look for approximate values with the hope that the exact value will emerge. We get the approximate values by partition the piece of land into smaller strips and approximate each strip with an appropriate rectangle, (since we know how to calculate areas of rectangles). In the figure below, it is divided into ten strips of width 0.2 units each. Actually, the strips do not have to be of equal width; we chose equal width here to simplify the computations to follow.

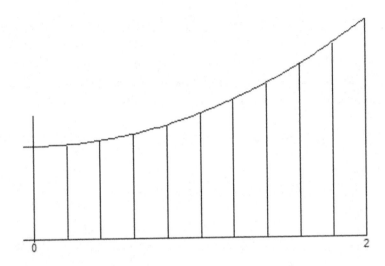

Each strip is then approximated with a rectangle. The figure below shows our choice of approximating rectangles.

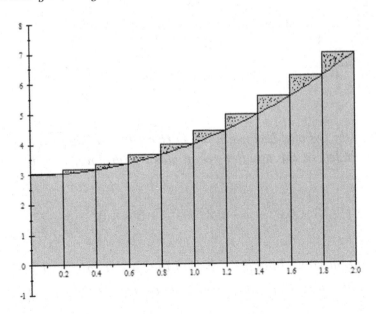

The area of the left-most rectangle is

$$(0.2)\left(3 + 0.2^2\right) \ square \ miles.$$

The next one has area

$$(0.2)\left(3+0.4^2\right)\ \ square\ miles.$$

Continuing in the same way, we compute the areas of all the remaining approximating rectangles then add up to get

$$0.2\left[3+0.2^2+3+0.4^2+3+0.6^2+\cdots+3+1.8^2+3+2^2\right]\ \ square\ miles.$$

Therefore, if the exact area of the land is A square miles then

$$A\simeq 0.2\left[3+0.2^2+3+0.4^2+3+0.6^2+\cdots+3+1.8^2+3+2^2\right]=9.08$$

The error in this estimate is the total area of the regions, (dotted in the above figure), that are above the parabola. We can reduce it by approximating the region with smaller rectangles as shown below where we have partitioned the land into 20 smaller strips.

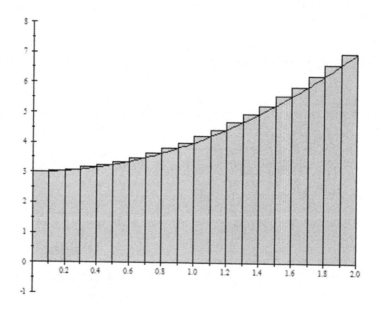

The error is definitely reduced because a big part of the region included in the first estimate is not included in the smaller rectangles. Therefore, a better approximation is

$$A\simeq 0.1\left[3+0.1^2+3+0.2^2+3+0.3^2+\cdots+3+1.9^2+3+2^2\right]=8.867$$

We now take a general step: Imagine dividing the region into n strips of width $\dfrac{2}{n}$ each then approximate each strip with a rectangle as above. We obtain the following estimate:

$$A\simeq \frac{2}{n}\left[3+\left(\frac{2}{n}\right)^2+3+\left(\frac{4}{n}\right)^2+3+\left(\frac{6}{n}\right)^2+\cdots+3+\left(\frac{2n-2}{n}\right)^2+3+\left(\frac{2n}{n}\right)^2\right]$$

This may be written, using the summation symbol, as

$$A \simeq \frac{2}{n}\left[\sum_{k=1}^{n}\left(3+\left(\frac{2k}{n}\right)^2\right)\right]$$

This sum includes n three's. They add up to 3n. Also, note that $\left(\frac{2}{n}\right)^2 = \frac{4}{n^2}$,

$\left(\frac{4}{n}\right)^2 = \frac{4}{n^2}(2)^2,\ \left(\frac{6}{n}\right)^2 = \frac{4}{n^2}(3^2),\ \ldots,\ \left(\frac{2n}{n}\right)^2 = \frac{4}{n^2}(n)^2.$ *Therefore*

$$A \simeq \frac{2}{n}\left[3n + \frac{4}{n^2}(1)^2 + \frac{4}{n^2}(2)^2 + \frac{4}{n^2}(3)^2 + \cdots + \frac{4}{n^2}(n)^2\right]$$

$$= 6 + \frac{8}{n^3}\left[1^1 + 2^2 + 3^2 + \cdots + n^2\right] = 6 + \frac{8}{n^3}\sum_{k=1}^{n}k^2$$

There is a formula, derived on page 237, for the sum of the squares of the first n positive integers. It is

$$\sum_{k=1}^{n}k^2 = 1^1 + 2^2 + 3^2 + \cdots + n^2 = \frac{n(n+1)(2n+1)}{6}$$

Applying it to the above sum gives

$$A \simeq 6 + \frac{8n(n+1)(2n+1)}{6n^3} = 6 + \frac{4}{3}\left(\frac{n+1}{n}\right)\left(\frac{2n+1}{n}\right) = 6 + \frac{4}{3}\left(1+\frac{1}{n}\right)\left(2+\frac{1}{n}\right)$$

For all large positive integers n, the numbers $6 + \frac{4}{3}\left(1+\frac{1}{n}\right)\left(2+\frac{1}{n}\right)$ *are close to*

$6 + \frac{8}{3} = \frac{26}{3}.$ *This must be the exact area of the piece of land in square miles.*

Example 178 *For another example consider the region R enclosed by the x-axis and the graph of* $f(x) = 2 + x^3$ *on the interval* $[0, 1]$. *Imagine dividing* $[0, 1]$ *into n equal subintervals*

$$\left[0, \tfrac{1}{n}\right],\ \left[\tfrac{1}{n}, \tfrac{2}{n}\right],\ \ldots,\ \left[\tfrac{n-1}{n}, \tfrac{n}{n}\right]$$

then approximate R with n rectangles as shown in the figure below.

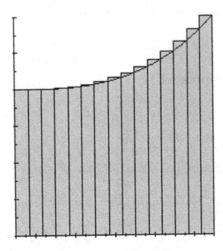

1. *Calculate the total area of the n rectangles.*

2. *Use the formula* $\displaystyle\sum_{k=1}^{n} k^3 = 1^3 + 2^3 + 3^3 + \cdots + k^3 = \left[\dfrac{k(k+1)}{2}\right]^2$, *(derived on page 237), to simplify your expression in (1)*

3. *Use the simplified expression in (2) to determine the exact area of R.*

You are probably thinking, (and worried), that we have to go through these elaborate computations every time we calculate an area. Fortunately, that is not the case. It is shown in the next section that if $f(x)$ has an antiderivative $F(x)$ then the limit of the above sums is $F(b) - F(a)$. In other words, the area enclosed by the x-axis and the graph of f on an interval $[a, b]$ is

$$F(b) - F(a) \tag{4.1}$$

To verify that this formula works, we apply it to the areas in the above examples.

1. The region in Example 174 is enclosed by the x-axis and the graph of $f(x) = 5$, on the interval $[-2, 4]$. An antiderivative for f is $F(x) = 5x$. By (4.1), the area of the region should be $F(4) - F(-2) = 20 - (-10) = 30$.

2. The region in Example 175 is enclosed by the x-axis and the graph of $f(x) = 4 - x$, on the interval $[-1, 3]$. An antiderivative for $f(x)$ is $F(x) = 4x - \frac{1}{2}x^2$. Therefore the area of the trapezium should be $F(3) - F(-1) = \left[12 - \frac{1}{2}(9)\right] - \left[-4 - \frac{1}{2}\right] = 12$.

3. In Example 176, triangle ACD is enclosed by the x-axis and the graph of $f(x) = \frac{1}{3}x + 1$ on $[-3, 3]$. An antiderivative of $f(x)$ is $F(x) = \frac{1}{6}x^2 + x$, hence ACD has area $F(3) - F(-3) = 6$. Triangle BCD is enclosed by the x-axis and the graph of $g(x) = 2x - 4$ on $[2, 3]$. An antiderivative for $g(x)$ is $G(x) = x^2 - 4x$, therefore the area of BCD should be $G(3) - G(2) = 1$. It follows that the shaded region has area $6 - 1 = 5$ units.

4. In Example 177, the region is enclosed by the x-axis and the graph of $f(x) = x^2 + 3$, on the interval $[0, 2]$. An antiderivative for $f(x)$ is $F(x) = \frac{1}{3}x^3 + 3x$, therefore the area under the parabola should be

$$F(b) - F(a) = F(2) - F(0) = \tfrac{8}{3} + 6 - 0 = \tfrac{26}{3}.$$

(We hope you are impressed.)

Exercise 179

1. *Find an antiderivative F of $f(x) = 2 + x^3$ and use it to verify that the area of the region R in Example 178 is $F(1) - F(0)$.*

2. *Use an antiderivative to calculate the area of the region:*

   (a) *Enclosed by the x-axis and the graph of $f(x) = x^3 + 2$ on $[-1, 2]$.*

(b) Enclosed by the x-axis and the graph of $f(x) = (1-x)(x+3)$.

(c) Enclosed by the x-axis and the graph of $f(x) = \sin x$ on $[0, \pi]$.

(d) Enclosed by the x-axis and the graph of $f(x) = x^3 + 2x$ on $[1, 3]$.

(e) Enclosed by the x-axis and the graph of $f(x) = \cos x$ on $\left[0, \frac{\pi}{2}\right]$.

(f) Enclosed by the x-axis and the graph of $f(x) = 3 + e^x$ on $[0, 2]$.

(g) Enclosed by the x-axis and the graph of $f(x) = x - \dfrac{1}{x^2}$ on $[1, 5]$.

(h) Below the line $y = 5$, but above the graph of $f(x) = 1 + x^2$.

(i) Below the graph of $f(x) = x^2 + 2x + 5$ on $[-1, 3]$ but above the line $y = 1$

Justifying the Area Formula

Let $f(x)$ be a function with an antiderivative $F(x)$. Let R be the region, (assumed to be above the x-axis), enclosed by the x-axis and graph of f between $x = a$ and $x = b$.

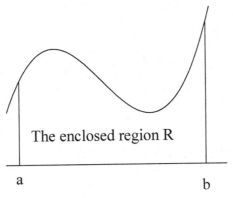

The enclosed region R

Denote its area by A. Following what we did in Example 177, we partition R into n strips of equal width $h = (b-a)/n$ as shown below. We then approximate each strip with an appropriate rectangle. A typical strip between $x = x_i$ and $x = x_{i+1}$ may be approximated with the rectangle whose base is the interval $[x_i, x_{i+1}]$ and height $f(x_i)$. Since the width of the strip is h, the area of the rectangle is $hf(x_i)$.

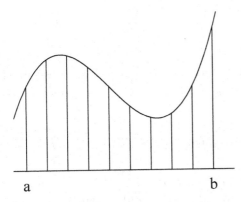

R partitioned into 9 strips

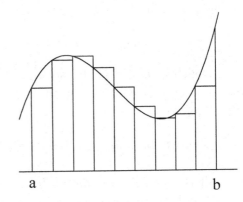

9 rectangles approximating R

The total area of all the n approximating rectangles is $hf(x_0) + hf(x_1) + hf(x_2) + \cdots + hf(x_{n-1})$, therefore

$$A \simeq hf(x_0) + hf(x_1) + hf(x_2) + \cdots + hf(x_{n-1})$$

We must reduce this sum to a simpler expression. This is where an antiderivative $F(x)$ of $f(x)$ comes into the picture, in the form of equation (1.12) on page 43. By definition, F satisfies the condition $F'(x) = f(x)$ for all x in the domain of f. By equation (1.12),

$$F(x + h) - F(x) = hF'(x) + e = hf(x) + e$$

where e is an error term that is close to 0 when h is "small". In particular,

$$F(x_1) - F(x_0) = F(x_0 + h) - F(x_0) = hf(x_0) + e_1$$

where e_1 is a very small error term. This enables us to express the term $hf(x_0)$, which appears in the approximation for A, in terms $F(x_1)$, $F(x_0)$, and the error term e_1. In fact

$$\begin{aligned} hf(x_0) &= F(x_1) - F(x_0) - e_1 \\ &= F(x_1) - F(a) - e_1 \end{aligned}$$

Likewise,

$$\begin{aligned} hf(x_1) &= F(x_2) - F(x_1) - e_2 \\[2mm] hf(x_2) &= F(x_3) - F(x_2) - e_3 \end{aligned}$$

After n such steps we get

$$\begin{aligned} hf(x_{n-1}) &= F(x_n) - F(x_{n-1}) - e_n \\ &= F(b) - F(x_{n-1}) - e_n \end{aligned}$$

When we add, many terms cancel in pairs and the result is

$$hf(x_0) + hf(x_1) + hf(x_2) + \cdots + hf(x_{n-1}) = F(b) - F(a) - (e_1 + e_2 + \cdots + e_n)$$

The exact value of A is the limit of the sums

$$hf(x_0) + hf(x_1) + hf(x_2) + \cdots + hf(x_{n-1})$$

as h approaches 0. It can be shown, using theorems we do not address in this course, that if f is continuous on $[a, b]$ then $(e_1 + e_2 + \cdots + e_n)$ approaches 0 as $h \to 0$. Therefore

$$\lim_{h \to 0} [hf(x_0) + hf(x_1) + hf(x_2) + \cdots + hf(x_{n-1})] = F(b) - F(a).$$

This result is called the *Fundamental Theorem of Calculus*. The title "*Fundamental Theorem*" is fitting because it reduces the difficult task of calculating the limit of sums to a mere evaluation of an antiderivative at two points. A more precise form of the theorem which does not mention areas under curves is the following:

Let $f(x)$ be a given function that is continuous on an interval $[a, b]$ and let $F(x)$ be any one of its antiderivatives. Take a positive integer n and divide $[a, b]$ into n equal subintervals $[x_0, x_1]$, $[x_1, x_2]$, ..., $[x_{n-1}, x_n]$ of length $h = \frac{1}{n}(b-a)$ each. Thus $x_0 = a$, $x_1 = a + h$, $x_2 = x_1 + h$, ..., $x_i = x_{i-1} + h$, ..., and $x_n = b = x_{n-1} + h$. Form the sums

$$hf(x_0) + hf(x_1) + hf(x_2) + \cdots + hf(x_{n-1}) = \sum_{i=0}^{n-1} hf(x_i).$$

Then $\displaystyle\lim_{h \to 0} \sum_{i=0}^{n-1} hf(x_i) = F(b) - F(a)$

Some Standard Terms and Notation

As above, consider a function f defined on a closed interval $[a, b]$. Let x_0, x_1, x_2, ..., x_n be points in $[a, b]$ such that $a = x_0 < x_1 < x_2 < \cdots < x_n = b$. The set $\{x_0, x_1, x_2, \ldots, x_n\}$ is called a **partition** of $[a, b]$ and is denoted by P, (for partition). It subdivides the interval $[a, b]$ into smaller subintervals

$$[x_0, x_1], \ [x_1, x_2], \ \ldots, \ [x_{n-1}, x_n]$$

Note that the points x_0, x_1, x_2, ..., x_n do not have to be equally spaced. We used equally spaced ones in the above examples to simplify the computations.

The length of the longest subinterval among $[x_0, x_1]$, $[x_1, x_2]$, ..., $[x_{n-1}, x_n]$ is called the **"norm"** or "fineness" of the partition P and will be denoted by $||P||$. Pick points θ_1 in $[x_0, x_1]$, θ_2 in $[x_1, x_2]$, ..., θ_n in $[x_{n-1}, x_n]$. (In Examples 177 and 178, we picked $\theta_1 = x_1$, $\theta_2 = x_2$, ..., $\theta_n = x_n$ but we do not have to.)

1. The expression

$$f(\theta_1)(x_1 - x_0) + f(\theta_2)(x_2 - x_1) + \cdots + f(\theta_n) = \sum_{i=1}^{n} f(\theta_i)(x_i - x_{i-1}) \quad (4.2)$$

   is called a **Riemann sum** of f on $[a, b]$ corresponding to the points x_0, x_1, x_2, ..., x_n. (Visualize it as an approximate value of the area enclosed by the graph of f, the x-axis and the graph of f on the interval $[a, b]$.)

2. The limit of the Riemann sums as the norm $||P||$ of the partitions shrinks to 0 is denoted by

$$\int_a^b f(x)dx \quad \text{or simply} \quad \int_a^b f \, dx$$

   and it is called the **definite integral** of f on $[a, b]$. (Visualize it as the exact value of the area enclosed by the x-axis and the graph of f on the interval $[a, b]$.) We have therefore verified, (non-rigorously), that if $f(x)$ has an antiderivative $F(x)$ then the limit of its Riemann sums on $[a, b]$ is $F(b) - F(a)$. More precisely,

$$\lim_{||P|| \to 0} \sum_{i=1}^{n} f(\theta_i)(x_i - x_{i-1}) = F(b) - F(a)$$

The symbol $\int$ is called an integral sign and the function f is called an **integrand**. The numbers a and b on the integral sign are called, respectively, the **lower limit** and the **upper limit** of integration. For the moment, regard dx as a symbol that specifies the independent variable of the function f. When the variable is x, we say that we are integrating f over $[a, b]$ with respect to x. If the variable is a different letter, say u, we would say that f is integrated over $[a, b]$ with respect to u and write the definite integral as

$$\int_a^b f(u)\,du \quad \text{or simply} \quad \int_a^b f\,du$$

3. Determining the limit of the Riemann sums of f on $[a, b]$ is called **integrating** f **over** $[a, b]$.

4. An antiderivative of a given function $f(x)$, (actually, the most general antiderivative of $f(x)$), which we have so far denoted by the corresponding upper case $F(x)$, is formally denoted by $\int f(x)\,dx$ or simply $\int f\,dx$, and is called the **indefinite integral,** (or simply the integral), of f. Determining an antiderivative of $f(x)$ is called **integrating** f **with respect to** x.

5. The expression $(x_i - x_{i-1})$ is often denoted by Δx, (pronounced "delta x"). Consequently, the Riemann sum (4.2) may be written as $\sum_{i=1}^n f(\theta_i)\Delta x$.

6. A convenient way of evaluating $\int_a^b f(x)\,dx$ is to put an antiderivative of $f(x)$, (any one will do), in square brackets that have the lower and upper limit at the end as $\left[F(x)\right]_a^b$, evaluate what is in the brackets at b and a then subtract. For example

   (a) $\displaystyle\int_0^1 (x^2 + 1)\,dx = \left[\frac{x^3}{3} + x\right]_0^1 = \left(\tfrac{1}{3} + 1\right) - (0 + 0) = \frac{4}{3}$.

   (b) $\displaystyle\int_1^2 \left(x + \frac{1}{x}\right)dx = \left[\frac{x^2}{2} + \ln x\right]_1^2 = \left(\tfrac{1}{2}\cdot 4 + \ln 2\right) - \left(\tfrac{1}{2} + 0\right) = \frac{3}{2} + \ln 2$.

Since antiderivatives are useful, it is necessary to address systematic methods of evaluating them.

Integration by Inspection

Recall that determining antiderivatives is the reverse of differentiating functions. Thus given a function $h(x)$, we have to answer the question: "*what is the most general function f whose derivative is $h(x)$?*" Equivalently, we imagine a table similar to the one below, giving the derivatives of various functions. Then determining an antiderivative means reading the table from right to left and add a constant to what you get.

Function	Derivative
1. x^n	nx^{n-1}
2. $\sin x$	$\cos x$
3. $\cos x$	$-\sin x$
4. $\tan x$	$\sec^2 x$
5. $\csc x$	$-\csc x \cot x$
6. $\sec x$	$\sec x \tan x$
7. $\cot x$	$-\csc^2 x$
8. e^x	e^x
9. $\ln|x|$	$\dfrac{1}{x}$
10. $\arcsin x$	$\dfrac{1}{\sqrt{1-x^2}}$
11. $\arctan x$	$\dfrac{1}{1+x^2}$
12. $f(x) \pm g(x)$	$f'(x) \pm g'(x)$
13. $kf(x)$, k a constant	$kf'(x)$

$$(4.3)$$

For example, to determine $\int \sec x \tan x\, dx$ we look for $\sec x \tan x$ in the column for derivatives. We find it in formula number 6 paired with $\sec x$, therefore

$$\int \sec x \tan x\, dx = \sec x + c.$$

The reality though, is that it is impossible to construct a table containing the derivative of every conceivable differentiable function because there are infinitely many of them. In table (4.3), we listed the derivatives of the building blocks x^n, $\sin x$, etc, and two rules for calculating derivatives. As the examples below demonstrate, it is possible to determine integrals of a number of more complicated functions by using the integrals of these elementary functions and the listed rules.

Example 180 *Consider* $\int \left(2\sin x - \frac{4}{x}\right) dx$. *Even though $h(x) = 2\sin x - \frac{4}{x}$ is not listed in table (4.3), its elementary components $2\sin x$ and $-\frac{4}{x}$ are indirectly listed. Indeed formulas 3 and 13 imply that $2\sin x$ is the derivative of $-2\cos x$ while 9 and 13 imply that $-\frac{4}{x}$ is the derivative of $-4\ln x$. It follows from formula 12 that $2\sin x - \frac{4}{x}$ is the derivative of $-2\cos x - 4\ln x$, therefore*

$$\int \left(2\sin x - \frac{4}{x}\right) dx = -2\cos x - 4\ln x + c$$

144

Example 181 *To determine* $\int (x^2 + 3x - 1) \sqrt{x} dx$, *remove parentheses to get*

$$\int (x^2 + 3x - 1) \sqrt{x} dx = \int \left(x^{5/2} + 3x^{3/2} - x^{1/2} \right) dx$$

Now formulas 1 and 13 in the table imply that $x^{5/2}$ *is the derivative of* $\frac{2}{7} x^{7/2}$, $3x^{3/2}$ *is the derivative of* $\frac{6}{5} x^{5/2}$ *and* $-x^{1/2}$ *is the derivative of* $-\frac{2}{3} x^{3/2}$. *Therefore*

$$\int (x^2 + 3x - 1) \sqrt{x} dx = \frac{2}{7} x^{7/2} + \frac{6}{5} x^{5/2} - \frac{2}{3} x^{3/2} + c$$

Example 182 *Consider* $\int \dfrac{3x + 1}{2\sqrt{x}} dx$. *Unlike derivatives, there is no quotient rule for integrals, therefore divide as much as possible then look for antiderivatives. The result is*

$$\int \left(\frac{3x}{2\sqrt{x}} + \frac{1}{2\sqrt{x}} \right) dx = \int \left(\tfrac{3}{2}\sqrt{x} + \frac{1}{2\sqrt{x}} \right) dx = \int \left(\tfrac{3}{2} x^{1/2} + \tfrac{1}{2} x^{-1/2} \right) dx.$$

Formula 1 of the table implies that $\frac{3}{2}\sqrt{x}$ *is the derivative of* $x^{3/2}$ *and* $\frac{1}{2} x^{-1/2}$ *is the derivative of* $x^{1/2}$. *Therefore*

$$\int \frac{3x + 1}{2\sqrt{x}} dx = x^{3/2} + x^{1/2} + c$$

Example 183 *To determine* $\int x^2 \cos x^3 dx$, *observe that the integrand is a product of the terms* $\cos x^3$ *and* x^2. *The chain rule suggests that it was obtained by differentiating an expression involving* $\sin x^3$, *because*

The Derivative of $\sin ($ $)$ *is* $\cos ($ $) \times$ *Derivative of what is in* $($ $)$

Since the derivative of $\sin x^3$ *is* $(\cos x^3)(3x^2)$, *which is not quite* $x^2 \cos x^3$, *the choice* $\sin x^3$ *is off the target by a **constant** 3. But that is easy to fix; we simply divide* $\sin x^3$ *by 3. Therefore*

$$\int x^2 \cos x^3 dx = \tfrac{1}{3} \sin x^3 + c$$

You can easily check that the derivative of $F(x) = \frac{1}{3} \sin x^3 + c$ *is* $F'(x) = x^2 \cos x^3$.

Example 184 *To determine* $\int x\sqrt{3x^2 + 4} dx$, *we also note that the presence of the terms* $\sqrt{3x^2 + 4} = (3x^2 + 4)^{1/2}$ *and* x *suggest that the integrand must be the result of differentiating an expression involving* $(3x^2 + 4)^{3/2}$. *By the chain rule, the derivative of* $(3x^2 + 4)^{3/2}$ *is* $\frac{3}{2} \cdot (3x^2 + 4)^{1/2} (6x) = 9x\sqrt{3x^2 + 4}$ *which is off what we want by the constant factor 9. We fix this by dividing by 9. Therefore*

$$\int x\sqrt{3x^2 + 4} dx = \tfrac{1}{9} \left(3x^2 + 4 \right)^{3/2} + c$$

It should be easy to verify that the derivative of $f(x) = \frac{1}{9} \left(3x^2 + 4 \right)^{3/2} + c$ *is* $x\sqrt{3x^2 + 4}$.

The Trial-And-Error method we used in the above examples is called **integration by inspection** (or integration by guessing wisely).

Exercise 185

1. *Integrate each function by inspection. Check your answer by differentiating your indefinite integral.*

(1) $\displaystyle\int \sqrt{x}\,dx$

(2) $\displaystyle\int x^{10}\,dx$

(3) $\displaystyle\int \frac{1}{x^4}\,dx$

(4) $\displaystyle\int x^{-1}\,dx = \int \frac{1}{x}\,dx$

(5) $\displaystyle\int x^n\,dx,\ n \neq -1$

(6) $\displaystyle\int \frac{3}{\sqrt{x}}\,dx$

(7) $\displaystyle\int e^{4x}\,dx$

(8) $\displaystyle\int e^{0.5x+6}\,dx$

(9) $\displaystyle\int e^{-x}\,dx$

(10) $\displaystyle\int \left(\frac{4}{x} - \frac{x}{4}\right)\,dx$

(11) $\displaystyle\int \left(\pi^2 + \frac{8}{x}\right)\,dx$

(12) $\displaystyle\int \left(\frac{2}{\sqrt{1-x^2}}\right)\,dx$

(13) $\displaystyle\int \frac{3}{x^2+1}\,dx$

(14) $\displaystyle\int \left(x^3 - \frac{1}{4x^3}\right)\,dx$

(15) $\displaystyle\int \left(\sqrt{x} - \sqrt{2}\right)\,dx$

(16) $\displaystyle\int 3\csc x \cot x\,dx$

(17) $\displaystyle\int \left(7x^{\frac{5}{2}} - x^{\frac{3}{2}}\right)\,dx$

(18) $\displaystyle\int \left(1 + x\sqrt{x}\right)\,dx$

(19) $\displaystyle\int \left(\pi x - \sec^2 x\right)\,dx$

(20) $\displaystyle\int \left(\frac{3x+4}{\sqrt{x}}\right)\,dx$

(21) $\displaystyle\int \left(\frac{x^2+1}{x^3}\right)\,dx$

(22) $\displaystyle\int \left(\frac{2}{\pi x^2} - 3\right)\,dx$

(23) $\displaystyle\int (2x+1)^2\,dx$

(24) $\displaystyle\int x\left(3x^2+1\right)^3\,dx$

(25) $\displaystyle\int x^2\left(4x^3+1\right)^4\,dx$

(26) $\displaystyle\int x^3\left(5x^4+1\right)^5\,dx$

(27) $\displaystyle\int \left(\sqrt{8x+1}\right)\,dx$

(28) $\displaystyle\int \left(x\sqrt{8x^2+1}\right)\,dx$

(29) $\displaystyle\int \left(x^2\sqrt{x^3+1}\right)\,dx$

(30) $\displaystyle\int \frac{1}{\sqrt{8x+1}}\,dx$

(31) $\displaystyle\int \frac{x}{\sqrt{8x^2+1}}\,dx$

(32) $\displaystyle\int \frac{x^2}{\sqrt{8x^3+1}}\,dx$

(33) $\displaystyle\int (2x-3)^8\,dx$

(34) $\displaystyle\int x\left(2x^2-3\right)^8\,dx$

(35) $\displaystyle\int x^2\left(\frac{x^3}{2}-3\right)^8\,dx$

(36) $\displaystyle\int x^3\left(2x^4-3\right)^8\,dx$

(37) $\displaystyle\int \sin 5x\,dx$

(38) $\displaystyle\int \sin(5x-4)\,dx$

(39) $\displaystyle\int \sin(\tfrac{2}{3}x+1)\,dx$

(40) $\displaystyle\int \sin(ax+b)\,dx$

(41) $\displaystyle\int x\sec^2\left(x^2\right)\,dx$

(42) $\displaystyle\int x^2\sec^2\left(x^3\right)\,dx$

(43) $\displaystyle\int x^3\sec^2\left(x^4\right)\,dx$

(44) $\displaystyle\int x^4\sec^2\left(x^5\right)\,dx$

(45) $\displaystyle\int xe^{x^2}\,dx$

(46) $\displaystyle\int x^2e^{x^3}\,dx$

(47) $\displaystyle\int x^3e^{x^4}\,dx$

(48) $\displaystyle\int x^4e^{x^5}\,dx$

2. *(In this exercise, you obtain a result that is used, (see page 211), to derive Simpson's rule.) Let $q(x) = Ax^2 + Bx + C$ be a quadratic function and $[a,b]$ be an interval.*

(a) *Show that* $\int_a^b q(x)dx = \frac{(b-a)}{6}[2A(b^2+ab+a^2)+3B(a+b)+6C]$. *(Hint:* b^3-a^3 *factors as* $(b-a)(b^2+ab+a^2)$.*)*

(b) *Show that* $2A(b^2+ab+a^2)+3B(a+b)+6C$ *may be written as*

$$(Aa^2+Ba+C)+(Ab^2+Bb+C)+A(a+b)^2+2B(a+b)+4C$$

(c) *Note that* $A(a+b)^2+2B(a+b)+4C = 4\left[A\left(\frac{a+b}{2}\right)^2+B\left(\frac{a+b}{2}\right)+C\right]$.

Use this to deduce that $\int_a^b q(x)dx = \frac{(b-a)}{6}\left[q(a)+4q\left(\frac{a+b}{2}\right)+q(b)\right]$.

Integration by Substitution

There are integrals which may not yield to integration by inspection. An example is

$$\int_3^9 \frac{x^3}{(x+1)^2}dx.$$

In such a case, we change to a new variable, a process called integration by substitution, to convert the unfamiliar integral into a familiar one. To motivate this process, consider the problem of determining the area of the piece of land in Example 177 on page 134. Say it is given to an individual who is used to measuring distances in kilometers, and he insists on measuring distances along the road in kilometers. Then the land is between the 0 and the $\frac{16}{5} = 3.2$ kilometer marks, (because 1 mile is equal to $\frac{8}{5}$ kilometers). Say he divides it into small strips using subintervals $[0, u_1], [u_1, u_2], \ldots, [u_{n-2}, u_{n-1}], [u_{n-1}, \frac{16}{5}]$ of $[0, \frac{16}{5}]$ and approximates each strip with a rectangle as shown below.

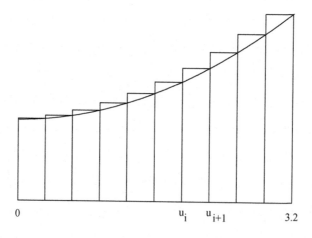

$\begin{array}{ccc} 0 & u_i \quad u_{i+1} & 3.2 \end{array}$

The strip on a typical interval $[u_{i-1}, u_i]$ is approximated with a rectangle of width $(u_i - u_{i-1})$ kilometers and height $\left[\left(\frac{5u_i}{8}\right)^2+3\right]$ miles. (To get the correct height of the rectangle, the distance u_i in kilometers has to be converted into miles before using the formula x^2+3 for the height.) To compute the area of the rectangle in square miles, we have to convert the $(u_i - u_{i-1})$ kilometers into miles, thus we multiply it by $\frac{5}{8}$. Then its area is

$$\left[\left(\frac{5u_i}{8}\right)^2+3\right]\left(\frac{5}{8}\right)(u_i - u_{i-1}) \text{ square miles.}$$

It follows that the area of the piece of land is the limit of the Riemann sums

$$\sum_{i=1}^{n} \left[\left(\tfrac{5}{8}u_i\right)^2 + 3 \right] \tfrac{5}{8} \left(u_i - u_{i-1}\right)$$

Take particular note of the factor $\tfrac{5}{8}$. We introduced it in order to convert the length $(u_i - u_{i-1})$, which is in kilometers, into miles. Therefore the area of the land is

$$\int_0^{16/5} \left[\left(\tfrac{5}{8}u\right)^2 + 3 \right] \tfrac{5}{8} du$$

We say that we have made a substitution $x = \tfrac{5}{8}u$. There is no standard term for the factor $\tfrac{5}{8}$. We will simply call it a *scaling factor* for the substitution.

Note that if we define the function $g(u) = \tfrac{5}{8}u$, then the expression $\left(\tfrac{5}{8}u\right)^2 + 3$ is the composition $f \circ g(u) = f(g(u))$ and the scaling factor $\tfrac{5}{8}$ is the derivative of g. Furthermore, $\left[0, \tfrac{16}{5}\right]$ is the interval that g maps onto $[0, 2]$. Therefore we have verified that

$$\int_0^2 f(x)dx = \int_0^{16/5} f(g(u))g'(u)du$$

This little exercise reveals the following basic steps we take to determine a given integral $\int_a^b f(x)dx$ by substitution.

1. *We replace the independent variable x with a new variable u. This amounts to composing f with some suitable function g to get a new function $f(g(u))$.*

2. *We multiply $f(g(u))$ by a scaling factor which happens to be $g'(u)$.*

3. *We determine the interval $[c, d]$ that g maps onto $[a, b]$. Then*

$$\int_a^b f(x)dx = \int_c^d f(g(u))g'(u)du \tag{4.4}$$

This is the formula for integrating by substitution and a sketch of its proof is given in the next section. To apply it successfully, one should choose $g(u)$ such that $\int_c^d f(g(u))g'(u)du$ can be determined by inspection. A bad choice for $g(u)$ may give another unfamiliar integral $\int_c^d f(g(u))g'(u)du$.

Example 186 *Say we wish to integrate $\int_0^4 \dfrac{x}{\sqrt{2x+1}}dx$ by substitution: The integrand is*

$$f(x) = \frac{x}{\sqrt{2x+1}}.$$

We compose it with a suitable function g with the property that $f(g(u))g'(u)$ can be integrated by inspection. To this end, note that if the denominator of the integrand were $\sqrt{u}$ and the numerator was a polynomial in u then we would divide each term in the numerator by $\sqrt{u}$ and proceed to integrate by inspection, as in Example 182 *on page 144. Therefore it is reasonable to look for a function g with the property that when we form the composition f(g(u)), the expression $\sqrt{2x+1}$ becomes $\sqrt{u}$. The choice $g(u) = \left(\frac{u-1}{2}\right)$ does precisely that. (We get it by setting $u = 2x+1$ then solve for x.) The scaling factor is $g'(u) = \frac{1}{2}$, therefore*

$$f(g(u))g'(u) = \frac{1}{\sqrt{u}}\left(\frac{u-1}{2}\right) \cdot \frac{1}{2} = \frac{1}{4}\left(u^{1/2} - u^{-1/2}\right)$$

which we can integrate by inspection. We need the interval that g maps onto $[0, 2]$. Since g is increasing, we have to find numbers c and d such that $g(c) = 0$ and $g(d) = 4$. They are $c = 1$ and $d = 9$, (obtained by solving $\frac{u-1}{2} = 0$ and $\frac{u-1}{2} = 4$). Therefore,

$$\int_0^4 \frac{x}{\sqrt{2x+1}}dx = \int_1^9 \frac{1}{4}\left(u^{1/2} - u^{-1/2}\right) du = \frac{1}{4}\left[\frac{2}{3}u^{3/2} - 2u^{1/2}\right]_1^9 = \frac{10}{3}.$$

The following are short-cuts that lead to the same result. They are defended ahead.

- Instead of defining $g(u) = \left(\frac{u-1}{2}\right)$ then form the composition $f(g(u))$, simply let $u = 2x+1$ and substitute it into the expression $\dfrac{x}{\sqrt{2x+1}}$ to get $\dfrac{x}{\sqrt{u}}$. (In fact if we define $g(u) = \left(\frac{u-1}{2}\right)$ and form the composition $f(g(u))$, $\sqrt{2x+1}$ is replaced by $\sqrt{u}$.)

- Differentiate u with respect to x. The result is $\dfrac{du}{dx} = 2$. Now regard $\dfrac{du}{dx}$ as fraction and solve to get $dx = \frac{1}{2}du$. This is the scaling factor. (We are changing to a new independent variable u hence the symbol du.) Indeed $g'(u) = \frac{1}{2}$ by direct verification.

- Replace the interval $[0, 2]$ with $[u(0), u(2)] = [1, 9]$.

- Then $\displaystyle\int_0^2 \frac{x}{\sqrt{2x+1}}dx = \int_1^9 \frac{x}{2\sqrt{u}}du = \frac{1}{2}\int_1^9 \frac{x}{\sqrt{u}}du$. Since $u = 2x+1$, replace x with $\frac{1}{2}(u-1)$. Therefore

$$\frac{1}{2}\int_1^9 \frac{x}{\sqrt{u}}du = \frac{1}{2}\int_1^9 \frac{\frac{1}{2}(u-1)}{\sqrt{u}}du = \frac{1}{4}\int_1^9 \left(u^{1/2} - u^{-1/2}\right) du.$$

The rest is similar to what we have done above.

In general, to determine an unfamiliar integral $\displaystyle\int_a^b f(x)dx$ by substitution, we do the following:

1. Pick an expression $h(x)$ in the formula for $f(x)$ and denote it by u. This is called substituting u for $h(x)$.

2. Determine $u' = \dfrac{du}{dx}$ then regard $\dfrac{du}{dx}$ as a fraction with numerator du and denominator dx and solve for dx to get $dx = \dfrac{1}{u'}du$. This is the scaling factor for the change of variable.

3. Replace the dx in the integral $\int_a^b f(x)dx$ with $\frac{1}{u'}du$ then use the relation $u = h(x)$ to get rid of the variable x in the expression $f(x)\frac{1}{u'}du$. The result should be an equivalent expression in the new variable u.

4. Replace the lower limit a of integration by $u(a)$, (the value of u when $x = a$) and the upper limit by $u(b)$, (the value of u when $x = b$). Then

$$\int_a^b f(x)dx = \int_{u(a)}^{u(b)} (\text{Equivalent expression invariable } u)du.$$

Example 187 *To integrate $\int_0^3 x^2\sqrt{5x+1}dx$ by substitution, using the short-cuts,*

we let $u = 5x + 1$. Then $x = \frac{1}{5}(u-1)$, $\frac{du}{dx} = 5$, and $dx = \frac{1}{5}du$. When $x = 0$, $u = 1$ and when $x = 3$, $u = 4$. Therefore

$$\int_0^3 x^2\sqrt{5x+1}dx = \frac{1}{5}\int_1^4 \left[\frac{1}{5}(u-1)\right]^2\sqrt{u}du = \frac{1}{125}\int_1^4 \left(u^{5/2} - 2u^{3/2} + u^{1/2}\right)du$$

$$= \frac{1}{125}\left[\frac{2}{7}u^{7/2} - \frac{4}{5}u^{5/2} + \frac{2}{3}u^{3/2}\right]_1^4 = 0.129 \text{ (to 3 decimal places)}.$$

Justifying the short-cuts

Let $\int_a^b f(x)dx$ be a given integral. Say we have picked an expression $h(x)$ in the formula for $f(x)$ and have defined $u = h(x)$.

- If we choose $g = h^{-1}$, (the inverse of h) and form the composition $f(g(u))$, the expression $h(x)$ is replaced by u.

- Since $h(g(u)) = u$, the chain rule implies that $h'(g(u))g'(u) = 1$. Therefore

$$g'(u) = \frac{1}{h'(x)} = \frac{1}{u'}$$

It follows that multiplying $f(g(u))$ by $dx = \frac{1}{u'}du$ amounts to introducing the scaling factor $g'(u)$ for the integral.

- We have to find numbers c and d such that $a = g(c)$ and $b = g(d)$. Since h is the inverse of g,

$$c = h(g(c)) = h(a) = u(a) \quad \text{and} \quad d = h(g(d)) = h(b) = u(b).$$

Therefore

$$\int_a^b f(x)dx = \int_{u(a)}^{u(b)} f(g(u))g'(u)du = \int_{u(a)}^{u(b)} f(g(u))\frac{1}{u'}du.$$

Example 188 *To determine* $\int_0^1 \dfrac{x^3}{(3x+2)^2}dx$, *let* $u = 3x + 2$. *Then* $\dfrac{du}{dx} = 3$ *and*

so $dx = \frac{1}{3}du$. *When* $x = 0$, $u = 2$ *and when* $x = 1$, $u = 5$. *Also,* $x = \frac{1}{3}(u-2)$, *therefore*

$$\int_0^1 \frac{x^3}{(3x+2)^2}dx = \int_2^5 \frac{(u-2)^3}{27u^2} \cdot \frac{1}{3}du = \frac{1}{81}\int_2^5 \left(u - 6 + \frac{12}{u} - \frac{8}{u^2}\right)du$$

$$= \frac{1}{81}\left[\frac{1}{2}u^2 - 6u + 12\ln u + \frac{8}{u}\right]_2^5 = \frac{1}{81}(12\ln 2.5 - 9.9)$$

Example 189 *To determine* $\int_0^1 \dfrac{x^5}{\sqrt{x^2+1}}dx$, *let* $u = x^2 + 1$. *Then* $\dfrac{du}{dx} = 2x$ *and*

so $dx = \dfrac{1}{2x}du$. *When* $x = 0$, $u = 1$ *and when* $x = 1$, $u = 2$. *Therefore*

$$\int_0^1 \frac{x^5}{\sqrt{x^2+1}}dx = \int_1^2 \frac{x^5}{\sqrt{u}} \cdot \frac{1}{2x}du = \frac{1}{2}\int_1^2 \frac{x^4}{\sqrt{u}}du$$

Since $x^2 = u-1$, *we may replace* x^4 *with* $(x^2)^2 = (u-1)^2 = u^2 - 2u + 1$. *Therefore*

$$\int_0^1 \frac{x^5}{\sqrt{x^2+1}}dx = \frac{1}{2}\int_1^2 \frac{u^2 - 2u + 1}{\sqrt{u}}du = \frac{1}{2}\int_1^2 \left(u^{3/2} - 2u^{1/2} + u^{-1/2}\right)du$$

$$= \frac{1}{2}\left[\frac{2}{5}u^{5/2} - \frac{4}{3}u^{3/2} + 2u^{1/2}\right]_1^2 = \frac{7\sqrt{2}-8}{15}.$$

Example 190 *To determine* $\int \left(x^2\sqrt{4x+1}\right)dx$, *let* $u = 4x + 1$. *Then* $\dfrac{du}{dx} = 4$, *hence* $dx = \frac{1}{4}du$. *Also* $x = \frac{1}{4}(u-1)$. *Therefore*

$$\int \left(x^2\sqrt{4x+1}\right)dx = \int \frac{1}{16}(u-1)^2 \sqrt{u} \cdot \frac{du}{4} = \frac{1}{64}\int \left(u^{5/2} - 2u^{3/2} + u^{1/2}\right)du$$

$$= \frac{1}{64}\left(\frac{2}{7}u^{7/2} - \frac{4}{5}u^{5/2} + \frac{2}{3}u^{3/2}\right) + c$$

Thus $\int \left(x^2\sqrt{4x+1}\right)dx = \frac{1}{64}\left(\frac{2}{7}(4x+1)^{7/2} - \frac{4}{5}(4x+1)^{5/2} + \frac{2}{3}(4x+1)^{3/2}\right) + c$

Example 191 *To find* $\int \tan x\sqrt{\sec x}dx$, *let* $u = \sec x$. *Then* $\dfrac{du}{dx} = \sec x \tan x$ *and* $dx = \dfrac{du}{\sec x \tan x}$. *The integral becomes*

$$\int \tan x\sqrt{\sec x}dx = \int \sqrt{u} \cdot \frac{du}{\sec x} = \int \sqrt{u} \cdot \frac{du}{u} = \int \frac{1}{\sqrt{u}}du$$

$$= 2u^{1/2} + c = 2\sqrt{\sec x} + c.$$

Example 192 *To determine* $\int \sin^3 x \cos^4 x\, dx$, *let* $u = \cos x$. *Then* $\dfrac{du}{dx} = -\sin x$

and $dx = -\dfrac{1}{\sin x} du$. *The integral becomes*

$$\int \sin^3 x \cos^4 x\, dx = -\int \left(\sin^3 x\right) u^4 \cdot \frac{1}{\sin x} du = -\int \left(\sin^2 x\right) u^4 du$$

To replace the term $\sin^2 x$ *by an expression involving* u, *we use the trigonometric identity* $\sin^2 x = 1 - \cos^2 x = 1 - u^2$. *Therefore*

$$\int \sin^3 x \cos^4 x\, dx = -\int \left(1 - u^2\right) u^4 x\, dx = \int \left(u^6 - u^4\right) du = \frac{u^7}{7} - \frac{u^5}{5} + c$$

$$= \tfrac{1}{7} \cos^7 x - \tfrac{1}{5} \cos^5 x + c.$$

You should memorize the results of the next two examples.

Example 193 *To determine* $\int \cot x\, dx = \int \dfrac{\cos x}{\sin x} dx$, *let* $u = \sin x$. *Then* $\dfrac{du}{dx} =$
$\cos x$ *and* $dx = \dfrac{1}{\cos x} du$. *Therefore*

$$\int \cot x\, dx = \int \frac{\cos x}{\sin x} dx = \int \frac{\cos x}{u} \cdot \frac{1}{\cos x} du = \int \frac{1}{u} du = \ln|u| + c = \ln|\sin x| + c$$

Example 194 *To determine* $\int \tan x\, dx = \int \dfrac{\sin x}{\cos x} dx$, *let* $u = \cos x$. *Then* $\dfrac{du}{dx} =$
$-\sin x$ *and* $dx = -\dfrac{1}{\sin x} du$. *Therefore*

$$\int \tan x\, dx = \int \frac{\sin x}{\cos x} dx = -\int \frac{\sin x}{u \sin x} du = -\int \frac{1}{u} du = -\ln|u| + c = \ln|\sec x| + c$$

Example 195 *To determine* $\int \dfrac{1}{\sqrt{1 - 9x^2}} dx$, *write* $\sqrt{1 - 9x^2}$ *as* $\sqrt{1 - (3x)^2}$. *Now*
the substitution $u = 3x$ *converts the integrand into the familiar function* $\dfrac{1}{\sqrt{1 - u^2}}$.

Therefore let $u = 3x$. *Then* $\dfrac{du}{dx} = 3$ *and* $dx = \dfrac{du}{3}$. *The integral becomes*

$$\int \frac{1}{\sqrt{1 - 9x^2}} dx = \tfrac{1}{3} \int \frac{1}{\sqrt{1 - u^2}} du = \tfrac{1}{3} \arcsin u + c = \tfrac{1}{3} \arcsin 3x + c$$

Example 196 *To determine* $\int \dfrac{1}{1 + 16x^2} dx$, *note that* $\dfrac{1}{1 + 16x^2} = \dfrac{1}{1 + (4x)^2}$ *and*
the substitution $u = 4x$ *converts this into the familiar integrand* $\dfrac{1}{1 + u^2}$. *Therefore*

let $u = 4x$. *Then* $\dfrac{du}{dx} = 4$ *and* $dx = \tfrac{1}{4} du$. *The integral becomes*

$$\int \frac{1}{1 + 16x^2} dx = \tfrac{1}{4} \int \frac{1}{1 + u^2} du$$

$$= \tfrac{1}{4} \arctan u + c = \tfrac{1}{4} \arctan 4x + c$$

Example 197 *To determine* $\int \dfrac{1}{5+x^2}dx$, *we look for a substitution that changes* x^2 *into* $5u^2$ *so that we may factor out the 5. The choice* $x^2 = 5u^2$, *which is equivalent to* $\sqrt{5}u = x$, *does it. We find that* $\dfrac{du}{dx} = \dfrac{1}{\sqrt{5}}$, *hence* $dx = \sqrt{5}du$, *and the integral becomes*

$$\int \frac{1}{5+x^2}dx = \int \frac{\sqrt{5}}{5(1+u^2)}du = \frac{\sqrt{5}}{5}\arctan u + c = \frac{1}{\sqrt{5}}\arctan\left(\frac{x}{\sqrt{5}}\right) + c$$

Example 198 *To determine* $\int \dfrac{1}{a^2+b^2x^2}dx$ *where* a *and* b *are non-zero constants, we need a substitution that changes* b^2x^2 *into* a^2u^2, *(so we can factor out the* a^2*). The choice* $b^2x^2 = a^2u^2$, *which is equivalent to* $u = \dfrac{bx}{a}$, *does precisely that. We get* $\dfrac{du}{dx} = \dfrac{b}{a}$, *hence* $dx = \dfrac{a}{b}du$, *and the integral becomes*

$$\int \frac{1}{a^2+b^2x^2}dx = \frac{a}{b}\int \frac{1}{a^2(1+u^2)}du = \frac{1}{ab}\arctan u + c = \frac{1}{ab}\arctan\left(\frac{bx}{a}\right) + c.$$

Example 199 *To determine* $\int \dfrac{1}{\sqrt{a^2-b^2x^2}}dx$ *where* a *and* b *are non-zero constants, we again use the substitution* $u = \dfrac{bx}{a}$ *to change* b^2x^2 *into* a^2u^2. *Then* $\dfrac{du}{dx} = \dfrac{b}{a}$ *and* $dx = \dfrac{a}{b}du$. *The integral becomes*

$$\int \frac{1}{\sqrt{a^2-b^2x^2}}dx = \frac{a}{b}\int \frac{1}{a\sqrt{1-u^2}}dx = \frac{1}{b}\arcsin u + c = \frac{1}{b}\arcsin\left(\frac{bx}{a}\right) + c$$

Exercise 200

1. *Use the suggested substitution to determine the given integral*

(a) $\displaystyle\int_{-1}^{2} \frac{x}{\sqrt{x+2}}dx$, $u = x+2$

(b) $\displaystyle\int_{0}^{8} x^3\sqrt{x+1}dx$, $u = x+1$

(c) $\displaystyle\int_{5}^{10} \frac{x+1}{\sqrt{x-1}}dx$, $u = x-1$

(d) $\displaystyle\int \frac{1}{3+x^2}dx$, $\sqrt{3}u = x$

(e) $\displaystyle\int_{0}^{\frac{\pi}{3}} (\tan x)(\sec x)^{\frac{3}{2}}\,dx$, $u = \sec x$

(f) $\displaystyle\int x^2\sqrt{x+4}dx$, $u = x+4$

(g) $\displaystyle\int_{0}^{1} \frac{1-\sqrt{x}}{1+\sqrt{x}}dx$, $u = 1+\sqrt{x}$

(h) $\displaystyle\int (x+1)\sqrt{x-2}dx$, $u = x-2$

(i) $\displaystyle\int \frac{1}{\sqrt{1-3x^2}}dx$, $u = \sqrt{3}x$

(j) $\displaystyle\int_{0}^{7} \frac{x^2}{\sqrt{x+1}}dx$, $u = x+1$

(k) $\displaystyle\int \frac{(1+\sqrt{x})^3}{\sqrt{x}}dx$, $u = \sqrt{x}+1$

(l) $\displaystyle\int_{-2}^{13} \frac{x^2}{\sqrt{x+3}}dx$, $u = x+3$

(m) $\int_0^5 \dfrac{x+5}{\sqrt{x+3}}dx,\ u = x+3$ (n) $\int \dfrac{1}{\sqrt{1-(x-3)^2}}dx,\ u = x-3$

(o) $\int \dfrac{1}{1+5x^2}dx\ \ u = \sqrt{5}x$ (p) $\int \dfrac{x^3}{\sqrt{x^2+4}}dx,\ u = x^2+4$

(q) $\int \dfrac{x^2+2}{x+2}dx,\ u = x+2$ (r) $\int \dfrac{\tan x}{\sec^3 x}dx,\ u = \sec x$

2. *Since* $x^2 + 4x + 7 = 3 + (x+2)^2$, $\displaystyle\int \dfrac{1}{x^2+4x+7} = \int \dfrac{1}{3+(x+2)^2}dx$. *The*

substitution $\sqrt{3}u = x+2$ *gives*

$$\int \dfrac{1}{x^2+4x+7}dx = \dfrac{\sqrt{3}}{3}\int \dfrac{1}{1+u^2}du = \dfrac{1}{\sqrt{3}}\arctan u + c = \dfrac{1}{\sqrt{3}}\arctan\left(\dfrac{x+2}{\sqrt{3}}\right) + c$$

Integrate the following in a similar way:

a) $\int \dfrac{1}{2+2x+x^2}dx$ b) $\int \dfrac{1}{x^2+4x+8}dx$ c) $\int \dfrac{1}{x^2+3x+10}dx$

3. *Since* $4x - x^2 - 3 = 1 - (x-2)^2$, $\displaystyle\int \dfrac{1}{\sqrt{4x-x^2-3}}dx = \int \dfrac{1}{\sqrt{1-(x-2)^2}}dx$.

Let $u = x - 2$. *Then*

$$\int \dfrac{1}{\sqrt{4x-x^2-3}}dx = \int \dfrac{1}{\sqrt{1-u^2}}du = \arcsin u + c = \arcsin(x-2) + c.$$

Integrate the following in a similar way

(a) $\int \dfrac{1}{\sqrt{6x-x^2-8}}dx$ (b) $\int \dfrac{1}{\sqrt{8-2x-x^2}}dx$

4. *Evaluate the following:* (a) $\displaystyle\int_3^{11} \dfrac{x+1}{\sqrt{2x+3}}dx$ (b) $\displaystyle\int_0^1 \dfrac{x^3}{x^2+1}dx$

Justifying the Integration by Substitution Formula

Let $f(x)$ be a given function, and assume that we wish to determine $\displaystyle\int_a^b f(x)dx$.

Let g be a differentiable function that maps some interval $[c, d]$ onto $[a, b]$. To simplify the argument, assume that g is increasing on $[c, d]$. In theory, to determine $\displaystyle\int_a^b f(x)dx$, we do the following:

- Select points $a = x_0 < x_1 < \cdots < x_n = b$ that divide $[a, b]$ into smaller subintervals $[x_0, x_1], [x_1, x_2], \ldots, [x_{n-1}, x_n]$.

- Pick points t_1 in $[x_0, x_1]$, t_2 in $[x_1, x_2]$, $\ldots$, and t_n in $[x_{n-1}, x_n]$, and

- Form Riemann sums $\sum_{i=1}^{n} f(t_i)(x_i - x_{i-1})$.

Then $\int_a^b f(x)dx$ is the limit of the above Riemann sums as the lengths of the subintervals shrink to 0.

Take points $c = u_0 < u_1 < \cdots < u_n = d$ in $[c, d]$, shown on the number line below,

that divide $[c, d]$ into smaller subintervals $[u_0, u_1]$, $[u_1, u_2]$, ..., $[u_{n-1}, u_n]$. We use them to select the points x_0, x_1, ..., x_n and t_0, t_1, ..., t_n in $[a, b]$, in the following special way: We apply the Mean Value Theorem to g on the interval $[u_0, u_1]$ to deduce that there is a number s_1 in the interval $[u_0, u_1]$, (see the figure below), such that

$$g(u_1) - g(u_0) = (u_1 - u_0)\, g'(s_1)$$

We then choose $x_0 = g(u_0)$, $x_1 = g(u_1)$ and $t_1 = g(s_1)$. We next apply the theorem to g on $[u_1, u_2]$ to obtain a number s_2 in the interval $[u_1, u_2]$ such that

$$g(u_2) - g(u_1) = (u_2 - u_1)\, g'(s_2),$$

then choose $x_2 = g(u_2)$ and $t_2 = g(s_2)$.

We continue in the same way till the nth step when we use the theorem to get a number s_n in the interval $[u_{n-1}, u_n]$ such that

$$g(u_n) - g(u_{n-1}) = (u_n - u_{n-1})\, g'(s_n),$$

then choose $x_n = g(u_n)$ and $t_n = g(s_n)$. Now the above Riemann sum may be written as

$$\sum_{i=1}^{n} f(t_i)(x_i - x_{i-1}) = \sum_{i=1}^{n} f(g(s_i))\,[g(u_i) - g(u_{i-1})] = \sum_{i=1}^{n} f(g(s_i))g'(s_i)(u_i - u_{i-1}).$$

Clearly, $\sum_{i=1}^{n} f(g(s_i))g'(s_i)(u_i - u_{i-1})$ is a Riemann sum of $f(g(u))g'(u)$ on the interval $[c, d]$. The limit of such sums as the lengths of the intervals $[u_0, u_1]$, $[u_1, u_2]$, ..., $[u_{n-1}, u_n]$ shrink to zero is $\int_c^d f(g(u))g'(u)du$. Therefore

$$\int_a^b f(x)dx = \lim \sum_{i=1}^{n} f(g(s_i))g'(s_i)(u_i - u_{i-1}) = \int_c^d f(g(u))g'(u)du.$$

The log integrals

These are integrals in which the integrand has two parts; a numerator and a denominator. Furthermore, the numerator should be the derivative of the denominator.

Examples: $(a) \int \frac{\cos x}{\sin x} dx$, $(b) \int \frac{2x}{x^2 + 5} dx$, $(c) \int \frac{1 + 2x}{x^2 + x - 7} dx$, $(d) \int \frac{1}{x + 3} dx$

If the numerator differs from the derivative of the denominator by a constant factor then the integrand can be modified by simply multiplying, (and then divide), by a **constant** to put it in the above form.

Example 201 $\int \frac{x}{x^2 + 1} dx$ *is not in the correct form because the derivative of the denominator is* $2x$ *which differs from the numerator by the constant factor* 2. *We modify the integrand as follows*

$$\int \frac{x}{x^2 + 1} dx = \int \frac{2x}{2(x^2 + 1)} dx = \frac{1}{2} \int \frac{2x}{x^2 + 1} dx$$

Now the integrand is of the correct form.

Example 202 $\int \frac{4}{6x - 5} dx$ *is not in the correct form because the derivative of the denominator is* 6 *which differs from the numerator by a constant factor. We may modify the integrand as follows:*

$$\int \frac{4}{6x - 5} dx = 4 \int \frac{6}{6(6x - 5)} dx = \frac{4}{6} \int \frac{6}{6x - 5} dx = \frac{2}{3} \int \frac{6}{6x - 5} dx$$

Example 203 $\int \frac{\sin x}{\cos x} dx$ *is not in the correct form because the derivative of* $\cos x$ *is* $-\sin x$, *NOT* $\sin x$. *We may modify it as* $- \int \frac{-\sin x}{\cos x} dx$

To determine $\int \frac{h'(x)}{h(x)} dx$, *make the substitution* $u = h(x)$. *Then* $dx = \frac{1}{h'(x)} du$ *and the integral becomes*

$$\int \frac{h'(x)}{h(x)} dx = \int \frac{h'(x)}{h(x)} \cdot \frac{1}{h'(x)} du = \int \frac{1}{u} du = \ln|u| + c = \ln|h(x)| + c.$$

In plain words, an antiderivative of $\frac{h'(x)}{h(x)}$ *is the logarithm of* $|h(x)|$, *hence the term "log integral".*

Example 204 *The derivative of* $\sin x$ *is* $\cos x$ *and the derivative of* $\cos x$ *is* $-\sin x$, *therefore*

$$\int \frac{\cos x}{\sin x}dx = \ln|\sin x| + c \quad and \quad \int \frac{\sin x}{\cos x}dx = -\ln|\cos x| + c = \ln|\sec x| + c.$$

We met these two integrals in Examples 193 and 194 on page 151.

Example 205

1. $\displaystyle\int \frac{x}{x^2+1}dx = \frac{1}{2}\int \frac{2x}{x^2+1}dx = \frac{1}{2}\ln|x^2+1| + c$

2. $\displaystyle\int \frac{4}{6x-5}dx = \frac{2}{3}\int \frac{6}{6x-5}dx = \frac{2}{3}\ln|6x-5| + c$

3. $\displaystyle\int \frac{3}{x+5}dx = 3\int \frac{1}{x+5}dx = 3\ln|x+5| + c$

There are many integrals, with a numerator and a denominator, that may not be reduced to the form

$$\int \frac{\text{(Derivative of Denominator)}}{\text{(Denominator)}}dx.$$

The above technique **does not apply to them.** Examples: (a) $\displaystyle\int \frac{1}{x^3+1}dx$, (b) $\displaystyle\int \frac{2}{\sin x+5}dx$.

There is no way of converting $\displaystyle\int \frac{1}{x^3+1}dx$ into the above form. You may try

$$\int \frac{1}{x^3+1}dx = \int \frac{3x^2}{3x^2(x^3+1)}dx = \int \frac{1}{3x^2}\cdot\frac{3x^2}{(x^3+1)}dx$$

but this is probably the farthest you may go since you cannot pull the term $\dfrac{1}{3x^3}$ out of the integral sign because it is not a constant. Therefore do not generalize irrationally by assuming that the integral of any quotient $\dfrac{p(x)}{q(x)}$ is $\ln|q(x)|$. Before you write $\displaystyle\int \frac{p(x)}{q(x)}dx = \ln|q(x)|$, make sure that the derivative of the denominator $q(x)$ is the numerator $p(x)$.

Exercise 206

1. *Determine each indefinite integral:*

(a) $\displaystyle\int \frac{\sin x}{1+4\cos x}dx$ (b) $\displaystyle\int \frac{\sin x - \cos x}{\sin x + \cos x}dx$ (c) $\displaystyle\int \frac{x+8}{x+4}dx$

(d) $\displaystyle\int \frac{1+x}{1+x^2}dx$ (*Hint:* $\dfrac{1+x}{1+x^2} = \dfrac{1}{1+x^2} + \dfrac{x}{1+x^2}$.)

(e) $\int \dfrac{3+x}{x^2+2x+2} dx$ (Hint: $\dfrac{3+x}{x^2+2x+2} = \dfrac{x+1}{x^2+2x+2} + \dfrac{2}{x^2+2x+2}$.)

(f) $\int \dfrac{(2x+3)\,dx}{x^2+2x+2}$, (g) $\int \dfrac{\sec^2 x + \tan x \sec x}{\sec x + \tan x} dx$, (h) $\int \dfrac{e^{-x} - e^x}{e^{-x} + e^x} dx$

2. Determine $\int \dfrac{1}{1+e^x} dx$. (Multiply the numerator and denominator by e^{-x}.)

3. We may determine $\int \dfrac{2\sin x + 3\cos x}{\cos x + 2\sin x} dx$ as follows: First determine constants A and B such that

$$2\sin x + 3\cos x = A\left(\cos x + 2\sin x\right) + B\left(-\sin x + 2\cos x\right)$$

Note that A is multiplied by the denominator of the integrand while B is multiplied by the derivative of the denominator. Now solve for A and B. You should get $A = \frac{7}{5}$ and $B = \frac{4}{5}$. Therefore

$$\int \dfrac{2\sin x + 3\cos x}{\cos x + 2\sin x} dx = \int \tfrac{7}{5} dx + \tfrac{4}{5}\int \dfrac{-\sin x + 2\cos x}{\cos x + 2\sin x}$$
$$= \tfrac{7}{5}x + \tfrac{4}{5}\ln\left|\cos x + 2\sin x\right| + c$$

Use a similar method to determine the following:

a) $\int \dfrac{\cos x}{\sin x + 2\cos x} dx$ b) $\int \dfrac{2\sin x + 3\cos x}{4\cos x + 5\sin x} dx$ c) $\int \dfrac{\sin x + \cos x}{3\cos x - 2\sin x} dx$

d) $\int \dfrac{\sin x}{3\cos x - 2\sin x} dx$ e) $\int_0^{\pi/3} \dfrac{1}{1+\tan x} dx$ f) $\int \dfrac{e^x + 3e^{-x}}{e^x + e^{-x}} dx$

4. We know the antiderivatives of $\sin x$, $\cos x$, $\tan x$ and $\cot x$. The remaining elementary trigonometric functions are $\sec x$ and $\csc x$. There are several ways of determining them but here is one we bet you could not have guessed: Write $\sec x$ as

$$\dfrac{\sec x \left(\sec x + \tan x\right)}{\tan x + \sec x} = \dfrac{\sec^2 x + \sec x \tan x}{\tan x + \sec x}.$$

It turns out that the derivative of $\tan x + \sec x$, (which is the denominator) is equal the numerator $\sec^2 x + \sec x \tan x$. Therefore

$$\int \sec x\,dx = \int \dfrac{\sec x \left(\sec x + \tan x\right)}{\tan x + \sec x} dx = \ln\left|\tan x + \sec x\right| + c.$$

Determine $\int \csc x\,dx$ in a similar way.

Integration by Parts

This is essentially the reverse of the product rule for derivatives. Recall that the derivative of a product $f(x)g(x)$ is

$$\frac{d}{dx}(f(x)g(x)) = g(x)\frac{df}{dx} + f(x)\frac{dg}{dx}.$$

We may re-arrange this as

$$g(x)\frac{df}{dx} = \frac{d}{dx}(f(x)g(x)) - f(x)\frac{dg}{dx} \qquad (4.5)$$

Now consider antiderivatives of the terms in (4.5).

- An antiderivative of $g(x)\dfrac{df}{dx}$ is written as $\displaystyle\int g(x)\frac{df}{dx}dx$.

- An obvious antiderivative of $\dfrac{d}{dx}(f(x)g(x))$ is $f(x)g(x)$.

- An antiderivative of $-f(x)\dfrac{dg}{dx}$ is written as $-\displaystyle\int f(x)\frac{dg}{dx}dx$.

Therefore

$$\int g(x)\frac{df}{dx}dx = f(x)g(x) - \int f(x)\frac{dg}{dx}dx \qquad (4.6)$$

This is the formula for integrating by parts. To apply it to an integral $\displaystyle\int h(x)dx$, do the following:

(a) Write $h(x)$ as a product of two functions. Call one of them $g(x)$ and the other one $\dfrac{df}{dx}$.

(b) Determine $f(x)$ from your knowledge of $\dfrac{df}{dx}$ and differentiate g to get $\dfrac{dg}{dx}$.

(c) Substitute $f(x)$, $g(x)$, $\dfrac{df}{dx}$ and $\dfrac{dg}{dx}$ into (4.6).

Your choice of $f(x)$ and $g(x)$ should be such that $\displaystyle\int f(x)\frac{dg}{dx}dx$ is easier to evaluate than $\displaystyle\int g(x)\frac{df}{dx}dx$.

Example 207 *To determine* $\displaystyle\int x\sin x\,dx$.

We choose $g(x) = x$ *and* $\dfrac{df}{dx} = \sin x$. *Then* $f(x) = -\cos x$, *(there is no need to introduce a constant of integration at this stage), and* $\dfrac{dg}{dx} = 1$. *Substituting these into (4.6) gives*

$$\int x\sin x\,dx = -x\cos x + \int \cos x\,dx = -x\cos x + \sin x + c$$

Note that the choice $g(x) = \sin x$ and $\dfrac{df}{dx} = x$ leads to

$$\int x \sin x dx = \tfrac{1}{2}x^2 \sin x - \int \tfrac{1}{2}x^2 \cos x dx$$

which is no easier to evaluate than $\int x \sin x dx$.

It may be necessary to apply the technique more than once as the next example shows:

Example 208 *To determine* $\int x^2 e^{2x} dx$, *let* $g(x) = x^2$ *and* $\dfrac{df}{dx} = e^{2x}$. *Then* $\dfrac{dg}{dx} = 2x$ *and* $f(x) = \tfrac{1}{2}e^{2x}$. *Substituting into (4.6) gives*

$$\int x^2 e^{2x} dx = \tfrac{1}{2}x^2 e^{2x} - \int x e^{2x} dx.$$

We have reduced the power of x under the integral sign by 1. To evaluate $\int x e^{2x} dx$, *we integrate by parts a second time. Take* $g(x) = x$ *and* $\dfrac{df}{dx} = e^{2x}$. *(Hopefully, you do not confuse the $g(x)$ in this part of the solution with the $g(x)$ in the first part.) Then* $\dfrac{dg}{dx} = 1$ *and* $f(x) = \tfrac{1}{2}e^{2x}$. *Substituting into (4.6) gives*

$$\int x e^{2x} dx = \tfrac{1}{2}x e^{2x} - \int \tfrac{1}{2}e^{2x} dx = \tfrac{1}{2}x e^{2x} - \tfrac{1}{4}e^{2x} + c$$

Therefore $\int x^2 e^{2x} dx = \tfrac{1}{2}x^2 e^{2x} - \left(\tfrac{1}{2}x e^{2x} - \tfrac{1}{4}e^{2x}\right) + c = \tfrac{1}{2}e^{2x}\left(x^2 - x + \tfrac{1}{2}\right) + c.$

In the next example, we write the integrand $h(x)$ as $h(x) \cdot 1$ then choose $g(x) = h(x)$ and $\dfrac{df}{dx} = 1$.

Example 209 *To determine* $\int \arcsin x dx$, *choose* $g(x) = \arcsin x$ *and* $\dfrac{df}{dx} = 1$. *Then* $f(x) = x$ *and* $\dfrac{dg}{dx} = \dfrac{1}{\sqrt{1 - x^2}}$. *Therefore*

$$\int \arcsin x dx = x \arcsin x - \int \dfrac{x}{\sqrt{1 - x^2}} dx.$$

You should be able to determine $\int \dfrac{x}{\sqrt{1 - x^2}} dx = \int x \left(1 - x^2\right)^{-1/2} dx$ *by inspection. The result is*

$$\int \arcsin x dx = x \arcsin x + \sqrt{1 - x^2} + c$$

Exercise 210

1. Integrate $\int x \cos x dx$ by parts.

2. Integrate $\int x^2 \ln x dx$ by parts.

3. Show that if $k \neq -1$ and it is a constant then $\int x^k \ln x dx = \frac{x^{k+1}}{k+1} \left(\ln x - \frac{1}{k+1} \right)$.

4. Use the substitution $u = \ln x$ to show that $\int \frac{\ln x}{x} dx = \frac{1}{2} (\ln x)^2 + c$.

5. Use the result of Exercises 3 to determine $\int \ln x dx$.

6. Determine $\int \arctan x dx$.

7. Determine $\int x \arctan x dx$. (Hint: $\frac{x^2}{1+x^2} = \frac{1+x^2-1}{1+x^2} = 1 - \frac{1}{1+x^2}$.)

8. In some instances, two or more integrations by parts may lead one to an integral involving the given integrand . One is then able to form an equation involving the given integral and solve for it. For an example, consider

$$\int e^{2x} \cos x dx.$$

Take $g(x) = e^{2x}$ and $\frac{df}{dx} = \cos x$. Then $\frac{dg}{dx} = 2e^{2x}$, $f(x) = \sin x$ and

$$\int e^{2x} \cos x dx = e^{2x} \sin x - 2 \int e^{2x} \sin x dx \qquad (4.7)$$

The next step is to integrate $\int e^{2x} \sin x dx$ by parts. Take $g(x) = e^{2x}$ and $\frac{df}{dx} = \sin x$. Then $\frac{dg}{dx} = 2e^{2x}$ and $f(x) = -\cos x$, therefore

$$\int e^{2x} \sin x dx = -e^{2x} \cos x + \int 2e^{2x} \cos x dx.$$ Substituting into(4.7) gives

$$\int e^{2x} \cos x dx = e^{2x} \sin x + 2e^{2x} \cos x - 4 \int e^{2x} \cos x dx.$$ In other words,

$$\int e^{2x} \cos x dx = e^{2x} \sin x + 2e^{2x} \cos x - 4 \int e^{2x} \cos x dx \qquad (4.8)$$

Note that the given integral $\int e^{2x} \cos x dx$ appears in the right hand side. We now solve (4.8) for the integral and the result is

$$\int e^{2x} \cos x dx = \frac{e^{2x} (\sin x + 2 \cos x)}{5} + c$$

Let a and b be constants. Use a similar procedure to:

(a) *Determine* $\int e^{2x} \sin 3x dx$.

(b) *Show that* $\int e^{ax} \sin bx dx = \dfrac{e^{ax}(a \sin bx - b \cos bx)}{a^2 + b^2} + c$

(c) *Show that* $\int e^{ax} \cos bx dx = \dfrac{e^{ax}(b \sin bx + a \cos bx)}{a^2 + b^2} + c$

9. *Let n and b be constants. Show that*

$$\int x^n e^{bx} dx = \frac{x^n e^{bx}}{b} - \frac{n}{b} \int x^{n-1} e^{bx} dx. \qquad (4.9)$$

10. *Formula (4.9) is an example of a reduction formula. Using it enables one to reduce the exponent of x, hence the term "reduction formula". Let us apply it repeatedly to determine* $\int x^3 e^{4x} dx$. *Thus*

$$\int x^3 e^{4x} dx = \frac{x^3 e^{4x}}{4} - \frac{3}{4} \int x^2 e^{4x} dx, \qquad \textit{(first application)}$$

The second step is to apply it to $\int x^2 e^{4x} dx$ *and the result is*

$$\int x^2 e^{4x} dx = \frac{x^2 e^{4x}}{4} - \frac{1}{2} \int x e^{4x} dx.$$

Therefore

$$\int x^3 e^{4x} dx = \frac{x^3 e^{4x}}{4} - \frac{3}{4} \left(\frac{x^2 e^{4x}}{4} - \frac{2}{4} \int x e^{4x} dx \right) = \left(\frac{x^3}{4} - \frac{3x^2}{16} \right) e^{4x} + \frac{3}{8} \int x e^{4x} dx$$

Finally, we apply it to $\int x e^{4x} dx$ *to get*

$$\int x e^{4x} dx = \frac{x e^{4x}}{4} - \frac{1}{4} \int e^{4x} dx = \frac{x e^{4x}}{4} - \frac{e^{4x}}{16} + c$$

The last integral was determined by inspection. Therefore

$$\int x^3 e^{4x} dx = \left(\frac{x^3}{4} - \frac{3x^2}{16} + \frac{3x}{32} - \frac{3}{128} \right) e^{4x} + c$$

Determine $\int x^4 e^{2x} dx$ *in a similar way.*

11. *Use your answer to question 8c above to determine* $\int e^{5x} \cos 6x dx$.

12. *Use the substitution* $x = u^2$ *to show that* $\int \cos \sqrt{x} dx = 2 \int u \cos u du$. *Now integrate by parts to complete the integration.*

13. *Consider the integral* $\int \cos^n x dx$ *where n is a positive integer. We may write* $\cos^n x$ *as* $(\cos^{n-1} x)(\cos x)$. *Take* $g(x) = \cos^{n-1} x$ *and* $\dfrac{df}{dx} = \cos x$ *then integrate by parts to get*

$$\int \cos^n x dx = \cos^{n-1} x \sin x + (n-1) \int \cos^{n-2} x \sin^2 x dx$$

Use the identity $\sin^2 x = 1 - \cos^2 x$ *to deduce that*

$$\int \cos^n x dx = \frac{\cos^{n-1} x \sin x}{n} + \frac{n-1}{n} \int \cos^{n-2} x dx \qquad (4.10)$$

This is another reduction formula which we may use to integrate powers of $\cos x$. *For example,*

$$\int \cos^3 x dx = \frac{\cos^2 x \sin x}{3} + \frac{2}{3} \int \cos x dx = \frac{\cos^2 x \sin x}{3} + \frac{2 \sin x}{3} + c$$

Use it to determine $\int \cos^5 x dx$.

14. *Let n be a positive integer. Show that*

$$\int \sin^n x dx = -\frac{\sin^{n-1} x \cos x}{n} + \frac{n-1}{n} \int \sin^{n-2} x dx \qquad (4.11)$$

then use this reduction formula to determine $\int \sin^3 x dx$

15. *Complete the following exercise to derive the following reduction formula:*

$$\int \tan^n x dx = \frac{\tan^{n-1} x}{n-1} + \int \tan^{n-2} x dx \qquad (4.12)$$

We write $\tan^n x$ *as* $\tan^{n-2} x \tan^2 x$. *Then*

$$\begin{aligned} \int \tan^n x dx &= \int \tan^{n-2} x \tan^2 x dx = \int \tan^{n-2} x (\sec^2 x - 1) dx \\ &= \int \tan^{n-2} x \sec^2 x dx - \int \tan^{n-2} x dx \end{aligned}$$

Now integrate $\int \tan^{n-2} x \sec^2 x dx$ *by parts and complete the exercise.*

16. *Complete the following exercise to derive the following reduction formula:*

$$\int \sec^n x dx = \frac{\sec^{n-2} x \tan x}{n-1} + \frac{n-2}{n-1} \int \sec^{n-2} x dx. \qquad (4.13)$$

We write $\sec^n x$ *as* $\sec^{n-2} x \sec^2 x$. *Then* $\int \sec^n x dx = \int \sec^{n-2} x \sec^2 x dx$. *Now integrate by parts.*

Integrals Involving Trigonometric Functions

1. To determine $\int \cos ax\, dx$ and $\int \sin ax\, dx$ where a is a non-zero constant, note that the derivative of $\sin ax$ is $a\cos ax$ and the derivative of $\cos ax$ is $-a\sin ax$. It follows that

$$\int \cos ax\, dx = \tfrac{1}{a}\sin ax + c \qquad \text{and} \qquad \int \sin ax\, dx = -\tfrac{1}{a}\cos ax + c$$

2. To find $\int \sin^2 x\, dx$, $\int \cos^2 x\, dx$, $\int \tan^2 x\, dx$ and $\int \cot^2 x\, dx$ note that

$$\sin^2 x = \tfrac{1}{2} - \tfrac{1}{2}\cos 2x \qquad \cos^2 x = \tfrac{1}{2} + \tfrac{1}{2}\cos 2x$$

$$\tan^2 x = \sec^2 x - 1 \qquad \cot^2 x = \csc^2 x - 1$$

Therefore

- $\int \sin^2 x\, dx = \int \left(\tfrac{1}{2} - \tfrac{1}{2}\cos 2x\right) dx = \tfrac{1}{2}x - \tfrac{1}{4}\sin 2x + c,$

- $\int \cos^2 x\, dx = \int \left(\tfrac{1}{2} + \tfrac{1}{2}\cos 2x\right) dx = \tfrac{1}{2}x + \tfrac{1}{4}\sin 2x + c,$

- $\int \tan^2 x\, dx = \int \left(\sec^2 x - 1\right) dx = \tan x - x + c,$

- $\int \cot^2 x\, dx = \int \left(\csc^2 x - 1\right) = -\cot x - x + c$

3. We determined the integrals of $\tan x$ and $\cot x$ in Examples 193 and 194 on page 151. We obtained

- $\int \tan x\, dx = -\ln|\cos x| + c = \ln|\sec x| + c,$ and

- $\int \csc x\, dx = -\ln|\csc x + \cot x| + c$

4. By Exercise 4 on page 157,

- $\int \sec x\, dx = \ln|\sec x + \tan x| + c$ and

- $\int \csc x\, dx = -\ln|\csc x + \cot x| + c$

5. Let n be a positive integer bigger than 1.

- To get $\int \cos^n x\, dx$, use the reduction formula (4.10) in Exercise 13.

- To get $\int \sin^n x\, dx$, use the reduction formula (4.11) in Exercise 14.

- To get $\int \tan^n x\, dx$, use the reduction formula (4.12) in Exercise 15.

- To get $\int \sec^n x\, dx$, use the reduction formula (4.13) in Exercise 16.

6. Integrals of the form $\int (\text{Expression in } \sin x) \cos x\, dx$ may yield to the substitution $u = \sin x$.

Example 211 *To determine* $\int (\sin^5 x + 1) \cos x\, dx$, *set* $u = \sin x$. *Then*
$\dfrac{du}{dx} = \cos x$, *hence* $dx = \dfrac{1}{\cos x} du$ *and the integral becomes*

$$\int (\sin^5 x + 1) \cos x\, dx = \int (u^5 + 1)\, du = \frac{u^6}{6} + u + c = \frac{\sin^6 x}{6} + \sin x + c$$

7. Integrals of the form $\int (\text{Expression in } \cos x) \sin x\, dx$ may yield to the substitution $u = \cos x$

Example 212 *To determine* $\int \cos^2 x \sin^3 x\, dx$, *first write the integral as*

$$\int \cos^2 x \left(\sin^2 x\right) \sin x\, dx = \int \cos^2 x \left(1 - \cos^2 x\right) \sin x\, dx$$

Now a substitution $u = \cos x$ *gives*

$$\int \cos^2 x \left(1 - \cos^2 x\right) \sin x\, dx = \int \left(u^4 - u^2\right)\, du = \frac{\cos^5 x}{5} - \frac{\cos^3 x}{3} + c$$

8. Integrals of the type $\int (\text{Expression in } \tan x) \sec^2 x\, dx$ may yield to the substitution $u = \tan x$.

Example 213 *To determine* $\int (\tan^3 x + 1) \sec^2 x\, dx$, *set* $u = \tan x$. *Then*
$dx = \dfrac{1}{\sec^2 x} du$ *and the integral becomes*

$$\int (\tan^3 x + 1) \sec^2 x\, dx = \int (u^3 + +1)\, du = \frac{u^4}{4} + u + c = \frac{\tan^4 x}{4} + \tan x + c.$$

9. Integrals of the type $\int (\text{Expression in } \sec x) \sec x \tan x\, dx$ may yield to the substitution $u = \sec x$.

Example 214 *To determine* $\int \tan^5 x \sec^3 x dx$, *first write the integral as*

$$\int \left(\tan^4 x \sec^2 x\right) \sec x \tan x dx = \int \left[\left(\sec^2 x - 1\right)^2 \sec^2 x\right] \sec x \tan x dx.$$

Now set $u = \sec x$. *Then* $dx = \dfrac{1}{\sec x \tan x} du$ *and the integral becomes*

$$\int \tan^5 x \sec^3 x dx = \int \left(u^2 - 1\right)^2 u^2 du = \int \left(u^6 - 2u^4 + u^2\right) du$$

$$= \tfrac{1}{7}u^7 - \tfrac{2}{5}u^5 + \tfrac{1}{3}u^3 + c = \tfrac{1}{7}\sec^7 x - \tfrac{2}{5}\sec^5 x + \tfrac{1}{3}\sec^3 x + c$$

10. Since the derivative of $\arcsin x$ is $\dfrac{1}{\sqrt{1-x^2}}$, an integrand with the term $\sqrt{1-x^2}$ may be the result of taking the derivative of an expression involving $\arcsin x$. This suggests a substitution $u = \arcsin x$. We actually define it by $x = \sin u$. One result of this substitution is that $\sqrt{1-x^2}$ is replaced by

$$\sqrt{1 - \sin^2 u} = \sqrt{\cos^2 u} = \cos u,$$

thus getting rid of the square root

Example 215 *To determine* $\int \dfrac{x^2}{\sqrt{1-x^2}} dx$, *set* $x = \sin u$. *Then* $\dfrac{dx}{du} = \cos u$, *and* $dx = \cos u du$. *The integral becomes*

$$\int \frac{x^2}{\sqrt{1-x^2}} = \int \frac{\sin^2 u}{\sqrt{1 - \sin^2 u}} \cos u du = \int \frac{\sin^2 u}{\cos u} \cos u du = \int \sin^2 u du$$

$$= \int \tfrac{1}{2}\left(1 - \cos 2u\right) du = \tfrac{1}{2}u - \tfrac{1}{4}\sin 2u + c$$

Since $\sin 2u = 2 \sin u \cos u$ *and* $\cos u = \sqrt{1 - \sin^2 u} = \sqrt{1 - x^2}$, *we may write the above result as*

$$\int \frac{x^2}{\sqrt{1-x^2}} = \tfrac{1}{2}\arcsin x - \tfrac{1}{4}\left(2x\sqrt{1-x^2}\right) + c = \tfrac{1}{2}\left(\arcsin x - x\sqrt{1-x^2}\right) + c$$

Example 216 *Consider* $\int_{1/2}^{1} \dfrac{\sqrt{1-x^2}}{x^2} dx$. *Let* $x = \sin u$. *Then* $\dfrac{dx}{du} = \cos u$, *and* $dx = \cos u du$. *When* $x = \tfrac{1}{2}$, $u = \tfrac{\pi}{6}$ *and when* $x = 1$, $u = \tfrac{\pi}{2}$. *Therefore*

$$\int_{1/2}^{1} \frac{\sqrt{1-x^2}}{x^2} dx = \int_{\pi/6}^{\pi/2} \frac{\sqrt{1 - \sin^2 u}}{\sin^2 u} \cdot \cos u du = \int_{\pi/6}^{\pi/2} \cot^2 u du$$

$$= \int_{\pi/6}^{\pi/2} \left(\csc^2 u - 1\right) du = -\left[\cot u + u\right]_{\pi/6}^{\pi/2} = \sqrt{3} - \tfrac{1}{3}\pi.$$

11. Since the derivative of $\arctan x$ is $\dfrac{1}{1+x^2}$, an integrand with the term $\dfrac{1}{1+x^2}$ may be the result of differentiating an expression related to $\arctan x$. This suggests a substitution $u = \arctan x$, also defined by $x = \tan u$. One consequence is that $1 + x^2$ is replaced by a single term, namely $1 + \tan^2 u = \sec^2 u$.

Example 217 *To determine* $\displaystyle\int \dfrac{1}{x^2\sqrt{1+x^2}}dx$, *set* $x = \tan u$. *Differentiating gives* $\dfrac{dx}{du} = \sec^2 u$. *Therefore* $dx = \sec^2 u\, du$ *and the integral becomes*

$$\int \frac{\sec^2 u}{\tan^2 u\sqrt{1+\tan^2 u}}du = \int \frac{\sec^2 u}{\tan^2 u \sec u}du = \int \frac{\cos u}{\sin^2 u}du = -\frac{1}{\sin u} + c.$$

Since $x = \tan u$, $\sin u = \dfrac{x}{\sqrt{1+x^2}}$, *hence* $\displaystyle\int \dfrac{1}{x^2\sqrt{1+x^2}}dx = -\dfrac{\sqrt{1+x^2}}{x} + c.$

12. Since $\sec^2 u - 1 = \tan^2 u$, a substitution $x = \sec x$ may simplify an integral involving $\sqrt{x^2 - 1}$ by getting rid of the square root. (Note that $\sqrt{x^2 - 1}$ is different from $\sqrt{1 - x^2}$.)

Example 218 *Consider* $\displaystyle\int \dfrac{1}{x\sqrt{x^2-1}}dx$. *Let* $x = \sec u$. $dx = \sec u \tan u\, du$ *and the integral becomes*

$$\int \frac{\sec u \tan u}{\sec u\sqrt{\tan^2 u}}du = \int 1\, du = u + c.$$

To express this in terms of x, *note that* $x = \sec u$ *implies that* $\cos u = \frac{1}{x}$, *therefore* $u = \arccos\frac{1}{x}$, *and so* $\displaystyle\int \dfrac{1}{x\sqrt{x^2-1}}dx = \arccos\frac{1}{x} + c.$

Exercise 219

1. *Determine each integral*

(a) $\displaystyle\int \sin^3 x \cos^3 x\, dx$ (b) $\displaystyle\int \sin^2 x \cos^2 x\, dx$ (c) $\displaystyle\int \frac{\cos x}{\sin^3 x}dx$

(d) $\displaystyle\int \tan^4 x\, dx$ (e) $\displaystyle\int \sin^4 2x \cos^2 2x\, dx$ (f) $\displaystyle\int \sec^4 x \tan x\, dx$

(g) $\displaystyle\int \tan^5 x\, dx$ (h) $\displaystyle\int \sec^4 x\, dx$ (i) $\displaystyle\int \sec^4 x \tan^4 x\, dx$

(j) $\displaystyle\int \frac{1}{x\sqrt{1-x^2}}dx$ (k) $\displaystyle\int \sec^6 x\, dx$ (l) $\displaystyle\int \frac{\cos^3 x}{\sin^3 x}x\, dx$

(m) $\displaystyle\int \frac{1}{x^2\sqrt{1-x^2}}dx$ (n) $\displaystyle\int \frac{x^2}{\sqrt{1-x^2}}dx$ (o) $\displaystyle\int \left(\sqrt{1-x^2}\right)dx$

(p) $\displaystyle\int \frac{1}{x\sqrt{1+x^2}}dx$ (q) $\displaystyle\int \frac{1}{(1+x^2)^2}dx$ (r) $\displaystyle\int \frac{1}{(x^2+1)^{3/2}}dx$

(s) $\displaystyle\int \frac{x^3}{(1-x^2)^{3/2}} dx$ (t) $\displaystyle\int \frac{x^3}{\sqrt{1-x^2}} dx$ (u) $\displaystyle\int \frac{x^4}{\sqrt{1-x^2}} dx$

(v) $\displaystyle\int \frac{x+1}{(x^2+4)^2} dx$ (w) $\displaystyle\int \frac{x^3}{(1+x^2)^{3/2}} dx$ (x) $\displaystyle\int \frac{1}{\sqrt{1+x^2}} dx$

(y) $\displaystyle\int \frac{1}{x^2\sqrt{x^2-1}} dx$ (z) $\displaystyle\int \frac{1}{x^3\sqrt{x^2-1}} dx$ (a') $\displaystyle\int \frac{x^3}{\sqrt{x^2-1}} dx$

2. Use the substitution $x = \cos^2 u$ to show that

$$\int \frac{1+x^{3/2}}{\sqrt{x^3(1-x)}} dx = -2\left(\sqrt{\frac{1-x}{x}} + \sqrt{1-x}\right) + c$$

3. Use the substitution $x = \tan u$ to show that

$$\int \left(x^3\sqrt{x^2+1}\right) dx = \frac{(x^2+1)^{5/2}}{5} - \frac{(x^2+1)^{3/2}}{3} + c$$

4. Use the substitution $x = \sin^2 u$ to show that

$$\int_{1/4}^{1/2} \left(\frac{x}{1-x}\right)^{1/2} dx = \frac{\pi}{12} + \frac{\sqrt{3}}{4} - \frac{1}{2}$$

5. Use the substitution $u = x^2$ to show that $\displaystyle\int_0^1 \frac{x}{1+x^4} dx = \frac{\pi}{8}$

6. Use the reduction formula (4.13) on page 162 to show that

$$\int \sec^3 x dx = \frac{1}{2}\left[\sec x \tan x + \ln|\sec x + \tan x|\right] + c.$$

7. Integrate $\displaystyle\int x\sec^2 x dx$ by parts.

Integration by Partial Fractions

Say you have to determine $\displaystyle\int \left(\frac{1}{1-x} + \frac{1}{1+x}\right) dx$. You would immediately integrate each term:

$$\int \left(\frac{1}{1-x} + \frac{1}{1+x}\right) dx = -\ln|1-x| + \ln|1+x| + c = \ln\left|\frac{1+x}{1-x}\right| + c$$

However,

$$\frac{1}{1-x} + \frac{1}{1+x} = \frac{2}{1-x^2},$$

therefore you could have been asked, instead, to determine $\displaystyle\int \frac{2}{1-x^2} dx$. In this form, you would probably try a substitution like $x = \sin u$, to get

$$\int \frac{2}{1-x^2} dx = 2\int \frac{\cos u}{\cos^2 u} du = 2\int \sec u du = 2\ln|\sec u + \tan u| + c$$

You would have to do some extra work to express $2\ln|\sec u + \tan u| + c$ in terms of x. Example:

$$2\int \sec u\, du \;=\; 2\ln|\sec u + \tan u| + c = 2\ln\left|\frac{1}{\sqrt{1-x^2}} + \frac{x}{\sqrt{1-x^2}}\right| + c$$

$$= \; 2\ln\left|\frac{1+x}{\sqrt{1-x^2}}\right| + c = 2\ln\left|\frac{\sqrt{1+x}}{\sqrt{1-x}}\right| + c = \ln\left|\frac{1+x}{1-x}\right| + c$$

A more complicated integral like

$$\int \frac{4x+2}{(x^2-1)(x+2)}\,dx$$

may not yield to the familiar substitutions we have encountered. But

$$\frac{4x+2}{(x^2-1)(x+2)} = \frac{1}{x+1} - \frac{2}{x-1} + \frac{3}{x+2}$$

and if the integral is given, instead, as

$$\int \left(\frac{1}{x+1} - \frac{2}{x-1} + \frac{3}{x+2}\right)dx$$

you would easily handle it. These examples suggest that given an integral of a rational function, it may be a good idea to split the integrand into partial fractions then integrate. Techniques of decomposing rational functions into partial fractions are developed in most pre-calculus textbooks. Here we consider only a few typical examples.

Example 220 *To determine* $\displaystyle\int \frac{x}{(1-x)(x-3)(x+2)}\,dx$, *we split the integrand as*

$$\frac{x}{(1-x)(x-3)(x+2)} = \frac{A}{1-x} + \frac{B}{x-3} + \frac{C}{x+2}$$

where A, B and C are constants. Solving gives $A = -\frac{1}{6}$, $B = -\frac{3}{10}$ and $C = \frac{2}{15}$. Therefore

$$\int \frac{x}{(1-x)(x-3)(x+2)}\,dx \;=\; -\frac{1}{6}\int \frac{1}{1-x}\,dx - \frac{3}{10}\int \frac{1}{x-3}\,dx + \frac{2}{15}\int \frac{1}{x+2}\,dx$$

$$= \; \tfrac{1}{6}\ln|1-x| - \tfrac{3}{10}\ln|x-3| + \tfrac{2}{15}\ln|x+2| + c$$

Example 221 *To determine* $\displaystyle\int \frac{1}{(1-x)(x+2)^2}\,dx$, *we split the integrand as*

$$\frac{1}{(1-x)(x+2)^2} = \frac{A}{1-x} + \frac{B}{x+2} + \frac{C}{(x+2)^2}$$

where A, B and C are constants. Solving gives $A = \frac{1}{9}$, $B = \frac{1}{9}$ and $C = \frac{1}{3}$. Therefore

$$\int \frac{1}{(1-x)(x+2)^2}dx = \frac{1}{9}\int \frac{1}{1-x}dx + \frac{1}{9}\int \frac{1}{x+2}dx + \frac{1}{3}\int \frac{1}{(x+2)^2}dx$$

$$= \frac{1}{9}\ln\left|\frac{x+2}{1-x}\right| - \frac{1}{3(x+2)} + c$$

Example 222 To determine $\int \frac{6}{(2x^2+1)(x-1)}dx$, we split the integrand as

$$\frac{6}{(2x^2+1)(x-1)} = \frac{Ax+B}{2x^2+1} + \frac{C}{x-1}$$

where A, B and C are constants. Solving for A, B and C yields $A = B = -4$ and $C = 2$. Write $\frac{6dx}{(2x^2+1)(x-1)}$ as $-\frac{4x}{2x^2+1} - \frac{4}{2x^2+1} + \frac{2}{x-1}$. Then

$$\int \frac{6dx}{(2x^2+1)(x-1)} = -\int \frac{4xdx}{2x^2+1} - 4\int \frac{dx}{2x^2+1} + 2\int \frac{dx}{x-1}$$

$$= -\ln(2x^2+1) - \frac{4}{\sqrt{2}}\arctan\sqrt{2}x + 2\ln|x-1| + c$$

Exercise 223

1. Integrate by partial fractions. In part (b), k is a non-zero constant.

(a) $\int \frac{xdx}{(x-1)(x+2)}$ b) $\int \frac{dx}{x^2-k^2}$ (c) $\int \frac{dx}{x^2(x+1)}$

(d) $\int \frac{(4x^2+6x-3)dx}{x^3+2x^2-3x}$ e) $\int \frac{(x-2)dx}{x^2(x-1)^2}$ (f) $\int \frac{x^2dx}{x^4-16}$

(g) $\int \frac{dx}{x+\sqrt{x}-2}$, (Let $u=\sqrt{x}$.) h) $\int \frac{dx}{x^2(x^2+2)}$ (i) $\int \frac{x}{2x+3\sqrt{x}+1}$

2. If the degree of the numerator of a rational integrand is not less than the degree of the denominator, first do a long division to get a numerator of lower degree than that of the denominator. For example, given $\int \frac{x^3+x^2-3}{x^2-1}dx$, we first do a long division to get

$$\frac{x^3+x^2-3}{x^2-1} = x+1+\frac{x-2}{x^2-1}$$

We then split $\frac{x-2}{x^2-1}$ into partial fractions. The result is

$$\frac{x-2}{x^2-1} = \frac{x-2}{(x-1)(x+1)} = \frac{3}{2(x+1)} - \frac{1}{2(x-1)}$$

Therefore

$$\int \frac{x^3 + x^2 - 3}{x^2 - 1} dx \;=\; \int \left(x + 1 + \frac{3}{2(x+1)} - \frac{1}{2(x-1)} \right) dx$$

$$= \; \tfrac{1}{2}x^2 + x + \tfrac{3}{2}\ln|x+1| - \tfrac{1}{2}\ln|x-1| + c$$

Determine the following in a similar way

a) $\displaystyle\int \frac{x^3 + x + 1}{(x+1)(x-3)} dx$ b) $\displaystyle\int \frac{x^4}{x^3 - 1} dx$ c) $\displaystyle\int \frac{x^4 + 1}{x^4 - 1} dx$ d) $\displaystyle\int \frac{x^3 - 1}{x^3 + 1} dx$

Integrals Involving Hyperbolic Functions

On page 131, you were asked to show that the derivatives of $f(x) = \cosh^{-1} x$ is $f'(x) = \dfrac{1}{\sqrt{x^2 - 1}}$ and the derivative of $g(x) = \sinh^{-1} x$ is $g'(x) = \dfrac{1}{\sqrt{x^2 + 1}}$. (Note that the derivative of $\sinh^{-1} x$ is different from the derivative of $h(x) = \arcsin x$, which is $h'(x) = \dfrac{1}{\sqrt{1 - x^2}}$.) These observations suggest that an integral involving $\sqrt{x^2 - 1}$, may yield to the substitution $u = \cosh^{-1} x$, and an integral involving $\sqrt{x^2 + 1}$ may yield to the substitution $u = \sinh^{-1} x$. We actually define these substitutions in the equivalent forms $\cosh u = x$, and $\sinh x = u$ respectively.

Example 224 *To determine* $\displaystyle\int \frac{x^2}{\sqrt{x^2 - 1}} dx$, *we set* $x = \cosh u$. *Then* $dx = \sinh u\, du$ *and the integral becomes*

$$\int \frac{x^2}{\sqrt{x^2 - 1}} dx \;=\; \int \frac{\cosh^2 u}{\sinh u} \cdot \sinh u\, du = \int \cosh^2 u\, du = \int \frac{1}{2}(\cosh 2u + 1)\, du$$

$$= \; \tfrac{1}{4}\sinh 2u + u + c = \tfrac{1}{2}\sinh u \cosh u + u + c$$

$$= \; \tfrac{1}{2}x\sqrt{x^2 - 1} + \ln\left(x + \sqrt{x^2 - 1} \right) + c$$

We used the following facts which were verified on page 131:

- *If* $x = \cosh u$ *then* $\sinh u = \sqrt{x^2 - 1}$,

- *If* $x = \cosh u$ *then* $u = \cosh^{-1} x = \ln\left(x + \sqrt{x^2 - 1} \right)$.

Example 225 *To determine* $\displaystyle\int_0^1 x^2\sqrt{x^2 + 1}\,dx$, *set* $x = \sinh u$. *Then* $dx = \cosh u\, du$. *When* $x = 0$, $\sinh u = 0$, *hence* $u = \sinh^{-1}(0) = \ln(1) = 0$. *When* $x = 1$, $\sinh u = 1$, *hence* $u = \sinh^{-1}(1) = \ln(1 + \sqrt{2})$. *For convenience, denote* $\ln(1 + \sqrt{2})$ *by* b.

The integral becomes

$$\int_0^1 x^2\sqrt{x^2+1}\,dx \;=\; \int_0^b \sinh^2 u \cosh u \cosh u\,du \;=\; \int_0^b (\sinh u \cosh u)^2\,du$$

$$= \int_0^b \tfrac{1}{4}\sinh^2 2u\,du = \int_0^b \tfrac{1}{8}(\cosh 4u - 1)\,du$$

$$= \left[\tfrac{1}{32}\sinh 4u - \tfrac{1}{8}u\right]_0^b = \tfrac{1}{64}\left(e^{4b} - e^{-4b}\right) - \tfrac{1}{8}b$$

$$= \tfrac{1}{64}\left[\left(1+\sqrt{2}\right)^4 - \left(1+\sqrt{2}\right)^{-4}\right] - \tfrac{1}{8}\ln\left(1+\sqrt{2}\right).$$

Example 226 *To determine* $\displaystyle\int \sqrt{1+x^2}\,dx$, *set* $x = \sinh u$. *Then* $dx = \cosh u\,du$ *and the integral becomes*

$$\int \sqrt{1+x^2}\,dx \;=\; \int (\cosh u)\cosh u\,du = \int \cosh^2 u\,du = \int \left(\frac{\cosh 2u + 1}{2}\right)du$$

$$= \frac{\sinh 2u}{4} + \frac{u}{2} + c = \frac{\sinh u \cosh u}{2} + \frac{1}{2}\sinh^{-1}x + c$$

Now recall that if $x = \sinh u$ *then* $\cosh u = \sqrt{1+x^2}$, *(see (3.23) on page 131).* *Also* $\sinh^{-1}x = \ln\left(x+\sqrt{1+x^2}\right)$, *(see (3.21) on page 130). Therefore*

$$\int \sqrt{1+x^2}\,dx = \frac{1}{2}x\sqrt{1+x^2} + \frac{1}{2}\ln\left(x+\sqrt{1+x^2}\right) + c$$

Exercise 227

1. *Determine the following:*

a) $\displaystyle\int \frac{x^2}{\sqrt{x^2+1}}\,dx$ b) $\displaystyle\int \frac{1}{\sqrt{a^2+x^2}}\,dx$ c) $\displaystyle\int \frac{x^3}{\sqrt{x^2-1}}\,dx$ d) $\displaystyle\int_2^5 \sqrt{x^2-4}\,dx$.

2. *Show that* $\displaystyle\int \sqrt{a^2+x^2}\,dx = \tfrac{1}{2}x\sqrt{a^2+x^2} + \tfrac{1}{2}a^2\ln\left(x+\sqrt{a^2+x^2}\right) + c.$

3. *Show that* $\displaystyle\int \frac{1}{x^2\sqrt{x^2-1}}\,dx = \frac{\sqrt{x^2-1}}{x} + c$

Test your skills - 8,   (90 minutes)

1. Determine each indefinite integral   (a) $\displaystyle\int \left(\frac{3\sin x}{4} + 2\csc^2 x + 1\right)dx$

(b) $\displaystyle\int \left(3x - \frac{3}{5x} - \frac{2}{3x^2}\right)dx$ (c) $\displaystyle\int \left(\frac{5}{3\sqrt{x}} + 2e^x + 3\sin x\right)dx$

2. Evaluate each definite integral.

(a) $\int_0^1 (3 - 2x + 5x^3)\, dx$

(b) $\int_0^{\pi/4} (3 + \sec^2 x)\, dx$

(c) $\int_0^2 \dfrac{x^2}{x^3 + 3}\, dx$

(d) $\int_1^2 \left(\dfrac{3}{7x} - \dfrac{2}{x^2} \right)\, dx$

3. Let R be the region enclosed by the graphs of $f(x) = x^2 - 2x - 2$ and $g(x) = 2x + 3$. Sketch it then calculate its area.

4. Use the substitution $u = 3x + 1$ to evaluate $\int_0^5 \left(\dfrac{x}{\sqrt{3x + 1}} \right)\, dx$

5. Integrate $\int 2x \cos 3x\, dx$ by parts.

6. Split $f(x) = \dfrac{x + 5}{(x + 3)(x + 4)}$ into partial fractions then determine

$$\int \dfrac{x + 5}{(x + 3)(x + 4)}\, dx.$$

7. Use the substitution $x = 3 \sin \theta$ to determine $\int_0^3 \dfrac{x^2}{\sqrt{9 - x^2}}\, dx.$

8. Use the substitution $u = 9 + x^2$ to evaluate $\int_0^4 x^3 \sqrt{9 + x^2}\, dx.$

9. Use a suitable substitution to determine $\int \left(\cos x - \dfrac{3}{2 \cos x} - 4 \right) \sin x\, dx$

10. Use the substitution $x = \tan \theta$ to show that

$$\int \dfrac{2}{(1 + x^2)^2}\, dx = \arctan x + \dfrac{x}{1 + x^2} + c$$

Chapter 5 SOME APPLICATIONS OF ANTIDERIVATIVES

All the applications of antiderivatives we consider in this chapter come down to approximating some required quantity with Riemann sums then determine the limit of the Riemann sums to get the exact value of the quantity.

Area Between Curves

We outlined a solution to this problem in Example 177 on page 134. We repeat the steps here for generality. Thus let R be the region enclosed by the x-axis and the graph of a function f, on an interval $[a, b]$. Let its area be A square units. To determine an approximate value of A, we divide $[a, b]$ into n equal subintervals $[a_0, x_1], [x_1, x_2], \ldots, [x_{n-1}, x_n]$ of length $h = \frac{b-a}{n} = \triangle x$ each. This partitions R into n small strips. We approximate the strip on an interval $[x_i, x_{i+1}]$ with a rectangle whose base is the interval $[x_i, x_{i+1}]$, and height $f(x_i)$. Its area is

$$f(x_i) \cdot (x_{i+1} - x_i) = f(x_i)h = f(x_i)\triangle x$$

Therefore an approximate value of A is the Riemann sum

$$\sum_{i=0}^{n-1} f(x_i)h = \sum_{i=0}^{n-1} f(x_i)\triangle x$$

The exact value of A is the limit of such Riemann sums which is

$$\int_a^b f(x)dx$$

Example 228 *Let R be the region enclosed by the graph of $f(x) = \sin x$ on the interval $\left[\frac{\pi}{6}, \frac{3\pi}{4}\right]$ and the x-axis.*

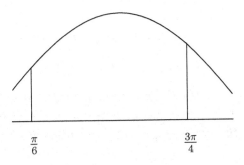

Its area is $A = \displaystyle\int_{\pi/6}^{3\pi/4} \sin x\, dx = \left[-\cos x \; \right]_{\pi/6}^{3\pi/4} = \frac{\sqrt{3}}{2} + \frac{\sqrt{2}}{2}.$

The area of the region enclosed by the graphs of functions f and g may be calculated in a similar way. Suppose the graph of f is above the graph of g on an interval $[a, b]$. Let R be the region they enclose on the interval $[a, b]$. To calculate its area, divide it into n small strips as before. A typical one determined by an interval

174

$[x_i, x_{i+1}]$ is shown below. It is approximated by a rectangle with width $(x_{i+1} - x_i)$, height $[f(x_i) - g(x_i)]$ and area $[f(x_i) - g(x_i)] \cdot (x_{i+1} - x_i) = [f(x_i) - g(x_i)]\, \triangle x$. Therefore the area of the strip is approximately equal to

$$[f(x_i) - g(x_i)] \cdot (x_{i+1} - x_i) = [f(x_i) - g(x_i)]\, \triangle x$$

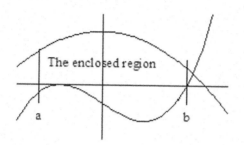
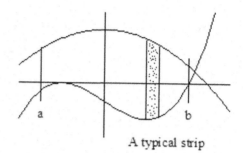

A typical strip

It follows that the area of the region is approximately equal to

$$\sum_{i=0}^{n-1} [f(x_i) - g(x_i)]\, \triangle x$$

Its exact area is the limit of such sums as $\triangle x \to 0$. That limit is

$$\int_a^b [f(x) - g(x)]\, dx$$

Example 229 *The graphs of $f(x) = 6x$ and $g(x) = x\,(x - 2)\,(x + 3)$ are shown below.*

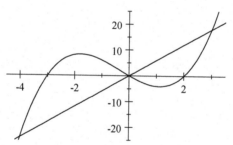

They intersect at points (x, y) where x satisfies the equation

$$6x = x\,(x - 2)\,(x + 3).$$

It has solution $x = -4$, or 0 or 3. To determine the area of the region enclosed by the two graphs, note that the graph of g is above the graph of f on the interval $(-4, 0)$, but it is below the straight line on the interval $(0, 3)$. Therefore the area they enclose is

$$\int_{-4}^0 [g(x) - f(x)]\, dx + \int_0^3 [f(x) - g(x)]\, dx$$

Since $f(x) - g(x) = 12x - x^3 - x^2$ and $g(x) - f(x) = x^3 + x^2 - 12x$, the area is

$$\int_{-4}^0 \left(x^3 + x^2 - 12x\right) dx + \int_0^3 \left(12x - x^3 - x^2\right) dx$$

Evaluate this. You should get the answer $78\frac{1}{12}$ square units.

Exercise 230

1. The figure shows the region R enclosed by the x-axis and graph of $f(x) = x^3 + x + 2$ on the interval $[-1, 2]$.

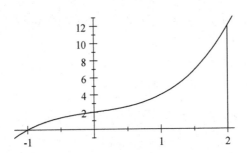

(a) Imagine dividing R into n strips. Draw a typical one and a rectangle that approximates it.

(b) Calculate the area of the approximating rectangle then determine a Riemann sum that approximates the area of R.

(c) Deduce the exact value of the area.

2. The figure below shows the graphs of $f(x) = -x^2$ and $g(x) = x^2 + x - 6$.

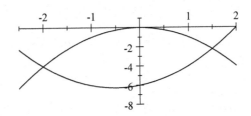

(a) Determine the coordinates of the points where the two graphs intersect.

(b) Let R be the region enclosed by the two graphs. Imagine dividing it into n strips. Draw a typical one and a rectangle that approximates it.

(c) Calculate the area of R.

3. Calculate the area of R, (shown below), enclosed by the graphs of $f(x) = \sin x$ and $g(x) = \cos x$ on the interval $\left[-\frac{\pi}{4}, \frac{5\pi}{4}\right]$.

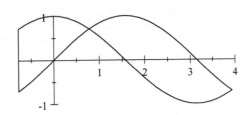

4. Calculate the area of the region enclosed by:

(a) The graphs of $f(x) = x^3$ and $g(x) = -x^3$ on the interval $[-1, 2]$.

176

(b) The x-axis and the graph of $f(x) = xe^x$ on the interval $[-3, 0]$. (First sketch the graph.)

(c) The x axis and the graph of $f(x) = x^3 - x^2 - 2x$. (First sketch the graph.)

(d) The graph of $f(x) = x^2 + 2x + 3$, (a parabola), and the graph of $g(x) = x + 9$, (a straight line).

(e) The graph of $f(x) = x^2 + 5x + 1$, (a parabola), and the graph of $g(x) = -x^2 + 2x + 3$, (also a parabola).

5. Let f be a function with an inverse f^{-1}. Consider the region R enclosed by the graph of f, the y-axis and the two lines $y = a$ and $y = b$. You may assume that R is in the first quadrant as depicted in the figure below. Subdivide it into smaller strips as shown in the second figure.

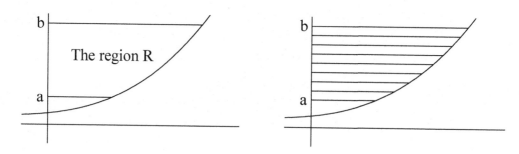

Approximate each strip with a rectangle then calculate an approximate value of the area of R.

(a) Use the approximation to deduce that the area of R is $\displaystyle\int_a^b f^{-1}(y)\,dy$.

(b) Use the formula in part (a) to calculate the area of the region R enclosed by the graph of $f(x) = 2x^3$, the y-axis and the two lines $y = 0$ and $y = 16$.

The First Mean Value Theorem for Integrals

Before stating the theorem, here is an example that throws light on the statement of the theorem:

Example 231 *Consider the function $f(x) = x^2 + 3x + 5$, $-3 \le x \le 2$. Let R be the region enclosed by the graph of f and the x-axis.*

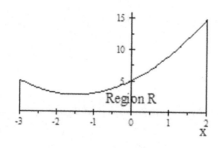

Its area is

$$A = \int_{-3}^{2} \left(x^2 + 3x + 5\right) dx = \left[\frac{x^3}{3} + \frac{3x^2}{2} + 5x\right]_{-3}^{2} = \frac{175}{6}$$

Surely, there is a rectangle with base $[-3, 5]$ and area $\dfrac{175}{6}$. It should be easy to figure out its height. Since its length is 5 units (because the interval $[-3, 2]$ has length 5), the height must be $\dfrac{175}{6} \div 5 = \dfrac{35}{6}$ units. The rectangle is shaded below.

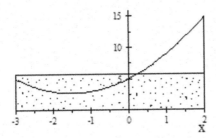

The first mean value theorem for integrals asserts that there is a number c between -3 and 2 such that $f(c) = \dfrac{35}{6}$, (i.e. the height of the rectangle is the value of f at some point c between -3 and 2). Since, (in this particular case), f is a quadratic, c is easy to determine. It is the solution of the equation

$$c^2 + 3c + 5 = \frac{35}{6}.$$

The quadratic formula gives $c = 0.26$ or $c = -3.26$ (to 2 decimal places). But -3.26 is not in the interval $[-3, 2]$, therefore c must be 0.26. The figure above supports this result.

Now to the general case: Let f be a *continuous* function defined on an interval $[a, b]$. Consider the area A of the region R enclosed by the graph of f on the interval $[a, b]$ and the x-axis. We know that

$$A = \int_{a}^{b} f(x) dx$$

It is possible to draw a rectangle with base $[a, b]$ and area A.

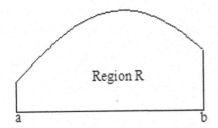

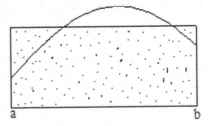

The region plus the rectangle

178

The First Mean Value Theorem for Integrals asserts that the height of the rectangle is a value $f(c)$ of the function f at some point c between a and b. More precisely, the theorem states that there is a number c between a and b such that

$$(b - a)\, f(c) = \int_a^b f(x)dx. \tag{5.1}$$

Here is a sketch of the proof: To begin with assume that the graph of f is above the x-axis on the interval $[a, b]$. Let m and M be, respectively, the smallest and the largest value of f on $[a, b]$.

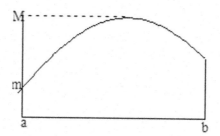

Then the rectangle with base $[a, b]$ and height m is inside the region R, but the rectangle with base $[a, b]$ and height M contains R.

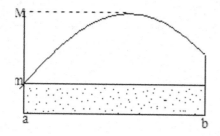
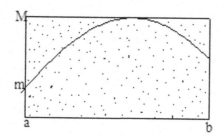

Small rectangle is in R Big rectangle contains R

The areas of the rectangles are $m\,(b - a)$ and $M\,(b - a)$ respectively. It follows that

$$m\,(b - a) \le \int_a^b f(x)dx \le M\,(b - a).$$

We may re-write this as

$$m \le \frac{1}{b - a}\int_a^b f(x)dx \le M.$$

Now we see that $\dfrac{1}{b - a}\displaystyle\int_a^b f(x)dx$ is a number between m and M. Since f is continuous, it does not skip any number between m and M, (because we can plot its graph without lifting the pencil from the paper). It follows that there is a number c between a and b such that $f(c) = \dfrac{1}{b - a}\displaystyle\int_a^b f(x)dx$. When we multiply both sides of this equation by $(b - a)$ we obtain the required equation:

$$(b - a)\, f(c) = \int_a^b f(x)dx.$$

If the graph of f is not completely above the x-axis, add a suitable positive constant k to $f(x)$ to shift the graph above the x-axis. Let $g(x) = f(x) + k$. It follows from what we did above that there is a number c between a and b such that

$$(b - a)\, g(c) = \int_a^b g(x)dx. \tag{5.2}$$

Deduce $(b - a)\, f(c) = \int_a^b f(x)dx$ from (5.2).

The value $f(c)$ in (5.1) is called the average value of f on the interval $[a, b]$.

Exercise 232

1. In each of the following cases, find a number c between a and b such that $(b - a)\, f(c) = \int_a^b f(x)dx$. (Follow the steps we took to get $c = 0.26$ in Example 231 above.)

   a) $f(x) = 1 + x^2$, $a = -3$, $b = 2$.   b) $f(x) = x^3$, $a = -2$, $b = 1$.

   c) $f(x) = x^3$, $a = -1$, $b = 3$ d) $f(x) = 2\sin x$, $a = 0$, $b = \frac{3\pi}{2}$.

   e) $f(x) = \cos x$, $a = 0$, $b = \frac{3\pi}{2}$ f) $f(x) = e^x$, $a = -3$, $b = 2$

2. Fill in the missing details in the following exercise to prove the Second Mean Value Theorem for Integrals which states that: **If the functions g and h are continuous on an interval $[a, b]$ and $g(x) \geq 0$ for all x in $[a, b]$ then there is a number c between a and b such that**

$$\int_a^b h(x)g(x)dx = h(c) \int_a^b g(x)dx.$$

   The first point to note is that $\int_a^b g(x)dx > 0$, unless $g(x) = 0$ for all x in $[a, b]$, and in that case any c in $[a, b]$ works. Let m be the smallest and M be the largest value of $h(x)$ on $[a, b]$. Then $mg(x) \leq h(x)g(x) \leq Mg(x)$ for all x in $[a, b]$. Use this to show that

$$m \int_a^b g(x)dx \leq \int_a^b h(x)g(x)dx \leq M \int_a^b g(x)dx$$

   Deduce that

$$m \leq \frac{1}{\int_a^b g(x)dx} \int_a^b h(x)g(x)dx \leq M.$$

   Now you are on familiar ground. Complete the proof then find such a number c given $h(x) = \sin x$, $g(x) = x$ and $[a, b] = \left[0, \frac{\pi}{2}\right]$. Also prove that the statement is true if $g(x) \leq 0$ for all x in $[a, b]$.

Volumes of Solids of Revolution by the Disc and Shell Methods

One gets a solid of revolution when one revolves a given region in the plane, about some given axis, through 4 right angles. The simplest one is a cylinder obtained by revolving, (about the x-axis), a rectangle with one edge on the x-axis. To sketch it, draw the reflection of the line segment in the x-axis as shown.

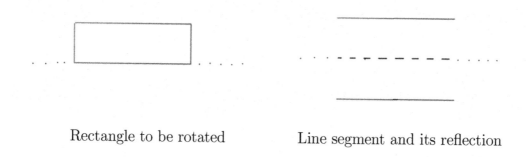

Rectangle to be rotated   Line segment and its reflection

Now join the ends with curves as shown below. To give the cylinder a more realistic look, dot what should be the invisible part of the curve at the back.

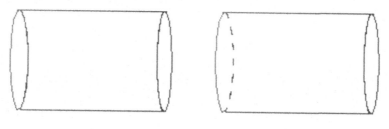

Join both ends with curves   The invisible edge is dotted

If you rotate, (about the x-axis), a rectangle that is above the x-axis, you get a solid called a **shell**. (It is the solid between two cylinders with the same axis.)

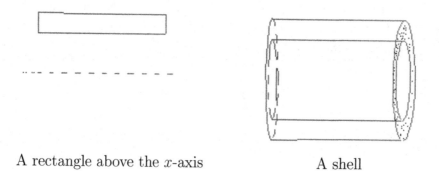

A rectangle above the x-axis   A shell

The next simple one to draw is a right circular cone obtained by revolving the region enclosed by a right triangle about one of its sides, (not the hypotenuse). The procedure for drawing it is similar; start by drawing a reflection of the hypotenuse.

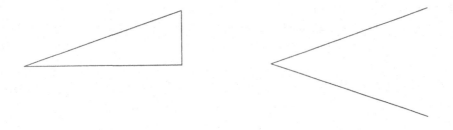

Region to be rotated   Hypotenuse and its reflection

Now join the two ends of the line segments with a curve as shown below.

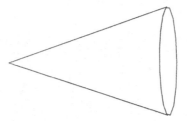

A right circular cone

The Disc Method

Consider the region R in the plane enclosed by the x-axis and the graph of some function f on an interval $[a, b]$. Imagine revolving it about the x-axis through 4 right angles to get a solid of revolution.

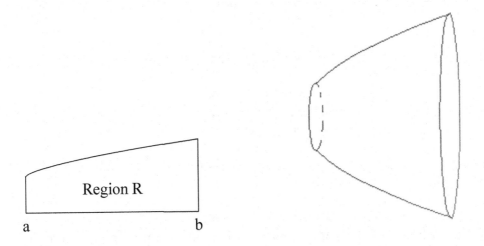

Region R

a   b

We wish to determine its volume. Let it be V. We may determine it as follows:

Divide $[a, b]$ into smaller subintervals $[x_0, x_1]$, $[x_1, x_2]$, $\ldots$, $[x_{n-1}, x_n]$ and use the subintervals to partition R into smaller strips. (In the figure below, $[a, b]$ is divided into 4 subintervals.)

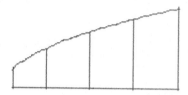

Note that each strip is perpendicular to the axis of rotation. When the graph is rotated about the x-axis, each strip determines a slice of the solid as shown in the figure below.

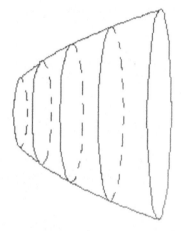

The slice determined by the interval $[x_i, x_{i+1}]$ is approximated with a disc of radius $f(x_i)$, thickness $\triangle x = (x_{i+1} - x_i)$ and volume $\pi \left[f(x_i) \right]^2 \triangle x$, (see the figure below).

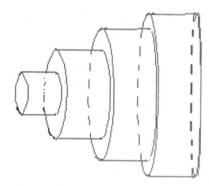

The sum of the volumes of the approximating discs is an approximate value of V. In other words,

$$V \simeq \pi \sum_{i=1}^{n} \left[f(x_i) \right]^2 \triangle x$$

The exact value of V is the limit of the Riemann sums $\pi \sum_{i=1}^{n} [f(x_i)]^2 \, \triangle x$ as $\triangle x \to 0$. That limit is

$$V = \pi \int_a^b [f(x)]^2 \, dx \qquad (5.3)$$

Because we obtained formula (5.3) by approximating the solid with appropriate discs, this method of computing volumes is called the **"disc method"**.

Example 233 *Take $f(x) = \frac{1}{2}x$, and let R be the triangle, shown below, enclosed by the graph of f on the interval $[0, 2]$ and the x-axis. Consider the right circular cone obtained by revolving R through 4 right angles about the x-axis.*

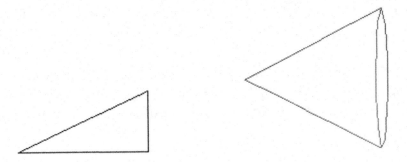

Its volume is

$$V = \int_0^2 \pi \left(\frac{1}{2}x\right)^2 dx = \pi \int_0^2 \frac{1}{4}x^2 dx = \frac{\pi}{4}\left[\frac{x^3}{3}\right]_0^2 = \frac{2\pi}{3}.$$

You probably know that the volume of a right circular cone with base radius r and height h is $\frac{1}{3}\pi r^2 h$. Use this formula to double check the above result.

Exercise 234

1. *Let R be the region enclosed by the graph of $f(x) = 2 + x^2$, $0 \le x \le 2$, (a parabola), and the x-axis. Show that the volume of the solid generated by revolving R about the x-axis through 4 right angles is $\frac{376\pi}{15}$ square units.*

2. *Let R be the region enclosed by the graph of $g(x) = \sin x$, $0 \le x \le \pi$ and the x-axis. It is revolved about the x-axis through 4 right angles. Show that the volume of the solid generated is $\frac{1}{2}\pi^2$ square units.*

3. *Consider the function $f(x) = \dfrac{x}{1 + x^2}$.*

   (a) *Sketch the graph of f.*

   (b) *Let R be the region enclosed by the x-axis and the graph of f between $x = \frac{1}{4}$ and $x = 4$. Show that its area is $\ln 4$ square units.*

   (c) *A wine bottle is obtained by revolving the region R about the x-axis through 4 right angles. Show that its volume is $\frac{\pi}{2}\left(\tan^{-1} 4 - \tan^{-2}\frac{1}{4}\right)$ cubic units.*

4. *The figure shows the graph of $f(x) = x^2 - 4$, $2 \le x \le 4$ and the three line segments; one of them joining $(4, 12)$ to $(0, 12)$, another one joining $(0, 12)$ to $(0, 0)$ and the third one joining $(0, 0)$ to $(2, 0)$. The region they enclose is*

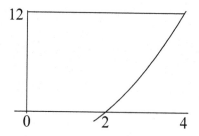

rotated about the x-axis through 4 right angles. Explain why the volume of the solid generated must be $\pi \int_0^4 144 dx - \pi \int_2^4 (x^2 - 4)^2 dx$ units then evaluate the integrals.

5. *Let $f(x) = x^2$ and $b > 0$. Consider the region enclosed by the x-axis and the graph of f on the interval $[0, b]$. It is rotated about the x-axis through 4 right angles. For what value of b is the volume of the solid generated equal to 2 cubic units?*

6. *Let f be a function with an inverse f^{-1}. Suppose the region R enclosed by the graph of f, the y-axis and the two lines $y = a$ and $y = b$, is in the first quadrant as shown in the figure below.*

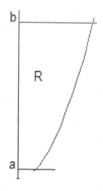

Solid of revolution

(a) *Show that the volume of the solid obtained by revolving R about the y-axis through 4 right angles is $\pi \int_a^b [f^{-1}(y)]^2 dy$.*

(b) *In particular, let R be the region enclosed by the graph of $f(x) = x^3$, the y-axis, the line $y = 1$ and the line $y = 8$. Use the result of part (a) to calculate the volume of the solid obtained by revolving R about the y axis through 4 right angles.*

7. Let $h > 0$ and R be the region enclosed by the x-axis and the graph of $f(x) = e^x$ on $[0, h]$.

   (a) Sketch the solid generated when R is rotated through 4 right angles about the x-axis, and show that its volume is $\frac{1}{2}\pi \left(e^{2h} - 1 \right)$ cubic meters. (Assume that x and $f(x)$ are measures in meters.)

   (b) Now imagine a container obtained by revolving the graph of $f(x) = e^x$, on the interval $[0, h]$, through 4 right angles about the x-axis. Turn the container counter-clockwise so that it is upright. Assume that it is closed at the bottom but open at the top. Water is poured into it at the rate of $\frac{1}{2}$ cubic meters per minute. At what rate is the water level rising when it is 1 meter deep?

8. A spherical water tank has radius R feet. Show that when the water level is H feet deep, then the volume of water in the tank is $\pi \left(RH^2 - \frac{H^3}{3} \right)$ cubic feet. (Imagine dividing the water into thin layers. A typical layer, shown in the figure below, which is y feet from the bottom of the tank has radius $\sqrt{R^2 - (R - y)^2}$ feet. Estimate its volume then form a Riemann sum.)

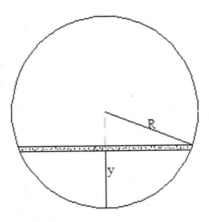

The Method of Shells

To introduce the method of shells, consider the region R enclosed by the graph of $f(x) = -\frac{5}{2}x + \frac{21}{2}$ in the interval $[1, 3]$ and the x-axis. Imagine revolving it about the y-axis. The result is the solid of revolution sketched below.

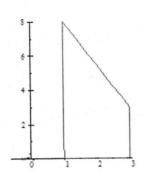

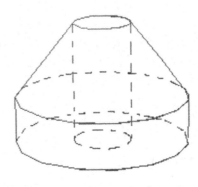

The region R The solid of revolution

To calculate its volume, partition the interval $[1, 3]$ into n smaller subintervals $[x_0, x_1]$, $[x_1, x_2]$, ..., $[x_{n-1}, x_n]$ of length $\triangle x = \frac{2}{n}$ each. This partitions R into n smaller strips. (In the left figure below, R is partitioned into 4 smaller strips.) Note that, this time, each strip is *parallel* to the axis of rotation. Approximate each strip with a suitable rectangle as shown in the right figure.

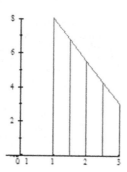
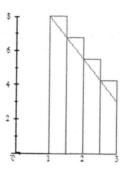

R partitioned into smaller strips   Rectangles that approximate the strips

Now revolve each rectangle about the y-axis. The result are n shells whose total volume approximates the volume of the solid. The figure below shows the four shells generated by the above rectangles. The solid of revolution can be seen inside the shells.

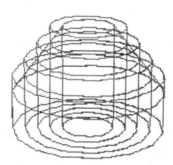

The shell generated by the rectangle on a typical interval $[x_i, x_{i+1}]$ has volume

$$\pi \left(x_{i+1}\right)^2 f(x_i) - \pi \left(x_i\right)^2 f(x_i) = \pi \left(x_{i+1} + x_i\right) \left(x_{i+1} - x_i\right) f(x_i)$$

Since $x_{i+1} = x_i + \triangle x$, it follows that the volume of the shell is $\pi \left(2x_i + \triangle x\right) f(x_i) \triangle x$, which is approximately equal to $2\pi x_i f(x_i) \triangle x$, because $2x_i + \triangle x \simeq 2x_i$ when $\triangle x$ is close to 0. Therefore the volume of the solid is approximately equal to

$$\sum_{i=1}^{n} 2\pi x_i \left[f(x_i)\right] \triangle x = 2\pi \sum_{i=1}^{n} x_i \left[f(x_i)\right] \triangle x$$

The exact volume is the limit as $\triangle x \to 0$ of the above Riemann sums, which is

$$2\pi \int_1^3 x f(x) dx = 2\pi \int_1^3 \left(-\frac{5x^2}{2} + \frac{21x}{2}\right) dx = 2\pi \left[-\frac{5x^3}{6} + \frac{21x^2}{4}\right]_1^3 = \frac{122\pi}{3}$$

In general, let R be a given region in the plane and V be the volume of the solid generated when R is revolved through 4 right angles about a given axis. To calculate V using the method of shells, (a) partition R into smaller strips **parallel** to the axis of rotation, (b) calculate the volume of the shell generated by each strip and sum up, (c) take limits.

Example 235 *Let R be the region enclosed by the graph of $f(x) = x^2 - 4x + 6$ on the interval $[0, 3]$ and the x-axis. Imagine revolving it about the y-axis through 4 right angles to generate a solid. Let the volume of the solid be V. To calculate it using the method of shells, divide $[0, 3]$ into smaller subintervals*

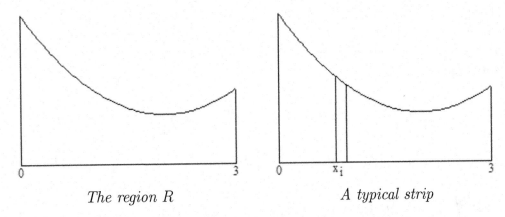

The region R A typical strip

$[x_0, x_1], [x_1, x_2], \ldots, [x_{n-1}, x_n]$. *These partition R into smaller strips parallel to the y-axis, (which is the axis of rotation). A typical one on some interval $[x_i, x_{i+1}]$ is shown above. Approximate it with a rectangle whose base is the interval $[x_i, x_{i+1}]$, and whose height is $f(x_i)$. The rectangle generates a shell with inner radius x_i and height $f(x_i)$ shown below.*

The shell with inner radius x_i and height $f(x_i)$

Its volume is approximately equal to

$$2\pi x_i \left[f(x_i)\right] \triangle x = 2\pi x_i \left[x_i^2 - 2x_i + 2\right] \triangle x.$$

Therefore V is approximately equal to the Riemann sum

$$2\pi \sum_{i=1}^{n} x_i \left[x_i^3 - 2x_i^2 + 2x_i\right] \triangle x.$$

The exact value of V is the limit of such Riemann sums as $\triangle x \to 0$, which is

$$2\pi \int_0^3 \left(x^3 - 2x^2 + 2x\right) dx = 2\pi \left[\frac{x^4}{4} - \frac{2x^3}{3} + x^2\right]_0^3 = \frac{45\pi}{2}$$

In the next example, the region is revolved about the x-axis

Example 236 *Let R be the region enclosed by the lines $y_1 = \frac{1}{3}x + 1$, $y_2 = 2x - 4$ and the x-axis, (see the figure in Example 176 on page 134). We calculate the volume of the solid generated by revolving R about the x-axis through 4 right angles. The region lies between $y = 0$ and $y = 2$. Partition the interval $[0, 2]$ into n subintervals. This partitions R into n strips. A typical strip determined by an interval $[y_i, y_{i+1}]$ generates a shell with inner radius radius y_i, approximate length $\frac{1}{2}(y_i + 4) - 3(y_i - 1)$ and thickness $\triangle y = (y_{i+1} - y_i)$. Its volume is approximately equal to $2\pi y_i \left[\frac{1}{2}(y_i + 4) - 3(y_i - 1)\right] \triangle y$. Therefore the volume of the solid is approximately*

$$2\pi \sum_{i=0}^{n-1} y_i \left[\frac{1}{2}(y_i + 4) - 3(y_i - 1)\right] \triangle y = 2\pi \sum_{i=0}^{n-1} y_i \left(5 - \tfrac{5}{2}y_i\right) \triangle y$$

Its exact volume is the limit of the above Riemann sums as $\triangle y \to 0$, which is

$$2\pi \int_0^2 y \left(5 - \tfrac{5}{2}y\right) dy = 10\pi \int_0^2 \left(y - \tfrac{1}{2}y^2\right) dy = 10\pi \left[\frac{y^2}{2} - \frac{y^3}{6}\right]_0^2 = \frac{20\pi}{3}$$

Exercise 237

1. *In the figure below*

   R *is the region enclosed by y-axis, the lines $y = 3x$ and $y = 4 - x$*

   S *is the region in the first quadrant enclosed by the x-axis, the line $y = 2x - 3$ and the curve $y = \sqrt{x}$.*

   T *is the region enclosed by the y-axis, the curve $y = \sqrt{4 - x^2}$ and the line $y = x$.*

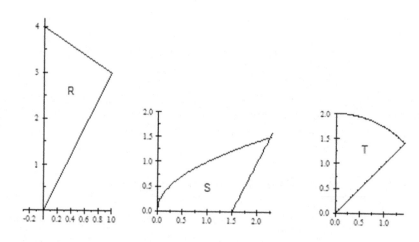

(a) *Use the method of shells to calculate the volume of the solid generated by revolving R about the y-axis through 4 right angles. (Draw a rectangle approximating a typical strip, draw the shell it generates, calculate its volume, add up to get a Riemann sum, then calculate the limit of the Riemann sums.)*

(b) *Use the method of shells to calculate the volume of the solid generated by revolving S about the x-axis through 4 right angles.*

(c) *Use the method of shells to calculate the volume of the solid generated by revolving T about the y-axis through 4 right angles.*

2. *Imagine a spherical ball of radius R units. A hole of radius r where $0 < r < R$, is drilled through the sphere. The result is the solid generated by revolving the area enclosed by the section of the circle shown below about the x-axis.*

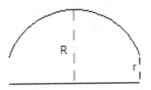

Use the shell method to show that its volume is $\frac{4\pi}{3}\left(R^2 - r^2\right)^{3/2}$.

Length of the Graph of a Function on an Interval

If f is a linear function then it is easy to determine the length of a section of its graph using the distance formula. For example, let $f(x) = 2x - 3$. Consider a section of the graph of f between $x = -1$ and $x = 4$. It is the line segment joining $(-1, -5)$ and $(4, 5)$.

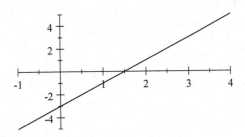

Its length is $\sqrt{(-1 - 4)^2 + (-5 - 5)^2} = \sqrt{5^2 + 10^2} = \sqrt{125} = 5\sqrt{5}$ units.

Now consider a non-linear function $f(x)$. Say we wish to determine the length L of the section of its graph between $x = a$ and $x = b$ with $a < b$. Since we have no direct means of calculating lengths of arbitrary curves, we resort to approximating the graph with straight line segments, then use the total length of the line segments

190

to approximate the length of the curve.

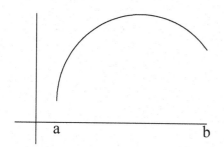 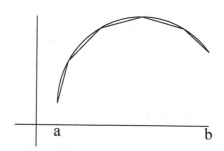

Graph between $x = a$ and $x = b$ Graph and five approximating line segments

In the above diagram, the curve is approximated by 5 line segments. In general we would divide the interval $[a, b]$ into n equal subintervals $[x_0, x_1]$, $[x_1, x_2]$, ..., $[x_i, x_{i+1}]$, ...,$[x_{n-1}, x_n]$ and approximate the curve with the line segments L_1 joining $(x_0, f(x_0))$ to $(x_1, f(x_1))$, L_2 joining $(x_1, f(x_1))$ to $(x_2, f(x_2))$, ..., L_n joining $(x_{n-1}, f(x_{n-1}))$ to $(x_n, f(x_n))$. The typical line segment joining $(x_{i-1}, f(x_{i-1}))$ to $(x_i, f(x_i))$ has length

$$\sqrt{(x_i - x_{i-1})^2 + [f(x_i) - f(x_{i-1})]^2}$$

Therefore the length of the curve is approximately equal to

$$\sum_{i=1}^{n} \sqrt{(x_i - x_{i-1})^2 + [f(x_i) - f(x_{i-1})]^2} \qquad (5.4)$$

We use the Mean Value theorem to write expression (5.4) in the form of a Riemann sum. The theorem asserts that if we consider any two values $f(x_{i-1})$ and $f(x_i)$ of f, we can find a number θ_i between x_{i-1} and x_i such that

$$f(x_i) - f(x_{i-1}) = (x_i - x_{i-1}) f(\theta_i)$$

It follows that the length of the curve is approximately equal to

$$\sum_{i=1}^{n} \sqrt{(x_i - x_{i-1})^2 + (x_i - x_{i-1})^2 (f'(\theta_i))^2} \qquad (5.5)$$

If we factor out $(x_i - x_{i-1})^2$ and denote $(x_i - x_{i-1})$ by $\triangle x$, we may write (5.5) as the Riemann sum

$$\sum_{i=1}^{n} \left(\sqrt{1 + (f'(\theta_i))^2} \right) (x_i - x_{i-1}) = \sum_{i=1}^{n} \left(\sqrt{1 + (f'(\theta_i))^2} \right) \triangle x$$

The exact length of the curve is the limit of the expression $\sum_{i=1}^{n} \left(\sqrt{1 + (f'(\theta_i))^2} \right) \triangle x$ as $\triangle x$ approaches 0. In other words,

$$L = \lim_{\triangle x \to 0} \sum_{i=1}^{n} \left(\sqrt{1 + (f'(\theta_i))^2} \right) \triangle x = \int_{a}^{b} \sqrt{1 + (f'(x))^2} dx$$

Example 238 *Let $f(x) = x^{3/2}$ and L be the length of the graph of f between $x = 1$ and $x = 8$. Then*

$$L = \int_1^8 \sqrt{1 + (f'(x))^2}\,dx = \int_1^8 \sqrt{1 + \tfrac{9x}{4}}\,dx = \int_1^8 \left(1 + \tfrac{9x}{4}\right)^{1/2}\,dx$$

We can integrate this by inspection. The result is

$$L = \left[\tfrac{2}{3}\left(1 + \tfrac{9x}{4}\right)^{3/2} \cdot \tfrac{4}{9}\right]_1^8 = \tfrac{8}{27}\left(19^{3/2} - \left(\tfrac{13}{4}\right)^{3/2}\right)$$

Example 239 *Let $f(x) = \tfrac{1}{4}x^2 - \tfrac{1}{2}\ln x$ and L be the length of the graph of f between $x = 1$ and $x = 9$. Then*

$$\begin{aligned}
L &= \int_1^9 \sqrt{1 + \left(\tfrac{x}{2} - \tfrac{1}{2x}\right)^2}\,dx = \int_1^9 \sqrt{\tfrac{x^2}{4} + \tfrac{1}{2} + \tfrac{1}{4x^2}}\,dx \\[2mm]
&= \int_1^9 \sqrt{\left(\tfrac{x}{2} + \tfrac{1}{2x}\right)^2}\,dx = \int_1^9 \left(\tfrac{x}{2} + \tfrac{1}{2x}\right)\,dx \\[2mm]
&= \left[\tfrac{x^2}{4} + \tfrac{1}{2}\ln x\right]_1^9 = 20 + \tfrac{1}{2}\ln 9 = 20 + \ln 3
\end{aligned}$$

Exercise 240

1. *Let $f(x) = \tfrac{1}{3}\left(x^2 + 2\right)^{3/2}$. Show that $1 + (f'(x))^2 = \left(1 + x^2\right)^2$. The length of the graph of f between $x = 0$ and $x = 2$ should be $\int_0^2 \sqrt{\left(1 + x^2\right)^2}\,dx$. Calculate it.*

2. *Let $f(x) = 2x^{3/2}$. Show that $\sqrt{1 + (f'(x))^2} = \sqrt{1 + 9x}$. Then the length of the graph of f between $x = 0$ and $x = 3$ should be $\int_0^3 \left(\sqrt{1 + 9x}\right)\,dx$. Calculate it.*

3. *Let $f(x) = \ln(\sec x)$. Show that $1 + (f'(x))^2 = \sec^2 x$, then calculate the length of the graph of f between $x = 0$ and $x = \tfrac{\pi}{3}$.*

4. *Let $f(x) = \cosh x$. Show that $1 + (f'(x))^2 = 1 + \sinh^2 x = \cosh^2 x$, then calculate the length of the graph of f between $x = -1$ and $x = 3$.*

5. *Calculate the length of the graph of $f(x) = \tfrac{1}{2}x^2$ between $x = 0$ and $x = \sqrt{3}$.*

6. *A curve in the plane may be described by stating its x and y components as functions of some variable t. (For example, the circle with center (a, b) and radius r may be described as the set of all points (x, y) such that $x = a + r\cos t$ and $y = b + r\sin t$, $0 \le t < 2\pi$.) Consider such a curve C consisting of all the points (x, y) such that $x = f(t)$ and $y = g(t)$, $a \le t \le b$, where f and g have derivatives on $[a, b]$. Let L be its length. Partition $[a, b]$ into n equal subintervals $[t_0, t_1]$, $[t_1, t_2]$, ..., $[t_{n-1}, t_n]$. The line segments joining $(f(t_{i-1}), g(t_{i-1}))$ to $(f(t_i), g(t_i))$ for $i = 1, \ldots, n$ may be used to approximate the length L of the curve.*

(a) *Show that* $L \simeq \sum_{i=1}^{n} \left(\sqrt{(f(t_i) - f(t_{i-1}))^2 + (g(t_i) - g(t_{i-1}))^2} \right)$ *then deduce that*

$$L = \int_{a}^{b} \sqrt{(f'(t))^2 + g'(t))^2} dt.$$

(b) *Use the above result to calculate the circumference of the circle* $x = a + r\cos t$, $y = b + r\sin t$, $0 \le t \le 2\pi$.

(c) *Use the result of part (a) to write down an expression for the length of an ellipse* $x = a\cos t$, $y = b\sin t$, $0 \le t \le 2\pi$. *What you get is an example of an elliptic integral.*

Area of a Surface of Revolution

Take the graph of a function f on an interval $[a, b]$ and revolve it about a given axis through 4 right angles. What you get is called a surface of revolution. This section addresses the problem of determining the areas of such surfaces.

Example 241 *When you revolve a line segment of length h about an axis parallel to the line segment, you get a cylinder. The radius of the cylinder is the distance r between the line segment and the axis of rotation.*

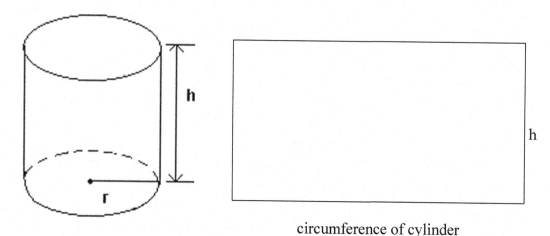

circumference of cylinder

To calculate the surface area of the above cylinder, imagine cutting the figure along a line in the surface parallel to the axis of rotation. When you open it out, you get a rectangle. One side of the rectangle is the circumference of the cylinder. The other side is its height. The area of the rectangle, which is $2\pi rh$ square units, is the surface area of the cylinder.

Example 242 *When you revolve a line segment of length s about an axis making a non-zero angle with the line segment, you get a right circular cone. The cone in the diagram below was obtained by revolving, about the y-axis, a line that makes a non-zero angle with the y-axis. The radius r of its base is called the **radius of the cone**. The distance h from the center of its base to its tip is called the **height of the cone**. The **slant height of the cone** is the shortest distance s from the tip of the cone to the circumference of the base. To calculate the surface area of the curved part of the cone, note that if you cut the cone along a slant line and open* *out, you get a sector of a circle, (shown below), with radius s and arc length $2\pi r$. Let the angle of the sector be θ. Since $s\theta = 2\pi r$, it follows that $\theta = \dfrac{2\pi r}{s}$.*

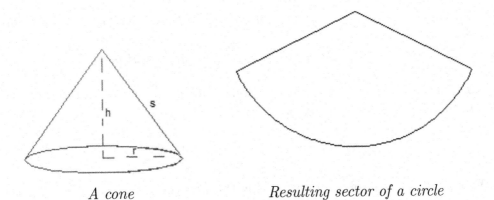

A cone   *Resulting sector of a circle*

The area of the sector, which is $\frac{1}{2}s^2\theta$, is the surface area of the curved part of the cone. It follows that the surface area of the cone is

$$\frac{1}{2}s^2\theta = \frac{1}{2}s^2 \cdot \frac{2\pi r}{s} = \pi rs \ \text{square units.} \tag{5.6}$$

To calculate the surface areas of a number of surfaces, we approximate them with **frustums of a cone**. A frustum of a cone is a figure, like the one below, obtained by chopping off a right circular cone from a given right circular cone.

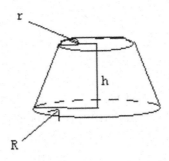

A frustum of a cone

We need an expression for its surface area. Let the radius of its bigger face be R and that of the smaller face be r. Let the slant height of the cone that is chopped off be l_1 and the slant height of the original cone (before the smaller cone is cut off) be l_2. Then the slant height l of the frustum is given by the equation $l_2 = l + l_1$, and the area of the curved part of the frustum is

$$\pi R l_2 - \pi r l_1 = \pi R(l_1 + l) - \pi r l_1 = \pi (R - r) l_1 + \pi R l$$

To get rid of the term $\pi (R - r) l_1$, we use the fact that $\dfrac{l_1}{r} = \dfrac{l_1 + l}{R}$. The result is $(R - r) l_1 = rl$, which implies that $\pi (R - r) l_1 = \pi rl$. Therefore the area of the curved part of the frustum is

$$\pi rl + \pi R l = \pi l (R + r) \ \text{square units.}$$

194

Example 243

1. *The area of the curved surface of a cone with base radius 3 cm and slant height 7 cm is 21π square centimeters.*

2. *The area of the curved surface of a cone with base radius 4 cm and height 6 cm is*
$$\pi(4)(\sqrt{4^2 + 6^2}) = 8\sqrt{13}\pi \text{ square centimeters.}$$

3. *The area of the curved surface of a frustum with slant height 10 cm, faces of radii 14 cm and 24 cm respectively is*
$$\pi(10)(14 + 24) = 380\pi \text{ cubic centimeters}$$

We can now derive an expression for the area of a surface of revolution. To this end, consider the graph of some function f on an interval $[a, b]$ and the surface generated by revolving the graph about the x-axis through 4 right right angles.

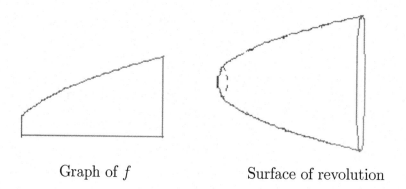

| Graph of f | Surface of revolution |

To calculate an approximate value of the area of revolution, we partition it into n segments by dividing $[a, b]$ into n equal subintervals $[x_0, x_1]$, $[x_1, x_2]$, ..., $[x_{n-1}, x_n]$ of length $\triangle x = h = \frac{1}{n}(b - a)$ each. Each subinterval gives a segment of the area. We approximate each segment by a frustum . The left figure below shows 2 segments that partition the surface. The figure to the right gives the frustums approximating the 2 segments.

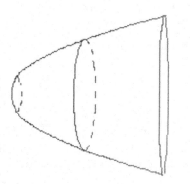 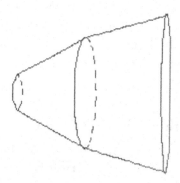

Surface partitioned into 2 segments   Approximating frustums

A typical segment on an interval $[x_{i-1}, x_i]$ is approximated by a frustum with faces of radii $f(x_{i-1})$ and $f(x_i)$. Its slant height is $s = \sqrt{(x_i - x_{i-1})^2 + [f(x_i) - f(x_{i-1})]^2}$. Therefore its area is

$$\pi \left(\sqrt{(x_i - x_{i-1})^2 + [f(x_i) - f(x_{i-1})]^2} \right) [f(x_i) + f(x_{i-1})] \tag{5.7}$$

There is a point θ_i between x_{i-1} and x_i such that $f(x_i) - f(x_{i-1}) = f'(\theta_i)(x_i - x_{i-1})$, (by the Mean Value Theorem for derivatives), hence we may write (5.7) as

$$\pi \left(\sqrt{1 + [f'(\theta_i)]^2} \right) [f(x_i) + f(x_{i-1})](x_i - x_{i-1})$$

Therefore an approximation of the required area is

$$\pi \sum_{i=1}^{n} \left(\sqrt{1 + [f'(\theta_i)]^2} \right) [f(x_i) + f(x_{i-1})](x_i - x_{i-1}) \tag{5.8}$$

Because x_i, x_{i-1} and θ_i are not the same point, (5.8) is not a Riemann sum. To get around this, assume that f is a continuous function. Then when $x_i - x_{i-1}$ is small, the numbers $f(x_i)$ and $f(x_{i-1})$ may be approximated by $f(\theta_i)$. Therefore the area of the surface of revolution is approximately equal to

$$2\pi \sum_{i=1}^{n} \left(\sqrt{1 + [f'(\theta_i)]^2} \right) f(\theta_i)(x_i - x_{i-1}) = 2\pi \sum_{i=1}^{n} \left(\sqrt{1 + [f'(\theta_i)]^2} \right) f(\theta_i)\triangle x$$

The exact area is the limit of these Riemann sums as $\triangle x \to 0$. In other words,

$$Area = 2\pi \int_a^b \left(\sqrt{1 + [f'(x)]^2} \right) f(x) dx \tag{5.9}$$

Example 244 *Let $f(x) = \cosh x$ and consider the surface generated when the graph of f between $x = 0$ and $x = 3$ is revolved about the x-axis through 4 right angles. Let its area be A. Since $1 + \sinh^2 x = \cosh^2 x$, and $\cosh^2 x = \cosh 2x + 1$,*

$$\begin{aligned} A &= 2\pi \int_0^3 \left(\sqrt{1 + \sinh^2 x} \right) \cosh x dx = 2\pi \int_0^3 \cosh^2 x dx \\ &= 2\pi \int_0^3 \tfrac{1}{2} (\cosh 2x + 1) \, dx = \pi \left[\frac{\sinh 2x}{2} + x \right]_0^3 = \pi \left(\frac{\sinh 6}{2} + 3 \right) \end{aligned}$$

Exercise 245

1. *Let $f(x) = \tfrac{1}{2}x$. Consider the graph of f between $x = 0$ and $x = 2$. When we revolve this graph about the x-axis through 4 right angles, we get a cone. Calculate the area of its curved surface using formula (5.9) and verify that the answer you get is in agreement with the value given by formula (5.6) on page 193.*

2. *Let $f(x) = a\sqrt{x}$ where a is a positive constant. Show that:*

(a) $\left(\sqrt{1 + [f'(x)]^2}\right) f(x) = \dfrac{a\sqrt{a^2 + 4x}}{2}.$

(b) The area of the surface generated by revolving the graph of f between $x = 0$ and $x = a^2$, about the x-axis through 4 right angle is $\frac{\pi a^4}{6}(5^{3/2} - 1)$.

3. *Consider the graph of $f(x) = \sin x$ on the interval $[0, \pi]$. Let A be the area of the surface obtained by revolving the graph about the x-axis through 4 right angles. Show that $A = \displaystyle\int_0^\pi \left(\sqrt{1 + \cos^2 x}\right) \sin x\, dx$, then use the substitution $u = \cos x$ to deduce that $A = \displaystyle\int_{-1}^1 \left(\sqrt{1 + u^2}\right) du$. We have, so far, introduced two methods of evaluating this integral. Evaluate it using any one of them.*

Work Done by a Force

Work is done when a force is exerted on an object and causes it to move from one point to another. For example, work is done when:

1. A student lifts a backpack full of books from the floor and places it on his shoulder.

2. A weight-lifter lifts a barbell above his head.

3. A horse pulls a plow through a field.

By definition, if a constant force of *magnitude F* is applied to an object and causes it to move a *distance d in the direction of the force* then the work W done by the force is given by the formula

$$W = F \times d$$

The following examples illustrate how we use Riemann sums to calculate work done by non-constant forces:

Example 246 *Consider the swimming pool shown in the Figure 1 below. It has a shallow side that is 2 meter deep and a deep (opposite) side that is 8 meters deep. It is 7 meters wide and 12 meters long.*

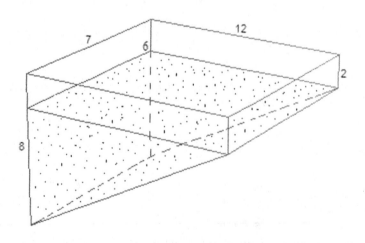

Figure 1.

It contains water that comes up to the 6 meter mark of the deep end of the pool. We wish to calculate the work done to pump all the water out. To this end, imagine partitioning the water into thin layers, by dividing the interval $[0, 6]$ into n equal subintervals of length $6/n$ meters each. Each interval determines a thin layer of the water. A typical one, determined by an interval $[z_i, z_{i+1}]$, is drawn in the Figure 2 below.

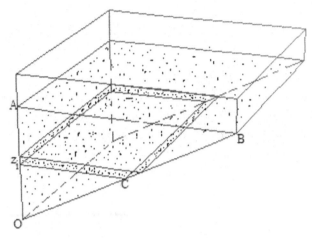

Figure 2.

It is also shown magnified in Figure 3 below.

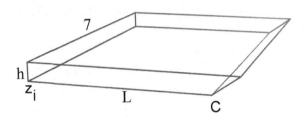

Figure 3.

We approximate it with a rectangular box of length L, width 7 and thickness $h = \frac{6}{n}$. Using the similar triangles Oz_iC and OAB in Figure 2, the length L of Cz_i is given by

$$\frac{L}{12} = \frac{z_i}{6}.$$

It follows that $L = 2z_i$ meters, therefore its volume is approximately

$$(7)(2z_i)(h) = 14z_i h \text{ cubic meters.}$$

Since a cubic meter of water weighs 1000 kilograms, the thin layer weighs approximately $14000z_i h$ kilograms. It has to be lifted through a distance of $(8 - z_i)$ meters to the surface of the pool. The work done is approximately

$$14000z_i h(8 - z_i) \text{ kilogram wt. meters}$$

An estimate of the total work done to empty the tank is obtained by adding the work done to move all the n layers. If the work is W then

$$W \simeq 14000z_0 h(8 - z_0) + \cdots + 14000z_{n-1} h(8 - z_{n-1}) = \sum_{i=0}^{n-1} 14000z_i h(8 - z_i)$$

This is a Riemann sum of the function

$$f(z) = 14000z(8 - z) = 112000z - 14000z^2.$$

The exact value of W is the limit of $\sum_{i=0}^{n-1} (112000z_i - 14000z_i^2) h$ as $h \to 0$. That limit is

$$\int_0^6 (112000z - 14000z^2)dz = \left[56000z^2 - \frac{14000}{3}z^3\right]_0^6 = 1008000 \ Kg. \ m$$

Example 247 *Imagine an elastic string on the x-axis with one end fixed to an object at the the origin $(0,0)$ and the other end free to move along the horizontal axis. Its length when it is not stretched is called the natural length of the string and we denote it by ℓ. Assume that when it is not stretched, the free end is at $(\ell, 0)$.*

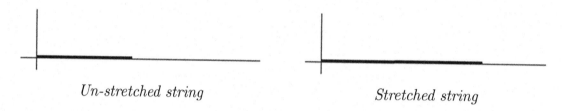

Un-stretched string   *Stretched string*

*It is known from experiments that when the free end is pulled a distance x to a point $(\ell + x, 0)$ the string exerts a force, (a pull), of magnitude kx where k is a constant, (called the constant of the string). Assume that force is measured in kilogram weight and distances in meters. Imagine pulling the free end from the point $(\ell, 0)$ all the way to a point $(\ell + b, 0)$ where $b > 0$. We wish to calculate the work done in doing so. The force you have to exert to pull it starts off small, when x is small, and increases as x increases. Therefore the work done it is **not** $(bk) \cdot (b) = kb^2$, which one obtains when one multiplies the force kb (when the string is stretched b units) by the distance b through which the free end of the string is pulled. It would be kb^2 if the force were constant and equal to bk all the way from the start to finish, but that is not the case. To deal with the changing force, we divide the interval $[0, b]$, (because the stretching changes from 0 to b), into n equal subintervals of length $h = \frac{b}{n} = \triangle x$ each. They are $[x_0, x_1] = \left[0, \frac{b}{n}\right]$, $[x_1, x_2] = \left[\frac{b}{n}, \frac{2b}{n}\right]$, $[x_2, x_3]$, ..., $[x_{n-1}, b]$.*

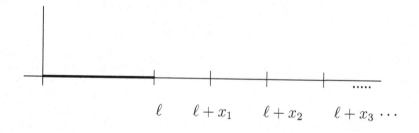

ℓ $\ell + x_1$ $\ell + x_2$ $\ell + x_3 \cdots$

Now imagine pulling the free end in stages: from $(\ell, 0)$ to $(\ell + x_1, 0)$, then from $(\ell + x_1, 0)$ to $(\ell + x_2, 0)$, and so on. In the last stage, we pull it from $(\ell + x_{n-1}, 0)$ to $(\ell + x_n, 0) = (\ell + b, 0)$. To estimate the work done in pulling it from $(\ell + x_i, 0)$ to $(\ell + x_{i+1}, 0)$, we assume that the force needed to pull it is constant and equal to $f(x_i) = kx_i$. In particular, we assume that the force needed to pull it from $(\ell, 0)$ to $(\ell + x_1, 0)$ is 0; the force needed to pull it from $(\ell + x_1, 0)$ to $(\ell + x_2, 0)$ is kx_1, and so on. Of course these are all approximations, because the force does not remain constant. But they are good approximations when h is very small. It follows that the work done to stretch it by a length of b units from its natural length is approximately equal to

$$hf(x_0) + hf(x_1) + \cdots + hf(x_{i-1}) = \sum_{i=0}^{n-1} hf(x_i) \qquad (5.10)$$

Let its exact value be W. Then W should be the limit of the sums (5.10) as $h \to 0$. In other words

$$W = \lim_{h \to 0} \sum_{i=0}^{n-1} hf(x_i) = \lim_{\triangle x \to 0} \sum_{i=0}^{n-1} f(x_i) \triangle x = \int_0^b f(x) dx$$

Since $f(x) = kx$ which has antiderivative $F(x) = \frac{1}{2}kx^2$, it follows that

$$W = \left[\frac{1}{2}kx^2 \right]_0^b = \frac{1}{2}kb^2.$$

Exercise 248

1. A chain weighs 1 kilogram per meter, and it hangs from a point on the roof of a house that is 20 meters from the ground.

   (a) What is the work done to pull such a 20 - meter chain to the roof?

   (b) What is the work done to pull such a 28 - meter chain to the roof?

2. The fuel tank for a tractor trailer is a cylinder lying on its curved side. It has radius r and length L meters. It contains fuel of density d kg per cubic meter. The fuel pump pumps the fuel to a point in the engine where it is ignited. The point is H meters above the highest point on the tank. Suppose the tank is half full.

   (a) Show that the work done, in kilogram meters, by the pump to pump all the fuel into the engine is

   $$2dL \int_0^r \left[r^2 - (r - z)^2 \right]^{1/2} \left[H + (r - z) \right] dz$$

   (b) Show that the substitution $u = (r - z)$ reduces the above integral to

   $$2dL \int_0^r \left(r^2 - u^2 \right)^{1/2} (H + u) \, du$$

   then evaluate it.

3. A water storage tank was made by welding a hemisphere onto each of the two ends of a cylinder of radius R and height H meters respectively. The tank is supported upright on a metal structure in such a way that its lowest point is L meters above the ground. It has to be filled with water pumped from the ground level. Given that 1 cubic meter of water weighs 1000 Kg:

(a) *Show that the work done, (the units are kilogram weight meters), to fill the bottom hemispherical part of the tank is*

$$1000\pi \int_0^R \left[R^2 - (R-z)^2\right] [L+z]\, dz = 1000\pi \int_0^R \left(2Rz - z^2\right)(L+z)\, dz$$

(b) *Calculate the work done to fill up the cylindrical part, and the work done to fill up the top hemispherical part then add up to get the total work to fill up the tank.*

Force Exerted by a Fluid

We use an example to show how forces may be calculated using Riemann sums:

Example 249 *Imagine a rectangular tank of length L and width W containing a fluid of uniform density u to height H units as shown in figure (i).*

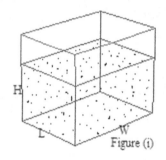
Figure (i)

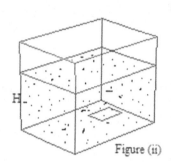
Figure (ii)

Imagine a rectangular hole, shown in figure (ii), in the base of the tank. The hole is plugged with a rectangular object of the same size. A force has to be exerted on the plug to stop it from popping out. It turns out, (these are experimental facts from physics), that the fluid force per unit area at any point on the plug is equal to

(Density of fluid) × (height H of column of fluid above point).

Therefore the total force on the plug, (we let you figure out the units), is

(Density of fluid) × (height H of column of fluid above point) × (area of plug).

This implies that the total force on the base of the tank is

(Density of fluid) × (height H of column of fluid above point) × (area of base).

Now imagine a hole in a side of the tank as shown in figure (iii) below.

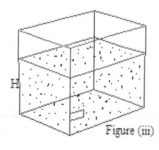
Figure (iii)

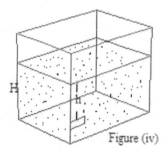
Figure (iv)

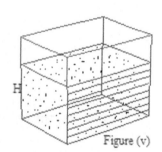
Figure (v)

It is plugged like the hole in the bottom. It is also an experimental fact that the force per unit area at a point on this plug is equal to

(Density of fluid) × (vertical distance h from the point to the surface of fluid).

The distance h is shown in figure (iv). This time the force per unit area is not the same at all the points of the plug because h may change from one point to another. To estimate the force on the plug, assume that it is very thin. Then it is reasonable to approximate the force per unit area at an arbitrary point on the plug with the force per unit area at the top end of the plug. Therefore the force needed to keep it in place is approximately equal to

(Density of fluid)×(distance h from top of plug to surface of fluid)×(area of plug).

Say we wish to calculate the force on the face in figure (iii) with the plug. We divide the face into n small strips of thickness $\triangle h$ each as shown in figure (v), estimate the force on each strip, add up to get an estimate of the total force on the face then take limits. The force on a strip whose center is h_i units from the surface of the fluid is $uh_iW\triangle h$. The total force on the face is approximately

$$\sum_{i=1}^{n} uh_iW\triangle h.$$

The exact force is the limit as $\triangle h \to 0$ of the above Riemann sum. That limit is

$$\int_0^H uWh\,dh = \frac{uWH^2}{2}.$$

Exercise 250

1. *A rectangular tank contains a liquid with density u kilogram per cubic meter to a depth of H meters. A circular plate with radius r meters, $(r < \frac{1}{2}H)$, is in one of the sides of the tank as shown in the figure below. Show that the fluid force acting on the plate is*

$$\int_0^{2r} \left[r^2 - (r-z)^2\right][H-z]\,dz$$

then evaluate the integral.

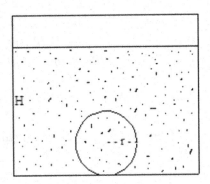

2. *A cylindrical tank with closed top and base has radius r and height H. It is half full with a fluid of density u. Imagine laying it on its curved side. What is the force on each of the two circular ends?*

3. *A dam is 300 m long and 40 m high. The density of water is 1000 kg per cubic meter. Calculate the force of water on the dam when the water level behind it is 25 m.*

Center of Gravity

Example 251 *Consider a uniform triangular lamina with vertices at $(0,0)$, $(3,4)$ and $(3,-4)$. It is the region enclosed by the lines $f(x) = \frac{4}{3}x$, $g(x) = -\frac{4}{3}x$ and the line $x = 3$. We wish to determine its center of gravity. This is the point where the lamina balances on a pin-head.*

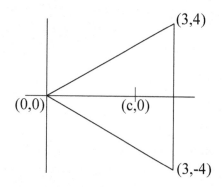
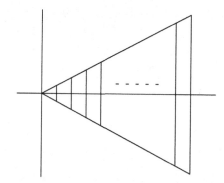

Given lamina　　　　　　　　*Lamina divided into small segments*

Since the lamina is symmetric about the x-axis, the center of gravity must be on the x-axis. Let it be at the point $(c,0)$. Then the moment of the lamina about the origin is Mc where M is its weight. Indeed $M = w \cdot 4 \cdot 3 = 12w$ where w is the weight per unit area of the lamina, therefore the moment of the lamina about the origin is

$$Mc = 12wc$$

We need another expression for the moment of the lamina about the origin in order to solve for c. To get it, we use the fact that the moment of a body is equal to the sum of the moments of its parts. We therefore divide the lamina into smaller segments, as shown above, estimate the moment of each segment and total up. Say we divide it into n segments of thickness $\triangle x = \frac{3}{n}$ each. A typical segment is between $x_i = i \cdot \frac{3}{n}$ and $x_{i+1} = (i+1) \cdot \frac{3}{n}$. Its area is approximately equal to $2f(x_i)\triangle x$, therefore its moment about the origin is approximately

$$2wf(x_i)\triangle x \cdot x_i = 2wf(x_i)x_i\triangle x$$

It follows that the moment of the lamina about the origin is approximately equal to

$$\sum_{i=0}^{n-1} 2wf(x_i)x_i\triangle x$$

The exact moment is $\displaystyle\lim_{\Delta x \to 0} \sum_{i=0}^{n-1} 2wf(x_i)x_i\Delta x = 2w\int_0^3 xf(x)dx = \frac{8w}{3}\int_0^3 x^2dx.$ *We*

have used the fact that $f(x) = \dfrac{4x}{3}$. *We easily evaluate the definite integral to get*

$$\frac{8w}{3}\int_0^3 x^2dx = \left[\frac{8w}{9}x^3\right]_0^3 = 24w$$

Equating the two expressions $12wc$ *and* $24w$ *for the total moment about the origin gives* $c = 2$. *Thus the lamina balances on a pin-head at* $(2,0)$.

Exercise 252

1. Find the position of the center of gravity of a uniform lamina in the shape of a half disc of radius r.

2. A uniform lamina has the shape of a trapezium with vertices at $(-3,0)$, $(3,0)$, $(2,4)$ and $(-2,4)$. Calculate its center of gravity. (Imagine dividing the lamina into small horizontal segments of width Δy. Estimate the moment of a typical segment about the x-axis then form a Riemann sum.)

3. Find the position of the center of gravity of a uniform lamina in the shape of the region enclosed by the graph of $f(x) = x^2$ and the line $y = 4$.

4. An isosceles triangle has vertices at $A(-a,0)$, $B(a,0)$ and $C(0,b)$ where $b > 0$. Let $O(0,0)$ be the origin. Show that its center of gravity is one third of the way up the line OC.

Improper Integrals

Consider a function $f(x)$ that is defined for all values of $x \geq a$. It may be meaningful to ask for the area of the region to the right of the line $x = a$, enclosed by the graph of f and the x-axis.

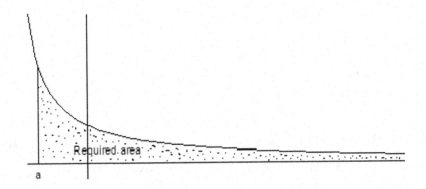

To evaluate it, do the following:

1. Take an arbitrary number $R > a$ and determine the area $\displaystyle\int_a^R f(x)dx$ of the region enclosed by the graph of f, the x-axis and the two lines $x = a$ and

204

$x = R.$

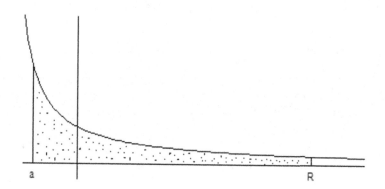

2. Check if $\lim\limits_{R \to \infty} \int_a^R f(x)dx$ exists. If it does and it is a finite number I then I should be the required area. The technical term for the limit is the **improper integral** of f on $[a, \infty)$. Its technical symbol is $\int_a^\infty f(x)dx$.

3. If $\lim\limits_{R \to \infty} \int_a^R f(x)dx$ does not exist then we say that the improper integral $\int_a^\infty f(x)dx$ diverges.

Example 253 *Let $f(x) = \dfrac{1}{x^3}$, $x \geq 1$. Its graph is given below together with the two lines $x = 1$ and $x = R$ where $R > 1$*

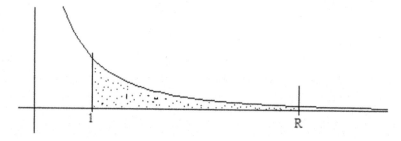

We find that

$$\int_1^R f(x)dx = \int_1^R \frac{1}{x^3}dx = \left[\frac{-1}{2x^2}\right]_1^R = \frac{1}{2} - \frac{1}{2R^2}$$

Clearly, when R is large and positive then $\int_1^R \dfrac{1}{x^3}dx$ is close to $\frac{1}{2}$, therefore we should define the area, to the right of $x = 1$, enclosed by the graph of f and the x-axis, to be $\frac{1}{2}$. We write $\int_1^\infty \dfrac{1}{x^3}dx = \frac{1}{2}$.

Example 254 *Let $f(x) = \dfrac{1}{x}$, $x > 1$. If $R > 1$ then $\int_1^R \dfrac{1}{x}dx = \ln R$. Since $\lim\limits_{R \to \infty} \ln R$ is infinite, the improper integral $\int_1^R \dfrac{1}{x}dx$ diverges.*

For a function g defined for all numbers $x \leq b$, the area to the left of b it encloses with the x-axis is denoted by

$$\int_{-\infty}^{b} f(x)dx$$

and it is defined by

$$\int_{-\infty}^{b} f(x)dx = \lim_{R \to \infty} \int_{-R}^{b} f(x)dx$$

provided the limit exists. If $\lim_{R \to \infty} \int_{-R}^{b} f(x)dx$ does not exist or is infinite then we say that $\int_{-\infty}^{b} f(x)dx$ diverges.

Example 255 *Let $f(x) = x^2 e^x$, $x \leq 0$. Then*

$$\int_{-R}^{0} x^2 e^x dx = 2 - \frac{R^2 + 2R + 2}{e^R}.$$

By L'Hopital's rule, $\lim_{R \to \infty} \left[2 - \frac{R^2 + 2R + 2}{e^R} \right] = 2$. Therefore $\int_{-\infty}^{0} x^2 e^x$ converges and we may write $\int_{-\infty}^{0} x^2 e^x = 2$

Finally if $f(x)$ is defined for all x then $\int_{-\infty}^{\infty} f(x)dx$ is defined by

$$\int_{-\infty}^{\infty} f(x)dx = \lim_{R \to \infty} \lim_{S \to \infty} \int_{-S}^{R} f(x)dx$$

provided the limit exists.

Example 256 *Consider $\int_{-\infty}^{\infty} \frac{1}{1 + x^2}dx$. The integrand is defined for all real numbers and*

$$\int_{-S}^{R} \frac{1}{1 + x^2}dx = \arctan R - \arctan(-S)$$

Since

$$\lim_{R \to \infty} \lim_{S \to \infty} \int_{-S}^{R} \frac{1}{1 + x^2}dx = \lim_{R \to \infty} \lim_{S \to \infty} (\arctan R - \arctan(-S)) = \frac{\pi}{2} - \left(-\frac{\pi}{2} \right) = \pi,$$

the improper integral converges and we may write $\int_{-\infty}^{\infty} \frac{1}{1 + x^2}dx = \pi$.

Exercise 257

1. Evaluate the improper integral (a) $\int_{-\infty}^{0} e^x dx$ and (b) $\int_{1}^{\infty} \frac{1}{x^p} dx, \ p > 1$

2. Use integration by parts to show that $\int xe^{-x} dx = -\frac{(1+x)}{e^x} + c$ then determine $\int_{0}^{\infty} xe^{-x} dx$.

3. Show that $\int \frac{dx}{x \, (\ln x)^2} = -\frac{1}{\ln x} + c$ then evaluate $\int_{2}^{\infty} \frac{dx}{x \, (\ln x)^2}$.

4. Integrate $\int \frac{1}{e^x + e^{-x}} dx$ by substituting $u = e^x$ then evaluate $\int_{-\infty}^{\infty} \frac{1}{e^x + e^{-x}} dx$.

5. Integrate $\int \frac{1}{\sqrt{x} \, (x+1)} dx$ by substituting $u = \sqrt{x}$ then find $\int_{1}^{\infty} \frac{1}{\sqrt{x} \, (x+1)} dx$.

6. Use partial fractions to show that $\int \frac{1}{x^2 + 4x} dx = \frac{1}{4} \ln \left(\frac{x}{x+4} \right) + c$ then find $\int_{1}^{\infty} \frac{1}{x^2 + 4x} dx$.

7. Evaluate $\int_{0}^{\infty} \frac{1}{4 + x^2} dx$ and $\int_{0}^{\infty} \frac{1}{81 + x^2} dx$.

8. Show that if t is a constant then $\int_{0}^{\infty} \frac{1}{t^2 + x^2} dx = \frac{\pi}{2t}$. *(This is an example of what is called **an integral depending on a parameter**. In this case the parameter is t.)*

9. Show that if y is a positive constant then $\int_{0}^{\infty} e^{-yx} \sin x dx = \frac{1}{1+y^2}$. *(By definition, the integral $\int_{0}^{\infty} e^{-yx} \sin x dx$ depending on the parameter y is called the Laplace transform of the function $f(x) = \sin x$.)*

10. Assume that y is a positive constant. Show that $\int_{0}^{\infty} e^{-yx} \cos x dx = \frac{y}{1+y^2}$. *(Likewise, $\int_{0}^{\infty} e^{-yx} \sin x dx$ is called the Laplace transform of $f(x) = \cos x$.)*

11. Let y be a positive constant. Determine $\int_{0}^{\infty} xe^{-yx} dx$, the Laplace transform of $f(x) = x$.

Chapter 6 APPROXIMATIONS OF DEFINITE INTEGRALS

We have, so far, relied on antiderivatives to evaluate areas under curves, work done by a variable force, volumes of revolution, etc. More precisely, whenever we have had to evaluate a definite integral $\int_a^b f(x)dx$, a standard procedure has been to look for an antiderivative $F(x)$ for the integrand $f(x)$ then use the fact that

$$\int_a^b f(x)dx = F(b) - F(a)$$

We have been fortunate in that every integrand $f(x)$ we have faced, to this point, has a "familiar" antiderivative. The bad news is that there are many useful functions with no such antiderivatives. An example is $f(x) = e^{-x^2}$, which is a very common function in probability theory. Also, there are functions that are given as tables, (from experimental data), with no formula describing the relationship between the variables. In cases like these, we have to settle for "good" approximate values of the definite integrals. To calculate an approximate value of $\int_a^b f(x)dx$, we visualize the integral as the area of the region enclosed by the x-axis and the graph of f on the interval $[a, b]$, then look for approximate values of the area. The standard procedure is to divide it into smaller strips by dividing the interval $[a, b]$ into smaller subintervals $[x_0, x_1]$, $[x_1, x_2]$, ..., $[x_{n-1}, x_n]$ of width $h = \frac{1}{n}(b-a)$ each. (Thus $x_0 = a$, $x_1 = a+h$, $x_2 = a+2h$, ..., $x_n = a+nh = b$.)

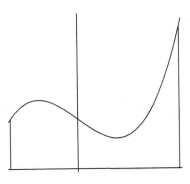

The region

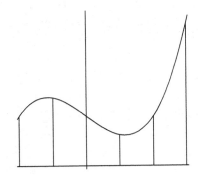

Region partitioned into strips

Each strip is then approximated with a geometric figure whose area is familiar. The standard ones are rectangles, trapeziums; in general, figures enclosed by polynomials. We start with rectangles.

Using Rectangles of Equal Width

We approximate the strip on an interval $[x_{i-1}, x_i]$ with a rectangle that has base $[x_{i-1}, x_i]$ and an appropriate height H_i. There are various choices for H_i:

1. We choose the height H_i to be the value $f(x_{i-1})$ of f at the left-end point of the interval $[x_{i-1}, x_i]$. Then $\int_a^b f(x)dx$ is approximated by the sum of the areas $hf(x_0)$, $hf(x_1)$, ..., $hf(x_{n-1})$. Since $h = \frac{1}{n}(b-a)$,

$$\int_a^b f(x)dx \simeq \frac{(b-a)}{n}\left[f(x_0) + f(x_1) + \cdots + f(x_{n-1})\right]$$

This is called the **left-endpoint rule**, because it results from approximating the strips with left-endpoint rectangles.

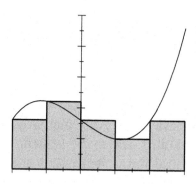

Left-endpoint rectangles

2. We choose H_i to be the value $f(x_i)$ of f at the right-end point of the interval $[x_{i-1}, x_i]$. Then $\int_a^b f(x)dx$ is approximated by the sum of the areas $hf(x_1)$, $hf(x_2)$, ..., $hf(x_n)$. Using $h = \frac{1}{n}(b-a)$,

$$\int_a^b f(x)dx \simeq \frac{(b-a)}{n}\left[f(x_1) + f(x_2) + \cdots + f(x_n)\right]$$

This is called the **right-endpoint rule**, because it results from approximating the strips with right-endpoint rectangles.

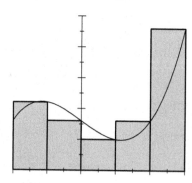

Right-endpoint rectangles

3. We choose H_i to be the value of f at the midpoint $\theta_i = \frac{1}{2}(x_{i-1} + x_i)$ of the interval $[x_{i-1}, x_i]$. Then $\int_a^b f(x)dx$ is approximated by the sum of the areas $hf(\theta_1)$, $hf(\theta_2)$, ..., $hf(\theta_n)$. Since $h = \frac{1}{n}(b-a)$

$$\int_a^b f(x)dx \simeq \frac{(b-a)}{n}\left[f(\theta_1) + f(\theta_2) + \cdots + f(\theta_n)\right]$$

This is called the **midpoint rule**, because it results from approximating the strips with midpoint rectangles.

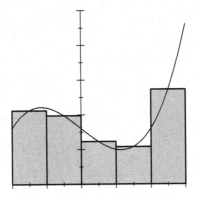

Midpoint rectangles

Example 258 *To approximate* $\int_0^1 e^{x^2}\,dx$ *using 15 rectangles of the types described above.*

Here the integrand is $f(x) = e^{x^2}$. *The interval* $[0,1]$ *is divided into the 15 subintervals* $\left[0, \frac{1}{15}\right]$, $\left[\frac{1}{15}, \frac{2}{15}\right]$, $\left[\frac{2}{15}, \frac{3}{15}\right]$, $\ldots$, $\left[\frac{14}{15}, 1\right]$.

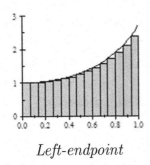

Left-endpoint

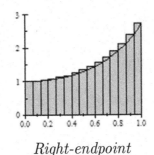

Right-endpoint

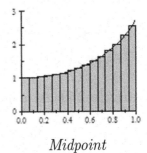

Midpoint

If we approximate the strips with left-endpoint rectangles then

$$\int_0^1 e^{x^2}\,dx \simeq \tfrac{1}{15} \cdot f(0) + \tfrac{1}{15} \cdot f\left(\tfrac{1}{15}\right) + \tfrac{1}{15} \cdot f\left(\tfrac{2}{15}\right) + \cdots + \tfrac{1}{15} \cdot f\left(\tfrac{14}{15}\right)$$

Pull out a calculator and evaluate $\left[\tfrac{1}{15}\left(e^0 + e^{1/15} + e^{2/15} + \cdots + e^{14/15}\right)\right]$. *The result should be 1.662, (to 3 decimal places).*

If we approximate the strips with right-endpoint rectangles we get

$$\int_0^1 e^{x^2}\,dx \simeq \tfrac{1}{15} \cdot f\left(\tfrac{1}{15}\right) + \tfrac{1}{15} \cdot f\left(\tfrac{2}{15}\right) + \tfrac{1}{15} \cdot f\left(\tfrac{3}{15}\right) + \cdots + \tfrac{1}{15} \cdot f\left(\tfrac{15}{15}\right) = 1.776$$

to 3 decimal places.

To approximate the strips with midpoint rectangles it is necessary to determine the mid-points of each interval. It is $\frac{1}{30}$ *for the interval* $\left[0, \frac{1}{15}\right]$, $\frac{3}{30}$ *for the interval* $\left[\frac{1}{15}, \frac{2}{15}\right]$, $\ldots$, $\frac{29}{30}$ *for the last interval. Therefore, to 3 decimal places,*

$$\int_0^1 e^{x^2}\,dx \simeq \tfrac{1}{15} \cdot f\left(\tfrac{1}{30}\right) + \tfrac{1}{15} \cdot f\left(\tfrac{3}{30}\right) + \tfrac{1}{15} \cdot f\left(\tfrac{5}{30}\right) + \cdots + \tfrac{1}{15} \cdot f\left(\tfrac{29}{30}\right) = 1.707$$

Using Trapeziums of Equal Width

This time we approximate the strip on an interval $[x_{i-1}, x_i]$ with the trapezium joining the four points $(x_{i-1}, 0)$, $(x_{i-1}, f(x_{i-1}))$, $(x_i, f(x_i))$ and $(x_i, 0)$. It has area $\frac{1}{2}(f(x_{i-1}) + f(x_i))h$. Therefore

$$\int_a^b f(x)dx \simeq \frac{1}{2}h\left[f(x_0) + f(x_1) + f(x_1) + f(x_2) + \cdots + f(x_{n-1}) + f(x_n)\right] \quad (6.1)$$

With the exception of $f(x_0)$ and $f(x_n)$, every other term inside the square brackets appears twice, therefore

$$\int_a^b f(x)dx \simeq \frac{(b-a)}{2n}\left[f(x_0) + 2f(x_1) + 2f(x_2) + \cdots + 2f(x_{n-1}) + f(x_n)\right]$$

This is called the **trapezoidal rule**.

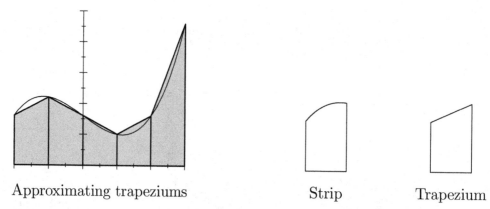

Approximating trapeziums Strip Trapezium

Example 259 *An approximation of* $\int_0^1 e^{x^2}dx$ *using 15 trapeziums is*

$$\int_0^1 e^{x^2}dx \simeq \frac{1}{2}\cdot\frac{1}{15}\left(e^0 + 2e^{1/15} + 2e^{2/15} + 2e^{3/15} + \cdots + 2e^{14/15} + e^{15/15}\right) = 1.719$$

to 3 decimal places.

Using Quadratic Curves

In this case we must partition the interval $[a, b]$ into an **even number** n of subintervals $[x_0, x_1]$, $[x_1, x_2]$, $\ldots$, $[x_{n-1}, x_n]$. We then approximate the two strips on an interval $[x_i, x_{i+2}]$ with the region enclosed by the quadratic curve $q_i(x)$ that passes through the three points $(x_i, f(x_i))$, $(x_{i+1}, f(x_{i+1}))$ and $(x_{i+2}, f(x_{i+2}))$, (see the figures below).

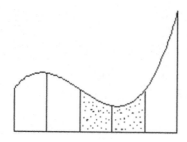

Two strips

Approximating quadratic

By Exercise 2 on page 145, the area under the quadratic curve is

$$\int_{x_i}^{x_{i+2}} q_i(x)dx = \frac{(x_{i+2} - x_i)}{6} [q(x_i) + 4q(x_{i+1}) + q(x_{i+2})]$$

$$= \frac{(x_{i+2} - x_i)}{6} [f(x_i) + 4f(x_{i+1}) + f(x_{i+2})].$$

Note that $(x_{i+2} - x_i) = \dfrac{2(b-a)}{n}$, hence $\dfrac{(x_{i+2} - x_i)}{6} = \dfrac{1}{3n}(b-a)$. Therefore

$$\int_{x_0}^{x_2} f(x)dx \simeq \int_{x_0}^{x_2} q_0(x)dx = \frac{1}{3n}(b-a)[f(x_0) + 4f(x_1) + f(x_2)].$$

Similarly, $\displaystyle\int_{x_2}^{x_4} f(x)dx \simeq \int_{x_2}^{x_4} q_2(x)dx = \frac{(b-a)}{3n}[f(x_2) + 4f(x_3) + f(x_4)].$

In general, $\displaystyle\int_{x_{2i}}^{x_{2i+2}} f(x)dx \simeq \int_{x_{2i}}^{x_{2i+2}} q_i(x)dx = \frac{(b-a)}{3n}[f(x_{2i}) + 4f(x_{2i+1}) + f(x_{2i+2})].$

Adding up gives

$$\int_a^b f(x)dx \simeq \frac{(b-a)}{3n}\{f(x_0) + 4f(x_1) + f(x_2) + f(x_2) + 4f(x_3) + f(x_4) + \cdots + f(x_{n-2}) + 4f(x_{n-1}) + f(x_n)\}$$

This simplifies to the following **Simpson's rule**.

$$\int_a^b f(x)dx \simeq \frac{(b-a)}{3n}\{f(x_0) + 2[f(x_2) + f(x_4) + \cdots + f(x_{n-2})] + 4[f(x_1) + f(x_3) + \cdots + f(x_{n-1})] + f(x_n)\}$$

This may look complicated but it is easy to remember. Add the values of f at the points $x_2, x_4, \ldots, x_{n-2}$ with even indexes then multiply the result by 2. This gives you the term $2[f(x_2) + f(x_4) + \cdots + f(x_{n-2})]$. Next add the values of f at the points $x_1, x_3, \ldots, x_{n-1}$ with odd indexes and multiply the result by 4. This gives the term $4[f(x_1) + f(x_3) + \cdots + f(x_{n-1})]$. To these two terms, add the value of f at x_0 and x_n, (i.e. the value of f at the end-points of the interval). Finally multiply the result by $\dfrac{(b-a)}{3n}$.

Exercise 260 *Let In each of the following questions, round off your answers to 2 decimal places.*

1. *Use the midpoint rule, with $n = 8$, to estimate $\displaystyle\int_0^1 \frac{1}{1 + x^4}dx$.*

2. *Use the trapezoidal rule, with $n = 8$, to estimate $\displaystyle\int_0^2 e^{-x^2}dx$.*

3. *Use the left-endpoint rule, with $n = 5$, to estimate* $\int_0^{\pi/2} \sin(x^2)dx$.

4. *Use the right-endpoint rule, with $n = 5$, to estimate* $\int_0^1 \sqrt{2 + x^3}dx$.

5. *Use Simpson's rule with $n = 8$ to estimate* $\int_0^1 e^{x^2} dx$.

6. *We know that* $\int_0^1 \dfrac{1}{1 + x^2}dx = \arctan 1 - \arctan 0 = \frac{1}{4}\pi$.

   (a) *Use the trapezoidal rule, with $n = 10$, to estimate π.*

   (b) *Use the midpoint rule, with $n = 10$, to get another estimate of π.*

   (c) *Use the Simpson's rule, with $n = 10$, to get yet another estimate of π.*

Errors in Estimates of Definite integrals

As before, we view $\int_a^b f(x)dx$ as the area of the region R enclosed by the x-axis and the graph of f on the interval $[a, b]$. Assume that f is differentiable on $[a, b]$. Partition the interval into n subintervals $[x_0, x_1]$, $[x_1, x_2]$, $\ldots$, $[x_{n-1}, x_n]$ where $x_0 = a$, $x_1 = x_0 + h$, $\ldots$, $x_n = x_{n-1} + h = b$, of length $h = \frac{1}{n}(b - a)$ each. Then R is partitioned into n strips with areas

$$\int_{x_0}^{x_1} f(x)dx, \quad \int_{x_1}^{x_2} f(x)dx, \quad \ldots, \quad \int_{x_{n-1}}^{x_n} f(x)dx$$

Say we approximate $\int_a^b f(x)dx$ with left-endpoint rectangles. This means that the area $\int_{x_i}^{x_{i+1}} f(x)dx$ of the ith strip, is approximated with the value $f(x_i)h$. Note that we may write this as $f(x_i)h = \int_{x_i}^{x_{i+1}} f(x_i)dx$. We wish to estimate the error in the approximation. The absolute value of the difference between the exact value and the approximation is

$$\left| \int_{x_i}^{x_{i+1}} [f(x) - f(x_i)]dx \right|.$$

For each x between x_i and x_{i+1}, the Mean Value Theorem assures us that there is a number θ_x between x and x_i such that

$$f(x) - f(x_i) = (x - x_i) f'(\theta_x)$$

Let M be the largest value of $|f'(x)|$ on the interval $[a, b]$. Then $|(x - x_i) f'(\theta_{x_i})| \leq M(x - x_i)$ and so

$$\left| \int_{x_i}^{x_{i+1}} [f(x) - f(x_i)]dx \right| = \left| \int_{x_i}^{x_{i+1}} (x - x_i) f'(\theta_x)dx \right|$$

$$\leq M \left| \int_{x_i}^{x_{i+1}} (x - x_i) \, dx \right| = M \left[\frac{(x - x_i)^2}{2} \right]_{x_i}^{x_i+h} = \frac{Mh^2}{2}.$$

Thus *the error in approximating* $\int_{x_i}^{x_{i+1}} f(x)dx$ *with* $f(x_i)h = \int_{x_i}^{x_{i+1}} f(x_i)dx$ *does not*

exceed $\frac{Mh^2}{2}$. Adding up, we conclude that *the error in approximating* $\int_a^b f(x)dx$
with the areas of the n left-endpoint rectangles does not exceed

$$\frac{Mh^2}{2} + \frac{Mh^2}{2} + \cdots + \frac{Mh^2}{2} = n\left(\frac{Mh^2}{2}\right) = \frac{Mn(b-a)^2}{2n^2} = \frac{M(b-a)^2}{2n}.$$

We have used the fact that $h = \frac{1}{n}(b-a)$.

Example 261 *Consider* $\int_0^1 \frac{1}{1+x^2}dx$. *The derivative of* $f(x) = \frac{1}{1+x^2}$ *is*

$$f'(x) = -\frac{2x}{(1+x^2)^2}.$$

On the interval $[0,1]$, the largest value of the numerator $2x$ is 2 and the smallest value of the denominator is 1. Therefore the largest value of $|f'(x)|$ on the interval does not exceed $\frac{2}{1} = 2$. *It follows that the error in approximating* $\int_0^1 \frac{1}{1+x^2}dx$ *with the areas of 10 left-endpoint rectangles does not exceed* $\frac{2 \cdot 1^2}{2 \cdot 10} = 0.1$. *Since the total area of the 10 rectangles is approximately 0.80998, and* $\int_0^1 \frac{1}{1+x^2}dx = \frac{\pi}{4}$, *we conclude that* $\frac{\pi}{4} = 0.8 \pm 0.1$, *or* $\pi = 3.2 \pm 0.4$

We state the next three error estimates without proof:

1. Suppose $f''(x)$, the second derivative of $f(x)$, is continuous, and $|f''(x)| \le M$ for all x in $[a,b]$. Then the error in approximating $\int_a^b f(x)dx$ with n midpoint rectangles does not exceed $\frac{M(b-a)^3}{24n^2}$.

Example 262 *Consider* $\int_0^1 \frac{1}{1+x^2}dx$. *The second derivative of* $f(x) = \frac{1}{1+x^2}$ *is*

$$f''(x) = -\frac{2}{(1+x^2)^2} + \frac{8x^2}{(1+x^2)^3}.$$

The value of $|f''(x)|$ on $[0,1]$ does not exceed $2 + 8 = 10$, (apply the argument in example (261) to each term). Therefore the error in approximating $\int_0^1 \frac{1}{1+x^2}dx$ *with the area of 10 midpoint rectangles does not exceed*

$$\frac{10 \cdot 1^2}{24 \cdot 100} \simeq 0.0042.$$

The area of the 10 midpoint rectangles is approximately 0.7856, *it follows that*
$$\frac{\pi}{4} = 0.7856 \pm 0.0042, \ or \ \pi = 3.14 \pm 0.01$$

2. Suppose $f''(x)$, the second derivative of $f(x)$, is continuous, and $|f''(x)| \le M$ for all x in $[a, b]$. Then the error in approximating $\int_a^b f(x)dx$ with n trapeziums does not exceed $\dfrac{M(b-a)^3}{12n^2}$.

3. Suppose $f^{(4)}(x)$, the fourth derivative of f, is continuous, and $\left|f^{(4)}(x)\right| \le M$ for all x in $[a, b]$. Let $[a, b]$ be divided into n even subintervals of length $(b-a)/n$ each. Then the error in approximating $\int_a^b f(x)dx$ using Simpson's rule does not exceed $\dfrac{M(b-a)^5}{180n^4}$.

Exercise 263

1. Let $f(x) = \cos x^2$, $0 \le x \le 1$.

   (a) *Use the left-endpoint rule with* $n = 6$ *to estimate* $\int_0^1 \cos x^2 dx$. *Also determine the maximum error in your answer.*

   (b) *Use the mid-point rule with* $n = 6$ *to estimate* $\int_0^1 \cos x^2 dx$. *Also determine the maximum error in your answer.*

   (c) *Use the trapezoidal rule with* $n = 6$ *to estimate* $\int_0^1 \cos x^2 dx$. *Also determine the maximum error in your answer.*

   (d) *Use Simpson's rule with* $n = 6$ *to estimate* $\int_0^1 \cos x^2 dx$. *Also determine the maximum error in your answer.*

2. *Determine a value of* n *so that the mid-point rule approximates* $\int_0^1 \sqrt{1 + x^3}dx$ *with an error of less than* 0.02.

Chapter 7   SEQUENCES AND SERIES OF NUMBERS

Sequences

A sequence of real numbers is a well-defined "string of numbers". A very famous one is the string of prime numbers 2, 3, 5, 7, 11, 13, ... It has intrigued people for centuries. It may be interesting to know that the really large prime numbers play a major role in encrypting information transmitted electronically on the web.

Some strings of numbers arise in problems we solve. For example, consider using Newton's method, (see page 46), to determine the largest root of the equation $x^3 - 12x + 5 = 0$ accurately to a specified number of decimal places, starting with an approximate root like $a_1 = 3$. If we define $f(x) = x^3 - 12x + 5$ then according to the method,

$$a_2 = a_1 - \frac{f(a_1)}{f'a_1)} = 3 + \frac{4}{15} = \frac{49}{15} \simeq 3.2666667$$

is a better approximate solution than a_1. If it does not satisfy the required level of accuracy, then we compute

$$a_3 = a_2 - \frac{f(a_2)}{f'a_2)} = \frac{49}{15} - \frac{f(49/15)}{f'(49/15)} \simeq 3.2337405$$

which is a better approximate solution than a_2. In general, if a_n has been obtained and it does not satisfy the required level of accuracy, then we calculate a_{n+1} given by

$$a_{n+1} = a_n - \frac{f(a_n)}{f'a_n)}$$

Clearly, a_1, a_2, a_3, a_4, ... is a sequence of numbers.

By its very nature, a sequence of numbers has a *first number*, called the first term of the sequence. This is followed by a *second number*, called the second term of the sequence, and so on. It is convenient, (as we have already done), to denote the first term by a_1, the second one by a_2; in general, the nth one by a_n. (In the case of the sequence of prime numbers, $a_1 = 2$, $a_2 = 3$, $a_3 = 5$ and so on.) Then the sequence may be written as

$$\{a_1, a_2, a_3, a_4, \ldots\}.$$

It is often the case that there is a formula $a_n = f(n)$ expressing the nth term a_n in terms of n. If the sequence has N terms $a_1, a_2, a_3, a_4, \ldots, a_N$ then it may be written briefly as

$$\{f(n)\}_{n=1}^{N}.$$

If it has infinitely may terms then it may be written as

$$\{f(n)\}_{n=1}^{\infty}.$$

For example, the nth positive odd number is $2n - 1$, and the nth positive even number is $2n$. Therefore the sequence consisting of the first 10 positive odd numbers may be written as

$$\{1, 3, \ldots, 19\} \text{ or more briefly as } \{2n - 1\}_{n=1}^{10},$$

and the sequence of positive even numbers may be written as

$$\{2, 4, 6, \ldots\} \text{ or more briefly as } \{2n\}_{n=1}^{\infty}.$$

Some Special Sequences
Arithmetic Progressions

Imagine depositing 100 dollars into a savings account that pays *simple interest* at the rate of 4% per year. This means that each year, the bank literally writes you a cheque for $100 \times \dfrac{4}{100} = 4$ dollars and deposits it into the account. Thus

- At the end of the first year, you will have $100 + 4 = 104$ dollars in the account.

- At the end of the second year, you will have $100 + 4 \times 2 = 108$ dollars.

- At the end of the third year, you will have $100 + 4 \times 3 = 112$ dollars.

- In general, at the end of n years, you will have $100 + 4n$ dollars.

The string of numbers 104, 108, 112, ... is an example of what is called *an arithmetic progression*. To get its second term, you add 4 to the first one. To get the third, you add 4 to the second, ..., to get the $(n+1)$th term, you add 4 to the nth one.

In general, an arithmetic progression, (abbreviated to A.P.), is a sequence $\{a_1, a_2, a_3, \ldots\}$ in which every term, except the first one, is obtained by adding some constant number d to the previous term. Thus

$$a_2 = a_1 + d, \quad a_3 = a_2 + d = a_1 + 2d, \quad a_4 = a_3 + d = a_1 + 3d,$$

in general, $a_n = a_{n-1} + d = a_1 + (n - 1)\,d$, therefore the A.P. may be written as $\{a_1 + (n - 1)\,d\}_{n=1}^{\infty}$. Note that

$$d = a_2 - a_1 = a_3 - a_2 = a_4 - a_3 = \cdots = a_n - a_{n-1} =$$

For this reason, d is called the *common difference* for the A.P.

Example 264 *An A.P. with first term -3 and common difference 2 has terms $-3, -1, 1, 3, 5, 7, \ldots$. Its nth term is $a_n = -3 + 2(n - 1)$.*

Example 265 *An A.P. with first term 10 and common difference $-\frac{1}{2}$ has terms $10, 9.5, 9, 8.5, \ldots$. Its nth term is $a_n = 10 - \frac{1}{2}(n - 1)$.*

If you know one term and the common difference of an A.P. then you essentially know all its terms.

Example 266 *Suppose the 5th term of an A.P. is 7 and its common difference is $-\frac{3}{2}$. Denote its first term by a_1. Then its terms are a_1, $a_1 - \frac{3}{2}$, $a_1 - \frac{6}{2}$, $a_1 - \frac{9}{2}$, ... Since its 5th term is $a_1 - \frac{12}{2} = a_1 - 6$ and we are given that it is 7, it follows that*

$$a_1 - 6 = 7$$

therefore $a_1 = 13$ and the sequence is $\{13, 11.5, 10, \ldots\} = \left\{13 - \frac{3}{2}(n-1)\right\}_{n=1}^{\infty}$

Also, if you know any two terms of an A.P. then you essentially know all its terms.

Example 267 *Suppose the 10th and 19th terms of an A.P. are known to be 24 and 27 respectively. Let its first term be a_1 and its common difference be d. Since the nth term of an A.P. with first term a_1 and common difference d is $a_1 + (n-1)d$, it follows that*

$$a_1 + 9d = 24 \qquad and \qquad a_1 + 18d = 27$$

Solving these two simultaneous equations gives $d = \frac{1}{3}$ and $a_1 = 21$, hence the sequence is $\left\{21 + \frac{1}{3}(n-1)\right\}_{n=1}^{\infty}$.

Exercise 268

1. James is hired by Great Motors with a starting salary of \$30,000 per year. His contract stipulates that he will receive a salary increase at the end of every year amounting to 5% of his starting salary. Calculate what he will earn in: (a) the second year of his employment, (b) the nth year of his employment. Also explain why his yearly salaries form an A.P. and give its common difference.

2. The 6th term of an A.P. is 68 and its common difference is $\frac{3}{4}$. Calculate:

   (a) Its first term   (b) Its nth term

3. The 9th and 25th terms of an A.P. are 25 and 7 respectively. Determine the nth term.

4. Let $\{a_n\}$ be an arithmetic progression. Show that for any $n \geq 2$, $a_n = \dfrac{a_{n-1} + a_{n+1}}{2}$. In other words, a_n is the (arithmetic) mean of the terms a_{n-1} and a_{n+1} that come before and after it respectively.

Geometric Progressions

Amanda is also hired by Great Motors in Question 1 of the above Exercise under the following terms: Her starting salary is \$32,000 per year and at the end of every year, she will receive a salary increase equal to 4% of what she earned in the ending year. This means that:

- In the second year of her employment, she will earn an extra 4% of what she earned in the first year, therefore she will earn a total of

$$32000 + \tfrac{4}{100} \times 32000 = 32000 \left(1 + \tfrac{4}{100}\right) = 32000 \left(\tfrac{104}{100}\right) \text{ dollars.}$$

It is tempting to multiply the 32000 by $\tfrac{104}{100}$ to get 33280, but when you do so you miss the pattern we are aiming for.

- In the third year, she will earn an extra 4% of $32000 \left(\tfrac{104}{100}\right)$, hence a total of

$$
\begin{aligned}
32000 \left(\tfrac{104}{100}\right) + \tfrac{4}{100} \times 32000 \left(\tfrac{104}{100}\right) &= 32000 \left(\tfrac{104}{100}\right) \left(1 + \tfrac{4}{100}\right) \\
&= 32000 \left(\tfrac{104}{100}\right)^2 \text{ dollars.}
\end{aligned}
$$

- In the fourth year, she will earn an extra 4% of $32000 \left(\tfrac{104}{100}\right)^2$, hence a total of

$$
\begin{aligned}
32000 \left(\tfrac{104}{100}\right)^2 + \tfrac{4}{100} \times 32000 \left(\tfrac{104}{100}\right)^2 &= 32000 \left(\tfrac{104}{100}\right)^2 \left(1 + \tfrac{4}{100}\right) \\
&= 32000 \left(\tfrac{104}{100}\right)^3 \text{ dollars.}
\end{aligned}
$$

- Continuing in this fashion, it is seen that in the nth year, she will earn a total of $32000 \left(\tfrac{104}{100}\right)^{n-1}$ dollars.

The string of numbers 32000, $32000 \left(\tfrac{104}{100}\right)$, $32000 \left(\tfrac{104}{100}\right)^2$, $32000 \left(\tfrac{104}{100}\right)^3$, ..., is an example of a geometric progression, abbreviated to G.P. It has the property that every term, except the first one, is obtained by multiplying the previous term by the constant $\left(\tfrac{104}{100}\right)$.

In general, a geometric progression is a sequence $\{a_1, a_2, a_3, \ldots\}$ in which every term, except the first one, is obtained by multiplying the previous term by some constant number r. Thus

$$a_2 = a_1 r, \quad a_3 = a_2 r = a_1 r^2, \quad a_4 = a_3 r = a_1 r^3, \tag{7.1}$$

in general, $a_n = r a_{n-1} = a_1 r^{n-1}$, so the sequence may be written as $\{a_1 r^{n-1}\}_{n=1}^{\infty}$. We may solve for r in (7.1) and the result is

$$r = \frac{a_2}{a_1} = \frac{a_3}{a_2} = \frac{a_4}{a_3} = \cdots = \frac{a_n}{a_{n-1}} = \cdots$$

Because of this, r is called the *common ratio* for the G.P.

Example 269 *By the end of every year, the value of a certain brand of car has depreciated by $\tfrac{1}{6}$th of its value at the beginning of the year. Suppose its value when new is $20,000. Then its value at the end of the first year will be*

$$20000 - 20000 \times \tfrac{1}{6} = 20000 \left(1 - \tfrac{1}{6}\right) = 20000 \left(\tfrac{5}{6}\right)$$

dollars. The following table shows its value, in dollars, as the years go by:

Year	Value of vehicle at end of the year
0	$20,000$ *dollars, (the value when brand new).*
1	$20000 - 20000 \times \frac{1}{6} = 20000\left(1 - \frac{1}{6}\right) = 20000\left(\frac{5}{6}\right)$
2	$20000\left(\frac{5}{6}\right) - 20000\left(\frac{5}{6}\right) \times \frac{1}{6} = 20000\left(\frac{5}{6}\right)\left(1 - \frac{1}{6}\right) = 20000\left(\frac{5}{6}\right)^2$
3	$20000\left(\frac{5}{6}\right)^2 - 20000\left(\frac{5}{6}\right)^2 \times \frac{1}{6} = 20000\left(\frac{5}{6}\right)^2\left(1 - \frac{1}{6}\right) = 20000\left(\frac{5}{6}\right)^3$
4	$20000\left(\frac{5}{6}\right)^3 - 20000\left(\frac{5}{6}\right)^3 \times \frac{1}{6} = 20000\left(\frac{5}{6}\right)^3\left(1 - \frac{1}{6}\right) = 20000\left(\frac{5}{6}\right)^4$
n	$20000\left(\frac{5}{6}\right)^{n-1} - 20000\left(\frac{5}{6}\right)^{n-1} \times \frac{1}{6} = 20000\left(\frac{5}{6}\right)^{n-1}\left(1 - \frac{1}{6}\right) = 20000\left(\frac{5}{6}\right)^n$

Clearly, the string of numbers $20000,\ 20000\left(\frac{5}{6}\right),\ 20000\left(\frac{5}{6}\right)^2,\ 20000\left(\frac{5}{6}\right)^3,\ \ldots$ *is a geometric progression. Its first term is* 20000 *and its common ratio is* $\left(\frac{5}{6}\right)$.

Exercise 270

1. Determine the 3rd term and the nth term of a G.P. with first term 7 and common ratio $\frac{5}{4}$.

2. The 4th and 7th terms of a G.P. are 27 and 8 respectively. Determine its first term and common ratio. (Hint: If the first term is a_1 and the common ratio is r then the 4th and 7th terms are $a_1 r^3$ and $a_1 r^6$ respectively. This implies that $a_1 r^3 = 27$ and $a_1 r^6 = 8$. Solve for a_1 and r.)

3. Let $\{a_n\}_{n=1}^{\infty}$ be a geometric progression whose terms are all positive. Show that if $n \geq 2$, then a_n is given by $a_n = \sqrt{a_{n-1}a_{n+1}}$. In other words, a_n is the geometric mean of the terms a_{n-1} and a_{n+1} that come before and after it respectively. (If you are not familiar with the **geometric mean** of a set of numbers, Google the expression **geometric mean**.)

Many sequences are neither arithmetic nor geometric progressions. Examples:

(a) $\{1, 4, 9, 16, 25, 36, \ldots\} = \{n^2\}_{n=1}^{\infty}$ (b) $\{0, 2, 0, 2, 0, 2, \ldots\}$

(c) $\left\{\frac{1}{2}, \frac{2}{3}, \frac{3}{4}, \frac{4}{5}, \ldots\right\} = \left\{\frac{n}{n+1}\right\}_{n=1}^{\infty}$ (d) $\left\{\frac{-1}{2}, \frac{1}{3}, \frac{-1}{4}, \frac{1}{5}, \ldots\right\} = \left\{\frac{(-1)^n}{n+1}\right\}_{n=1}^{\infty}$

Limit of a Sequence

Consider the sequence $\{a_n\}_{n=1}^{\infty} = \{20000, 20000\left(\frac{5}{6}\right), 20000\left(\frac{5}{6}\right)^2, 20000\left(\frac{5}{6}\right)^3, \ldots\}$ in Example 269. It is clear that the value of the car shrinks to 0 as the years pass by. For example, eleven years after it is purchased, its value has shrunk to $20000\left(\frac{5}{6}\right)^{10}$ which is about 3230 dollars. Thirty one years after the purchase, it will be down to $20000\left(\frac{5}{6}\right)^{30}$ which is less than 85 dollars. After sixty one years, it will be worth $20000 \left(\frac{5}{6}\right)^{60}$ which is close to 40 cents! If you plot the terms of the sequence on the number line, you will find that they eventually settle close to 0.

For this reason, we say that the sequence $\left\{20000 \left(\frac{5}{6}\right)^{n-1}\right\}_{n=1}^{\infty}$ has limit 0 as n approaches infinity.

With the above example in mind, let $\{a_n\}_{n=1}^{\infty}$ be a given sequence. We say that it has limit L as n approaches infinity if its terms eventually settle close to L. To be more precise than this, we have to introduce the idea of a neighborhood of a given number and we define it as follows:

Definition 271 *Let L be a given number. Pick a positive number ε. Then the interval $(L - \varepsilon, L + \varepsilon)$ is called a **neighborhood of L of radius** ε.*

To say that the terms of a sequence $\{a_n\}_{n=1}^{\infty}$ eventually settle close to a number L simply means that any neighborhood $(L - \varepsilon, L + \varepsilon)$ of L, however small, misses only a finite number of the terms. In other words, if you draw any neighborhood $(L - \varepsilon, L + \varepsilon)$ of L and then plot the terms $a_1, a_2, a_3, \ldots$, of the sequence on the number line, sooner or later, every term you plot falls inside the interval $(L - \varepsilon, L + \varepsilon)$. A neat way of saying this is that you can always find an integer N such that all the terms $a_N, a_{N+1}, a_{N+2}, \ldots$, are in the interval $(L - \varepsilon, L + \varepsilon)$. This suggests the following definition of a limit:

Let $\{a_n\}_{n=1}^{\infty}$ be a sequence. We say that it has limit L as n approaches infinity, which is abbreviated to $\lim\limits_{n \to \infty} a_n = L$, if given any neighborhood $(L - \varepsilon, L + \varepsilon)$ of L, it is possible to find an integer N such that all the terms $a_N, a_{N+1}, a_{N+2}, \ldots$ are in $(L - \varepsilon, L + \varepsilon)$.

In the case of $\left\{20000 \left(\frac{5}{6}\right)^{n-1}\right\}_{n=1}^{\infty}$, suppose we are given a neighborhood

$$(0 - \tfrac{1}{2}, 0 + \tfrac{1}{2}) = (-\tfrac{1}{2}, \tfrac{1}{2})$$

of 0. We show that it is possible to find an integer N such that all the terms a_N, $a_{N+1}, a_{N+2}, \ldots$ are in $(-\tfrac{1}{2}, \tfrac{1}{2})$. The first step is to find the integers n such that

$$-\tfrac{1}{2} < 20000 \left(\tfrac{5}{6}\right)^{n-1} < \tfrac{1}{2}.$$

We do not have to worry about the condition $-\tfrac{1}{2} < 20000 \left(\tfrac{5}{6}\right)^{n-1}$ because $20000 \left(\tfrac{5}{6}\right)^{n-1}$ is positive for all integers n. Turning to $20000 \left(\tfrac{5}{6}\right)^{n-1} < \tfrac{1}{2}$, we note that it is equivalent to

$$\left(\tfrac{5}{6}\right)^{n-1} < \tfrac{1}{40000}$$

Take logarithms to base e then divide by $\ln\left(\tfrac{5}{6}\right)$, which is a negative number, to get

$$n - 1 > -\ln(40000)/\ln\left(\tfrac{5}{6}\right) \simeq 58.1$$

It follows that if $N > 59$ then all the terms $a_N, a_{N+1}, a_{N+2}, \ldots$ are in $\left(-\frac{1}{2}, \frac{1}{2}\right)$.

If we are given a different interval, we may need a different integer. For instance, suppose we are given the interval $(-0.01, 0.01)$. Then we will have to find an integer N such that

$$-0.01 < 20000 \left(\tfrac{5}{6}\right)^{n-1} < 0.01.$$

Similar computations reveal that any $N > \frac{-\ln(2000000)}{\ln\left(\frac{5}{6}\right)} + 1 \simeq 81$ will do

In general, given any interval $(-\varepsilon, \varepsilon)$, we can find an integer N such that such that all the terms $a_N, a_{N+1}, a_{N+2}, \ldots$ are in $(-\varepsilon, \varepsilon)$. We trust that you can verify that any $N > \dfrac{\ln\left(\frac{\varepsilon}{20000}\right)}{\ln\left(\frac{5}{6}\right)} + 1$ works.

- A sequence $\{a_n\}_{n=1}^{\infty}$ that has a limit is called a **convergent sequence**.

- If a sequence $\{a_n\}_{n=1}^{\infty}$ has limit a number L, then we say that "$\{a_n\}_{n=1}^{\infty}$ **converges to L as n approaches ∞**".

Remark 272 *Every number x in the interval $(L - \varepsilon, L + \varepsilon)$ satisfies the condition $|x - L| < \varepsilon$. Therefore an equivalent way of verifying that a sequence $\{a_n\}_{n=1}^{\infty}$ has limit L is to show that given any positive number ε, it is possible to find an integer N such that $|a_n - L| < \varepsilon$ for all $n \geq N$.*

Here are several examples of proofs of limits using the precise definition of a limit:

Example 273 *The sequence $\left\{1, \frac{1}{2}, \frac{1}{3}, \ldots\right\} = \left\{\frac{1}{n}\right\}_{n=1}^{\infty}$ has limit 0. (Plot some terms on the number line. You will find that when n is large, then $\frac{1}{n}$ is close to 0.) To prove it using the more precise definition of a limit, take any neighborhood $(0 - \varepsilon, 0 + \varepsilon) = (-\varepsilon, \varepsilon)$ of 0. Then we have to show that it is possible to find an integer N such that all the terms $a_N, a_{N+1}, a_{N+2}, \ldots$ are in $(-\varepsilon, \varepsilon)$. Since the terms get smaller as n increases and they are all positive, it suffices to show that there is an integer N such that*

$$\frac{1}{N} < \varepsilon.$$

Clearly, any N bigger than $\frac{1}{\varepsilon}$ will do. In other words, if we choose any $N > \frac{1}{\varepsilon}$ then all the terms $a_N, a_{N+1}, a_{N+2}, \ldots$ are in $(-\varepsilon, \varepsilon)$.

Example 274 *The sequence $\left\{\frac{1}{2}, \frac{2}{3}, \frac{3}{4}, \frac{4}{5}, \ldots\right\} = \left\{\frac{n}{n+1}\right\}_{n=1}^{\infty}$ has limit 1. To prove it using the precise definition of a limit, take any neighborhood $(1 - \varepsilon, 1 + \varepsilon)$ of 1. Then we have to show that it is possible to find an integer N such that all the terms $a_N, a_{N+1}, a_{N+2}, \ldots$ are in $(1 - \varepsilon, 1 + \varepsilon)$. To do this, we look for the integers n such that*

$$1 - \varepsilon < \frac{n}{n+1} < 1 + \varepsilon \tag{7.2}$$

Subtract 1 from each of the three terms in (7.2) and simplify to get

$$-\varepsilon < \frac{-1}{n+1} < \varepsilon$$

These inequalities are satisfied by any $n > \frac{1}{\varepsilon} - 1$. Therefore if we choose any $N > \frac{1}{\varepsilon} - 1$ then all the terms $a_N, a_{N+1}, a_{N+2}, \ldots$ are in $(1 - \varepsilon, 1 + \varepsilon)$.

Example 275 *Let k be a constant. Then the sequence $\{a_n\}_{n=1}^{\infty} = \{k, k, k, \ldots\}$ has limit k. To prove it, take any neighborhood $(k - \varepsilon, k + \varepsilon)$ of k. We have to find an integer N such that all the terms a_N, a_{N+1}, a_{N+2}, $\ldots$ are in $(-\varepsilon, \varepsilon)$. Since all the terms a_1, a_2, a_3, $\ldots$ are in $(k - \varepsilon, k + \varepsilon)$, any positive integer N will do.*

Example 276 *Consider the sequence $\left\{\left(\frac{3}{4}\right)^n\right\}_{n=1}^{\infty} = \left\{\frac{3}{4}, \frac{9}{16}, \frac{27}{64}, \ldots\right\}$. If you plot a number of its terms you soon find that they settle close to 0, therefore it has limit 0. To prove it rigorously, we take $\varepsilon > 0$ and verify that there is an integer N such that all the terms $\left(\frac{3}{4}\right)^N$, $\left(\frac{3}{4}\right)^{N+1}$, $\left(\frac{3}{4}\right)^{N+2}$, $\ldots$ are in $(-\varepsilon, \varepsilon)$. To this end, we look for the integers n such that*

$$-\varepsilon < \left(\frac{3}{4}\right)^n < \varepsilon.$$

Since $\left(\frac{3}{4}\right)^n$ is positive, we only have to look for integers n such that $\left(\frac{3}{4}\right)^n < \varepsilon$. Taking logarithms, we reduce the problem to determining n such that $n \log \frac{3}{4} < \log \varepsilon$. When we divide both sides by the negative number $\log \frac{3}{4}$ we get $n > \frac{\log \varepsilon}{\log \frac{3}{4}}$. Therefore any $N > \frac{\log \varepsilon}{\log \frac{3}{4}}$ will do.

Sequences Without Limits

Consider the sequence $\{a_n\}_{n=1}^{\infty} = \left\{\frac{1}{n} + \cos \frac{n\pi}{2}\right\}_{n=1}^{\infty} = \left\{1, -\frac{1}{2}, \frac{1}{3}, \frac{5}{4}, \frac{1}{5}, \ldots\right\}$. You can easily confirm that $a_n = \frac{1}{n}$ if n is odd, it is $\frac{1}{n} + 1$ if n is a multiple of 4, and it is $\frac{1}{n} - 1$ if n is even but not a multiple of 4. The odd terms a_1, a_3, a_5, $\ldots$ settle close to 0. The terms a_4, a_8, a_{12}, $\ldots$ settle close to 1, and the rest settle close to -1. Therefore $\left\{\frac{1}{n} + \cos \frac{n\pi}{2}\right\}_{n=1}^{\infty}$ has no limit because there is no single "meeting point" for its terms. Here is a more rigorous proof:

Let L be any real number. Choose a positive number ε that is less than $\frac{1}{3}$. If $L \leq 0$ then the neighborhood $(L - \varepsilon, L + \varepsilon)$ misses the infinitely many terms a_4, a_8, a_{12}, $\ldots$ because they are all positive and bigger than 1. If $L > 0$ then $(L - \varepsilon, L + \varepsilon)$ misses the infinitely many terms a_2, a_6, a_8, $\ldots$ because they are all negative and less than or equal to $-\frac{1}{2}$. Therefore it is impossible to find an integer N such that a_N, a_{N+1}, a_{N+2}, $\ldots$ are all in $(L - \varepsilon, L + \varepsilon)$. This shows that L is not a limit of $\left\{\frac{1}{n} + \cos \frac{n\pi}{2}\right\}_{n=1}^{\infty}$. Since L was an arbitrary real number, this proves that $\left\{\frac{1}{n} + \cos \frac{n\pi}{2}\right\}_{n=1}^{\infty}$ has no limit.

Example 277 *The sequence $\{-1, 1, -1, 1, -1, 1 \ldots\} = \{(-1)^n\}$ has no limit. To see this, take any number L and choose a positive number ε that is smaller than 1. Consider the interval $(L - \varepsilon, L + \varepsilon)$. Since the distance between $L - \varepsilon$ and $L + \varepsilon$ is less than 2 and the distance between an odd term a_{2n-1} and an even term a_{2k} is 2, the following are the only possibilities:*

1. *$(L - \varepsilon, L + \varepsilon)$ does not contain any of the terms of the sequence. An example is $L = 5$ and $\varepsilon = \frac{1}{2}$*

2. *$(L - \varepsilon, L + \varepsilon)$ contains only the odd terms a_1, a_3, a_5, $\ldots$. An example is $L = -0.9$ and $\varepsilon = \frac{1}{2}$.*

3. *$(L - \varepsilon, L + \varepsilon)$ contains only the even terms a_2, a_4, a_6, $\ldots$. An example is $L = 1$ and $\varepsilon = \frac{1}{2}$.*

This implies that it is impossible to find an integer N such that all the terms $a_N, a_{N+1}, a_{N+2}, \ldots$ are in $(L - \varepsilon, L + \varepsilon)$, and proves that L is not a limit of the sequence. Since L was arbitrary, the sequence has no limit.

Remark 278 A sequence $\{a_n\}_{n=1}^{\infty}$ that has no limit is called a **divergent sequence**.

Exercise 279

1. Write the first 5 terms of the given sequence:

   (a) $\{3^n\}_{n=1}^{\infty}$ (b) $\left\{\frac{(-1)^n}{n}\right\}_{n=1}^{\infty}$ (c) $\left\{\frac{2^n}{n!}\right\}_{n=1}^{\infty}$

   (d) $\left\{\frac{n!}{(n+2)!}\right\}_{n=1}^{\infty}$ (e) $\{4/3^n\}_{n=1}^{\infty}$ (f) $\{1 + (-1)^n\}_{n=1}^{\infty}$

2. Let $\{a_n\}_{n=1}^{\infty}$ be an A.P. with first term a_1 and common difference d which is not zero. Let L be an arbitrary real number. Choose $\varepsilon < \frac{1}{3}|d|$ and show that the interval $(L - \varepsilon, L + \varepsilon)$ contains at most one term of $\{a_n\}_{n=1}^{\infty}$, then deduce that the A.P. has no limit.

3. Let $\{a_n\}_{n=1}^{\infty}$ be a G.P. with a non-zero first term a_1, and common ratio r. Show that if $|r| < 1$ then $\{a_n\}_{n=1}^{\infty}$ has limit 0. (Hint: study the proof in Example 276.)

Bounded Sequences

Let $\{a_n\}_{n=1}^{\infty}$ be a given sequence.

- We say that it is **bounded above** if there is a number K, (called an upper bound of the sequence), that is bigger than all its terms. For example, the sequence $\left\{\frac{1}{n}\right\}_{n=1}^{\infty}$ is bounded above because a number like $K = 2.7$ is bigger than all its terms.

- We say that it is **bounded below** if there is a number M, (called a lower bound of the sequence) that is smaller than all its terms. For example, the $\{2n - 1\}_{n=1}^{\infty} = \{1, 3, 5, \ldots\}$ of odd positive integers is bounded below. A number like $M = 0$ is smaller than all its terms.

- We say that it is **bounded** if it is bounded above and below. Thus there are numbers K and M such that $M < a_n < K$ for all n. For example, the sequence $\left\{\frac{1}{n}\right\}_{n=1}^{\infty}$ is bounded because we can find numbers like -3 and 5 such that $-3 < \frac{1}{n} < 5$ for all integers n.

Example 280 The sequence $\left\{20000, \ 20000\left(\frac{5}{6}\right), \ 20000\left(\frac{5}{6}\right)^2, \ 20000\left(\frac{5}{6}\right)^3, \ldots\right\}_{n=1}^{\infty}$

is bounded. For an upper bound, we may take $K = 20000$ and for a lower bound, we may take $M = 0$. Actually, any number bigger than 20000 is also an upper bound and any number smaller than 0 is a lower bound of the sequence.

Example 281 *The sequence $\left\{\frac{4}{n} + (-1)^n\right\}_{n=1}^{\infty}$ is bounded. For example, 5 is an upper bound and -2 is a lower bound of the sequence. Of course these are not the only bounds. There are infinitely many others, (give several examples).*

Remark 282 *If a sequence is bounded then, as we have seen in Examples 280 and 281, it has infinitely many upper bounds and infinitely many lower bounds. However, there is always a smallest upper bound, and a biggest lower bound, (an axiom of the real numbers guarantees this). In Example 280, the smallest upper bound is 20000 and the biggest lower bound is 0. In Example 281, the smallest upper bound is 3 and the biggest lower bound is -1.*

A sequence that is not bounded above, or below, is called an **unbounded sequence**. If it is not bounded above then it has arbitrarily large and positive terms. An example is $\{n^2\}_{n=1}^{\infty}$ which has arbitrarily large positive terms. If it is not bounded below then it has arbitrarily large negative terms. An example is $\{-6n\}_{n=1}^{\infty}$.

It is possible for a sequence to be unbounded above and below. An example is $\left\{(-n)^2\right\}_{n=1}^{\infty}$. It has both arbitrarily large and positive and arbitrarily large and negative terms. The following is a more precise definition of an unbounded sequence:

*A sequence $\{a_n\}_{n=1}^{\infty}$ is **unbounded** if given any positive number K it is possible to find a term a_m of the sequence such that $|a_m| > K$.*

Since the terms of a convergent sequence $\{a_n\}_{n=1}^{\infty}$ with limit L eventually settle close to L, the sequence must be bounded. We record this as a theorem.

Theorem 283 *If $\{a_n\}_{n=1}^{\infty}$ converges then it is bounded.*

For a proof, let its limit be L. Choose $\varepsilon = 1$, (actually, any positive number will do). Then we can find an integer N such that all the terms a_N, a_{N+1}, a_{N+2}, ... are in the interval $(L-1, L+1)$. Let K be the largest of the numbers a_1, a_2, ..., a_{n-1}, $L+1$. Since all the terms a_n with $n \geq N$ are smaller than $L+1$, $a_n \leq K$ for all n. Likewise, let M be the smallest of the numbers a_1, a_2, ..., a_{n-1}, $L-1$. Then $M \leq a_n$ for all n. It follows that $M \leq a_n \leq K$ for all n, and so the sequence is bounded.

One consequence of the above result is that if a sequence is not bounded then it has no limit. This simple observation is used in many arguments ahead. For example, the sequence $\{3^n\}_{n=1}^{\infty}$ has no limit because its terms get arbitrarily large and positive.

But you should not misinterpret the above statement to mean that every bounded sequence has a limit. That is far from true, and it is easy to produce divergent bounded sequences. Here are two:

$$\{(-1)^n\}_{n=1}^{\infty} = \{-1, 1, -1, 1, \ldots\} \quad \text{and} \quad \left\{2 + \tfrac{1}{n} + (-1)^n\right\}_{n=1}^{\infty} = \{2, \tfrac{7}{2}, \tfrac{4}{3}, \ldots\}$$

There are two exceptions; any bounded increasing sequence and every bounded decreasing sequence converges. These are defined as follows:

Definition 284 *Let* $\{a_n\}_{n=1}^{\infty}$ *be a sequence. It is **increasing** if* $a_n \leq a_{n+1}$ *for all positive integers* n. *It is **strictly increasing** if* $a_n < a_{n+1}$ *for all positive integers* n. *It is **decreasing** if* $a_n \geq a_{n+1}$ *for all positive integers* n. *It is **strictly decreasing** if* $a_n > a_{n+1}$ *for all positive integers* n.

Theorem 285 *If a sequence* $\{a_n\}_{n=1}^{\infty}$ *is increasing and bounded then it has a limit.*

(It follows that if a sequence is increasing and has no limit then it must be unbounded.) Here is a sketch of a proof: If $\{a_n\}_{n=1}^{\infty}$ is bounded then there are numerous numbers K, called upper bounds of the sequence, such that $a_n \leq K$ for all n. Take the smallest of them and denote it by u, (see Remark 282). Then $a_n \leq u$ for all n, (because u is an upper bound of $\{a_n\}_{n=1}^{\infty}$). Since the sequence is increasing,

$$a_1 \leq a_2 \leq \cdots \leq a_n \leq a_{n+1} \leq \cdots \leq u$$

therefore all its terms are squeezed between a_1 and u. We claim that they eventually settle close to u. To see this take any interval $(u - \varepsilon, u + \varepsilon)$. Because u is the smallest upper bound of $\{a_n\}_{n=1}^{\infty}$, $u - \varepsilon$ is not an upper bound of the sequence. Therefore there must be a term a_N that is bigger than $u - \varepsilon$. In other words, there is an integer N such that $u - \varepsilon < a_N$. Since the sequence is increasing,

$$u - \varepsilon < a_N \leq a_{N+1} \leq a_{N+2} \leq \cdots \leq u < u + \varepsilon.$$

In other words, all the terms $a_N, a_{N+1}, a_{N+2}, \ldots$ are in $(u - \varepsilon, u + \varepsilon)$, which proves that $\{a_n\}_{n=1}^{\infty}$ has limit u.

A similar argument may be used to show that if $\{a_n\}_{n=1}^{\infty}$ is bounded and decreasing then it has a limit.

Some Useful Properties of Limits

We start with some definitions. To this end, let $\{a_n\}_{n=1}^{\infty}$, $\{b_n\}_{n=1}^{\infty}$ be sequences of real numbers and k be a constant. Then:

1. $\{ka_n\}_{n=1}^{\infty}$ is the sequence whose terms are ka_1, ka_2, ka_3, $\ldots$

2. The sum of $\{a_n\}_{n=1}^{\infty}$ and $\{b_n\}_{n=1}^{\infty}$ is the sequence whose terms are $a_1 + b_1$, $a_2 + b_2$, $a_3 + b_3$, $\ldots$. It is denoted by $\{a_n + b_n\}_{n=1}^{\infty}$.

3. The product of $\{a_n\}_{n=1}^{\infty}$ and $\{b_n\}_{n=1}^{\infty}$ is the sequence whose terms are $a_1 b_1$, $a_2 b_2$, $a_3 b_3$, $\ldots$. It is denoted by $\{a_n b_n\}_{n=1}^{\infty}$.

4. If $b_n \neq 0$ for all n then the quotient of $\{a_n\}_{n=1}^{\infty}$ and $\{b_n\}_{n=1}^{\infty}$ is the sequence whose terms are $\frac{a_1}{b_1}$, $\frac{a_2}{b_2}$, $\frac{a_3}{b_3}$, $\ldots$. It is denoted by $\left\{\frac{a_n}{b_n}\right\}_{n=1}^{\infty}$.

The following statements should be easy to believe:

1. If $\{a_n\}_{n=1}^{\infty}$ has limit L and k is a constant then $\{ka_n\}_{n=1}^{\infty}$ has limit kL. This is called the constant multiple rule for limits.

2. If $\{a_n\}_{n=1}^{\infty}$ has limit L and $\{b_n\}_{n=1}^{\infty}$ has limit M then $\{a_n + b_n\}_{n=1}^{\infty}$ has limit $L + M$. (The sum of two numbers, one close to L and the other one close to M should be a number close to $L + M$.) This is called the sum rule for limits.

3. If $\{a_n\}_{n=1}^{\infty}$ has limit L and $\{b_n\}_{n=1}^{\infty}$ has limit M then $\{a_n b_n\}_{n=1}^{\infty}$ has limit LM. (The product of two numbers, one close to L and the other one close to M should be a number close to LM.) This is called the product rule for limits

4. If $\{a_n\}_{n=1}^{\infty}$ has limit L, $\{b_n\}_{n=1}^{\infty}$ has a non-zero limit M and all its terms are non-zero then $\left\{\frac{a_n}{b_n}\right\}_{n=1}^{\infty}$ has limit $\frac{L}{M}$. This is called the quotient rule for limits.

These properties enable us to calculate limits of a number of sequences if we know the limits of more elementary sequences. Here are some examples:

Example 286 *From Example 273, we know that the limit of $\left\{\frac{1}{n}\right\}_{n=1}^{\infty}$ is 0. It follows from the product rule for limits that the limit of $\left\{\frac{1}{n^2}\right\}_{n=1}^{\infty}$ is $0 \cdot 0 = 0$. In general, if p is a positive integer then $\left\{\frac{1}{n^p}\right\}_{n=1}^{\infty}$ has limit $0^p = 0$.*

Example 287 *If p is an integer and k is a constant then, by the constant multiple rule for limits, $\left\{\frac{k}{n^p}\right\}_{n=1}^{\infty}$ has limit $k \cdot 0^p = 0$.*

Example 288 *The sequence $\left\{4 + \frac{5}{n^3}\right\}_{n=1}^{\infty}$ is the sum of the constant sequence $\{4, 4, 4, \ldots\}$ with limit 4 and the sequence $\left\{\frac{5}{n^3}\right\}_{n=1}^{\infty}$ with limit 0. It follows from the sum rule for limits that $\left\{4 + \frac{5}{n^3}\right\}_{n=1}^{\infty}$ has limit $4 + 0 = 4$*

Example 289 *Consider the sequence $\{a_n\}_{n=1}^{\infty} = \left\{\frac{n^3+2n-1}{5-2n+n^2-3n^3}\right\}_{n=1}^{\infty}$. Divide the numerator and denominator of $\frac{n^3+2n-1}{5-2n+n^2-3n^3}$ by n^3, (the highest power of n in the denominator). The result is*

$$\left\{\frac{n^3+2n-1}{5-2n+n^2-3n^3}\right\}_{n=1}^{\infty} = \left\{\frac{1+\frac{2}{n^2}-\frac{1}{n^3}}{\frac{5}{n^3}-\frac{2}{n^2}+\frac{1}{n}-3}\right\}_{n=1}^{\infty}. \tag{7.3}$$

The numerator and denominator have limit 1 and -3 respectively, therefore $\{a_n\}_{n=1}^{\infty}$ has limit $-\frac{1}{3}$.

Exercise 290

1. *Let $a_n = \frac{3}{2n}$. Find an integer N such that all the terms $a_N, a_{N+1}, a_{N+2}, \ldots$ of the sequence $\left\{\frac{3}{2n}\right\}_{n=1}^{\infty}$ are in the interval $(0 - 0.03, 0 + 0.03) = (-0.03, 0.03)$*

2. *Let $a_n = \frac{3n}{2+n}$. Find an integer N such that all the terms $a_N, a_{N+1}, a_{N+2}, \ldots$ of the sequence $\left\{\frac{3n}{2+n}\right\}_{n=1}^{\infty}$ are in the interval $(3 - 0.1, 3 + 0.1) = (2.9, 3.1)$.*

3. *Write down the first six terms of the sequence $\left\{2\sin\frac{n\pi}{3}\right\}_{n=1}^{\infty}$ and prove that it is bounded.*

4. Let $\{a_n\}_{n=1}^{\infty}$ be a G.P. with first term a_1, (which is non-zero), and common ratio r. Suppose $|r| > 1$. Then we may write it as $|r| = 1 + b$ where $b > 0$. Show that $|r|^n > 1 + bn$ and deduce that $\{a_n\}_{n=1}^{\infty}$ has no limit.

5. Let $\{a_n\}_{n=1}^{\infty}$ be a G.P. with first term a_1, (which is non-zero), and common ratio r. Show that it diverges if $r = -1$ and converges if $r = 1$.

6. Use appropriate limit theorems to determine the limit of each sequence:

   a) $\left\{3 - \frac{4}{n} + \frac{7}{n^2}\right\}_{n=1}^{\infty}$ b) $\left\{4 + \frac{n}{n+15}\right\}_{n=1}^{\infty}$ c) $\left\{\frac{3+2n}{4-5n}\right\}_{n=1}^{\infty}$

   d) $\left\{\frac{3-4n^2}{10+4n^2}\right\}_{n=1}^{\infty}$ e) $\left\{\frac{2n^2+6n+5}{1-n^2}\right\}_{n=1}^{\infty}$ f) $\left\{\frac{5n^2+2n-3}{n^3+1}\right\}_{n=1}^{\infty}$

7. Show that if a sequence $\{a_n\}_{n=1}^{\infty}$ diverges and c is a nonzero constant then $\{ca_n\}_{n=1}^{\infty}$ also diverges. [Hint: Suppose $\{ca_n\}_{n=1}^{\infty}$ converges. Consider the sequence $\left\{\frac{1}{c} \cdot ca_n\right\}_{n=1}^{\infty}$. Explain why it must converge. Since we are given that $\{a_n\}_{n=1}^{\infty}$ diverges, what do you conclude?]

8. Show that the sequence $\{2^n\}_{n=1}^{\infty}$ is not bounded, and deduce that it diverges. (Hint: $2^n = (1+1)^n \geq 1 + n$ for all positive integers n.)

9. Let $b_n = \frac{2^n}{n!}$, $n = 1, 2, \ldots$. Note that

   $$b_3 = 2 \cdot 1 \cdot \frac{2}{3} = 2 \cdot \frac{2}{3} = \frac{4}{3}, \quad b_4 = 2 \cdot 1 \cdot \frac{2}{3} \cdot \frac{2}{4} < 2 \cdot \frac{2}{4} = \frac{4}{4}, \quad b_5 = 2 \cdot 1 \cdot \frac{2}{3} \cdot \frac{2}{4} \cdot \frac{2}{5} < 2 \cdot \frac{2}{5} = \frac{4}{5}$$

   Prove that in general, $b_n \leq \frac{4}{n}$, then find real numbers M and K such that $M < b_n < K$ for all positive integers n.

10. Let $a_n = \frac{n^n}{n!}$, $n = 1, 2, \ldots$. Note that

    $$a_2 = \left(\frac{2}{2}\right) \cdot \left(\frac{2}{1}\right) = 2, \qquad\qquad a_3 = \left(\frac{3}{3}\right) \cdot \left(\frac{3}{2}\right) \cdot \left(\frac{3}{1}\right) > 3,$$
    $$a_4 = \left(\frac{4}{4}\right) \cdot \left(\frac{4}{3}\right) \cdot \left(\frac{4}{2}\right) \cdot \left(\frac{4}{1}\right) > 4, \qquad a_5 = \left(\frac{5}{5}\right) \cdot \left(\frac{5}{4}\right) \cdot \left(\frac{5}{3}\right) \cdot \left(\frac{5}{2}\right) \cdot \left(\frac{5}{1}\right) > 5, \ldots$$

    Prove that $a_n \geq n$, and deduce that $\{a_n\}_{n=1}^{\infty}$ is not bounded.

11. Let $a_n = (n!)^{1/n}$. Fill in the missing steps in the following proof that $\{a_n\}_{n=1}^{\infty}$ diverges:

    $$n! \;=\; n \;\times\; (n-1) \;\times\; \cdots \;\times\; k \;\times\; \cdots \;\times\; 1$$

    $$=\; 1 \;\times\; 2 \;\times\; \cdots \;\times\; (n+1-k) \;\times\; \cdots \;\times\; n$$

    This implies that

    $$(n!)^2 = [n \times 1]\,[(n-1) \times 2] \cdots [(n+1-k) \times k] \cdots [2 \times (n-1)][1 \times n].$$

    Introduce the quadratic function

    $$q(x) = (n+1-x)\,x.$$

    Then $[n \times 1]\,[(n-1) \times 2] \cdots [(n+1-k) \times k] \cdots [2 \times (n-1)][1 \times n]$ is the product of $q(1)$, $q(2)$, $\ldots$, $q(n)$. Show that $n \leq q(x) \leq \left(\frac{n+1}{2}\right)^2$ for all $1 \leq x \leq n$ and deduce that $\sqrt{n} \leq (n!)^{1/n} \leq \left(\frac{n+1}{2}\right)$. Now complete the proof that $\{a_n\}_{n=1}^{\infty}$ diverges.

The sandwich Theorem

This is a useful theorem for solving a number of problems. It may be stated as follows:

*Let $\{a_n\}_{n=1}^{\infty}$, $\{b_n\}_{n=1}^{\infty}$ and $\{c_n\}_{n=1}^{\infty}$ be sequences such that $a_n \leq b_n \leq c_n$ for all n. If $\{a_n\}_{n=1}^{\infty}$ and $\{c_n\}_{n=1}^{\infty}$ converge to the **same** number L, then $\{b_n\}_{n=1}^{\infty}$ also converges to L.*

This should be easy to believe: the terms of $\{a_n\}_{n=1}^{\infty}$ and $\{c_n\}_{n=1}^{\infty}$ settle close to L. It stands to reason that the terms of $\{b_n\}_{n=1}^{\infty}$, which are in between, must also settle close to L. Here is a proof:

Fix a positive number ε. We must show that there is an integer N such that all the terms b_N, b_{N+1}, b_{N+2}, ... are in the interval $(L - \varepsilon, L + \varepsilon)$. Since $\{a_n\}_{n=1}^{\infty}$ converges to L, there is an integer K such that a_K, a_{K+1}, a_{K+2}, ... are all in the interval $(L - \varepsilon, L + \varepsilon)$. It is also the case that $\{c_n\}_{n=1}^{\infty}$ has limit L, therefore there is an integer M such that c_M, c_{M+1}, c_{M+2}, ... are all in the interval $(L - \varepsilon, L + \varepsilon)$. Let N be the larger of the two integers K and M. Then a_N, a_{N+1}, a_{N+2}, ... and c_N, c_{N+1}, c_{N+2}, ... are all in the interval $(L - \varepsilon, L + \varepsilon)$. This implies that

$$L - \varepsilon < a_n \leq b_n \leq c_n \leq L + \varepsilon$$

for all integers $n \geq N$, and proves that b_N, b_{N+1}, b_{N+2}, ... are all in the interval $(L - \varepsilon, L + \varepsilon)$. Therefore $\{b_n\}$ has limit L.

A famous limit

Let $a_n = \left(1 + \frac{1}{n}\right)^n$. With the aid of a calculator, evaluate $\left(1 + \frac{1}{n}\right)^n$ for different integers, including some really large ones. The values you get should convince you that $\{a_n\}_{n=1}^{\infty}$ must converge. Here we give an indirect proof that it converges by showing that it is bounded and increasing. By the Binomial Theorem, (see (3.10) on page 111),

$$(x + y)^n = x^n + nx^{n-1}y + \frac{n(n-1)}{2!}x^{n-2}y^2 + \frac{n(n-1)(n-2)}{3!}x^{n-3}y^3 + \cdots + \frac{n!}{n!}y^n$$

The general term in this expansion is $\frac{n(n-1)\cdots(n+1-k)}{k!}x^k y^{n-k}$. In particular, if we take $x = 1$ and $y = \frac{1}{n}$ then we get

$$a_n = \left(1 + \tfrac{1}{n}\right)^n = 1 + 1 + \frac{n(n-1)}{2!}\left(\tfrac{1}{n}\right)^2 + \frac{n(n-1)(n-2)}{3!}\left(\tfrac{1}{n}\right)^3 + \cdots + \frac{n!}{n!}\left(\tfrac{1}{n}\right)^n$$

If we re-write this as

$$\left(1 + \tfrac{1}{n}\right)^n = 1 + 1 + \frac{n(n-1)}{n^2}\frac{1}{2!} + \frac{n(n-1)(n-2)}{n^3}\frac{1}{3!} + \cdots + \frac{n!}{n^n}\frac{1}{n!}$$

then we see that

$$\left(1 + \tfrac{1}{n}\right)^n < 2 + \tfrac{1}{2!} + \tfrac{1}{3!} + \cdots + \tfrac{1}{n!}$$

Note that $3! = 3 \times 2 > 2^2$, $4! = 4 \times 3 \times 2 > 2^3$, ..., $n! > 2^{n-1}$. Therefore

$$\left(1 + \tfrac{1}{n}\right)^n < 2 + \tfrac{1}{2!} + \tfrac{1}{3!} + \cdots + \tfrac{1}{n!} < 2 + \tfrac{1}{2} + \tfrac{1}{2^2} + \tfrac{1}{2^3} + \cdots + \tfrac{1}{2^{n-1}}$$

We now have the sum $\frac{1}{2} + \frac{1}{2^2} + \frac{1}{2^3} + \cdots + \frac{1}{2^{n-1}}$ of the first n terms of a geometric progression. There is a neat way of simplifying such sums, (see Example 297):

$$\text{If} \quad S \; = \; \tfrac{1}{2} \; + \; \tfrac{1}{2^2} \; + \; \tfrac{1}{2^3} \; + \; \cdots \; + \; \tfrac{1}{2^{n-1}}$$

$$\text{then} \quad \tfrac{1}{2}S \; = \; \qquad \tfrac{1}{2^2} \qquad \tfrac{1}{2^3} \; + \; \cdots \; + \; \tfrac{1}{2^{n-1}} \; + \; \tfrac{1}{2^n}$$

Now subtract the second row from the first row and the result is

$$\tfrac{1}{2}S = \tfrac{1}{2} - \tfrac{1}{2^n} \quad \text{or} \quad S = 1 - \tfrac{2}{2^n} < 1$$

It follows that $\left(1 + \frac{1}{n}\right)^n < 2 + 1 = 3$ for all positive integers n. It is also true that $0 < \left(1 + \frac{1}{n}\right)^n$ for all n. Therefore

$$0 < \left(1 + \tfrac{1}{n}\right)^n < 2 + 1 = 3$$

hence the sequence is bounded. To prove that it is increasing, we compare $a_{n+1} = \left(1 + \frac{1}{n+1}\right)^{n+1}$ to $a_n = \left(1 + \frac{1}{n}\right)^n$. The general term in the expansion of a_n is

$$\frac{n(n-1)\cdots(n+1-k)}{k!} \left(\tfrac{1}{n}\right)^k \; = \; \tfrac{1}{k!}\left[\frac{n(n-1)\cdots(n+1-k)}{n^k}\right]$$

$$= \; \tfrac{1}{k!}\left[\left(\tfrac{n-1}{n}\right)\left(\tfrac{n-2}{n}\right)\cdots\left(\tfrac{n+1-k}{n}\right)\right]$$

This may be re-written as

$$\tfrac{1}{k!}\left[\left(1 - \tfrac{1}{n}\right)\left(1 - \tfrac{2}{n}\right)\cdots\left(1 - \tfrac{k}{n}\right)\right]$$

Similarly, the general term in the expansion of a_{n+1} may be written as

$$\tfrac{1}{k!}\left[\left(1 - \tfrac{1}{n+1}\right)\left(1 - \tfrac{2}{n+1}\right)\cdots\left(1 - \tfrac{k}{n+1}\right)\right]$$

Now consider the table below:

$$a_{n+1} = \; 2 \; + \; \tfrac{1}{2!}\left(1 - \tfrac{1}{n+1}\right) \; + \; \cdots \; + \; \cdots \; + \; \left(\tfrac{1}{n+1}\right)^{n+1}$$

$$a_n = \; 2 \; + \; \tfrac{1}{2!}\left(1 - \tfrac{1}{n}\right) \; + \; \cdots \; + \; \left(\tfrac{1}{n}\right)^n$$

The term $\frac{1}{2!}\left(1 - \frac{1}{n+1}\right)$ is bigger than the corresponding term $\frac{1}{2!}\left(1 - \frac{1}{n}\right)$

The term $\frac{1}{3!}\left(1 - \frac{1}{n+1}\right)\left(1 - \frac{2}{n+1}\right)$ is bigger than $\frac{1}{3!}\left(1 - \frac{1}{n}\right)\left(1 - \frac{2}{n}\right)$

$$\vdots$$

And $\frac{1}{n!}\left(1 - \frac{1}{n+1}\right)\left(1 - \frac{2}{n+1}\right)\cdots\left(1 - \frac{n-1}{n+1}\right)$ is bigger than $\left(\frac{1}{n}\right)^n = \frac{n!}{n!}\left(\frac{1}{n}\right)^n$

$\left[\text{To get the last line above, note that } \frac{n!}{n!}\left(\frac{1}{n}\right)^n = \frac{1}{n!}\left(\frac{n-1}{n}\right)\left(\frac{n-2}{n}\right)\cdots\left(\frac{n-(n-1)}{n}\right)\right].$

Therefore the terms you add to get $\left(1 + \frac{1}{n+1}\right)^{n+1}$ are bigger than or equal to the corresponding terms you add to get $\left(1 + \frac{1}{n}\right)^n$. This proves that $a_n \leq a_{n+1}$. We have now proved that $\left\{\left(1 + \frac{1}{n}\right)^n\right\}_{n=1}^{\infty}$ is increasing and bounded. By Theorem 285, it has a limit.

Remark 291 *The first few terms of this sequence are* 2, 2.25, 2.3704, 2.4414, *.... Its limit is denoted by e and is the base for the exponential function* $f(x) = e^x$. *You may use a scientific calculator to show that* $e \simeq 2.72$.

Infinite Limits

A sequence like $\{n^2\}_{n=1}^{\infty}$ whose terms get large and positive without bound is said to have limit ∞, (pronounced "infinity") and we write $\lim\limits_{n \to \infty} n^2 = \infty$. This is not to say that the terms of the sequence eventually settle close to some number ∞. There is no such a number. You should take the statement " $\lim\limits_{n \to \infty} n^2 = \infty$" to mean that when n is a large positive integer then n^2 is a very large positive number.

The sequences like $\left\{-\frac{1}{2}n\right\}_{n=1}^{\infty}$ whose terms get large and negative without bound are said to have limit $-\infty$. Again, there is no number $-\infty$. We write $\lim\limits_{n \to \infty}\left(-\frac{1}{2}n\right) = -\infty$ to convey the message that when n is a large positive integer then $-\frac{1}{2}n$ is a large negative number. The following are the formal definitions:

Definition 292 *Let* $\{a_n\}_{n=1}^{\infty}$ *be a given sequence.*

1. *We say that it has limit* ∞ *and write* $\lim\limits_{n \to \infty} a_n = \infty$ *if given any positive number* K *we can find an integer* N *such that* $a_n > K$ *for all* $n \geq N$.

2. *We say that it has limit* $-\infty$ *and write* $\lim\limits_{n \to \infty} a_n = -\infty$ *if given any positive number* K *we can find an integer* N *such that* $a_n < -K$ *for all* $n \geq N$.

Example 293 *We show that* $\{2n + 5\}_{n=1}^{\infty}$ *has limit* ∞. *Thus let* K *be a positive number. We have to show that there is an integer* N *such that* $2n + 5 > K$ *for all* $n \geq N$. *Clearly, the condition* $2n + 5 > K$ *is equivalent to* $n > \frac{K-5}{2}$. *Therefore if we take any* $N > \frac{K-5}{2}$ *then* $a_n > K$ *for all* $n \geq N$.

Exercise 294

1. Let $a_n = 0$, $b_n = \frac{2^n}{n!}$ and $c_n = \frac{4}{n}$. You showed in question 9 on page 227 that $b_n \leq c_n$. Use the sandwich theorem to prove that $\{b_n\}_{n=1}^{\infty}$ has limit 0.

2. Calculate $n^{1/n}$ for a number of integers n including some really big ones. Your answers should settle close to 1. To prove that $\lim\limits_{n \to \infty} n^{1/n} = 1$, follow the following steps: First, define $a_n = n^{1/n} - 1$. Then you have to prove that $\lim\limits_{n \to \infty} a_n = 0$. Clearly, if $n > 1$ then its nth root is bigger than 1, therefore $a_n > 0$ for all n. Write $a_n = n^{1/n} - 1$ as $n^{1/n} = 1 + a_n$. Then, by (3.10) on page 111,

$$n = (1 + a_n)^n = 1 + na_n + \frac{n(n-1)}{2}a_n^2 + \cdots + a_n^n \qquad (7.4)$$

All the terms in the right hand side of (7.4) are positive. This means that if we drop some of them, the right hand side becomes smaller. In particular, if we keep the third and drop the rest we get $n > \frac{n(n-1)}{2}a_n^2$. Use this to show that $0 < a_n < \sqrt{\frac{2}{n-1}}$, then use the sandwich theorem to prove that $\lim\limits_{n \to \infty} a_n = 0$. Finally, deduce that $\lim\limits_{n \to \infty} n^{1/n} = 1$.

3. Let $a_n = \frac{n^n}{n!}$. In Exercise 10 an page 227, you were asked to show that $a_n \geq n$ for all positive integers n. Use this result to prove that $\lim\limits_{n \to \infty} a_n = \infty$.

4. Fill in the missing steps in the following proof that the sequence $\left\{ \frac{2^n}{n^2} \right\}_{n=1}^{\infty}$ has limit ∞.

$$2^n = (1+1)^n = 1 + n + \frac{n(n-1)}{2!} + \frac{n(n-1)(n-2)}{3!} + \cdots + 1.$$

(by (3.10) on page 111). Use this to show that $\frac{2^n}{n^2} > \frac{n(n-1)(n-2)}{(3!)n^2}$ when $n > 3$. Now show that given any positive number K, it is possible to find an integer N such that $\frac{2^n}{n^2} > K$ for all $n \geq N$.

5. Let $a_n = \ln n$ and K be a given positive number. Show that if $n > e^K$ then $a_n > K$, and deduce that $\lim\limits_{n \to \infty} \ln n = \infty$.

6. Prove that $\lim\limits_{n \to \infty} \left(4 - \sqrt{n}\right) = -\infty$.

7. Let c be a constant. Fill in the missing details in the following proof that if $c > 1$ then $\lim\limits_{n \to \infty} \frac{c^n}{n} = \infty$. Since $c > 1$, we may write it as $c = 1 + \theta$ where $\theta > 0$. Therefore

$$c^n = (1 + \theta)^n = 1 + n\theta + \tfrac{1}{2}n(n-1)\theta^2 + \cdots + \theta^n$$

Deduce that $c^n > \tfrac{1}{2}n(n-1)\theta^2$ then complete the proof.

Series

One may want to add the terms of a given sequence. For example, if James worked for forty years at Great Motors, (see problem 1 on page 217), before retiring and you want to know how much he earned over the years, you would have to add up the forty numbers 30000, 31500, 33000, ..., 88500. With a calculator, that would not be a problem. Indeed, any finite string of real numbers can be added. The challenging problem is to add the terms of an infinite sequence like

$$\left\{ \frac{1}{n(n+1)} \right\}_{n=1}^{\infty} = \left\{ \frac{1}{1 \times 2}, \frac{1}{2 \times 3}, \frac{1}{3 \times 4}, \frac{1}{4 \times 5}, \cdots \right\}$$

The standard method of using a calculator to add $\frac{1}{1 \times 2}$ to $\frac{1}{2 \times 3}$ then add $\frac{1}{3 \times 4}$ to the result, and so on, does not work because it does not matter how long you spend doing this, at the end of any day, there will still be infinitely many terms that are untouched. The following seems to be a reasonable approach:

Add the first k terms for different integers k then ask the question: "Is there a single number around which these sums eventually settle?" If the answer is YES then surely that number is the best candidate to be the sum of the infinitely many terms of the sequence. To apply this to the above sequence, note that $\frac{1}{n(n+1)}$ splits into partial fractions as $\frac{1}{n(n+1)} = \frac{1}{n} - \frac{1}{n+1}$. In particular,

$$\frac{1}{1 \times 2} = \frac{1}{1(1+1)} = \frac{1}{1} - \frac{1}{2}, \quad \frac{1}{2 \times 3} = \frac{1}{2(2+1)} = \frac{1}{2} - \frac{1}{3}, \cdots, \quad \frac{1}{k(k+1)} = \frac{1}{k} - \frac{1}{k+1}$$

Therefore

$$\frac{1}{1\times 2} + \frac{1}{2\times 3} = \frac{1}{1} - \frac{1}{2} + \frac{1}{2} - \frac{1}{3} = 1 - \frac{1}{3}$$

$$\frac{1}{1\times 2} + \frac{1}{2\times 3} + \frac{1}{2\times 3} = \frac{1}{1} - \frac{1}{2} + \frac{1}{2} - \frac{1}{3} + \frac{1}{3} - \frac{1}{4} = 1 - \frac{1}{4}$$

In general, $\frac{1}{1\times 2} + \cdots + \frac{1}{k(k+1)} = \frac{1}{1} - \frac{1}{2} + \cdots + \frac{1}{k} - \frac{1}{k+1} = 1 - \frac{1}{k+1}$.

Now we see that the answer to the question; "*Is there a single number around which these sums eventually settle?*"; is YES. The number is 1, therefore we should define

$$\frac{1}{1\times 2} + \frac{1}{2\times 3} + \frac{1}{3\times 4} + \cdots = 1$$

The numbers $s_1 = \frac{1}{1\times 2}$, $s_2 = \frac{1}{1\times 2} + \frac{1}{2\times 3}$, $\ldots$, $s_k = \frac{1}{1\times 2} + \cdots + \frac{1}{k(k+1)}$, $\ldots$ are called, respectively, the first, the second, the kth partial sums of the sequence $\left\{\frac{1}{n(n+1)}\right\}_{n=1}^{\infty}$. We used them to construct the sequence $\{s_1, s_2, s_3, \ldots\}_{n=1}^{\infty}$ of partial sums. It is called the series with terms $\frac{1}{1\times 2}, \frac{1}{2\times 3}, \cdots$. Its limit is called the sum of the infinitely many numbers $\frac{1}{1\times 2}, \frac{1}{2\times 3}, \cdots$ and we write $\sum_{n=1}^{\infty} \frac{1}{n(n+1)} = 1$.

To generalize the above ideas, let $\{a_n\}_{n=1}^{\infty}$ be a given sequence of real numbers. Form the sums

$s_1 = a_1$, called the first partial sum of $\{a_n\}_{n=1}^{\infty}$

$s_2 = a_1 + a_2$, called the second partial sum of $\{a_n\}_{n=1}^{\infty}$

$s_3 = a_1 + a_2 + a_3$, called the third partial sum of $\{a_n\}_{n=1}^{\infty}$

$\vdots$

$s_k = a_1 + a_2 + \cdots + a_k$, called the kth partial sum of $\{a_n\}_{n=1}^{\infty}$

$\vdots$

The resulting sequence $\{s_k\}_{k=1}^{\infty}$ of partial sums is called the series with terms a_1, a_2, a_3, $\ldots$. If it converges then its limit s is called the sum of the infinitely many numbers a_1, a_2, a_3, $\ldots$ and we write $\sum_{n=1}^{\infty} a_n = s$. If $\{s_k\}_{k=1}^{\infty}$ has no limit then we say that the series diverges.

In general, one is not able to write down a simple formula for the kth partial sum $s_k = a_1 + a_2 + \cdots + a_k$ of a given series. For example if we take $a_n = \frac{n^2}{e^n}$ then there is no simple formula for the sum

$$s_k = \frac{1}{e} + \frac{4}{e^2} + \cdots + \frac{k^2}{e^k}.$$

Therefore, the notation $\{s_k\}_{k=1}^{\infty}$ is not very practical. The practical one is $\sum_{n=1}^{\infty} a_n$, the same notation for the sum of the terms a_1, a_2, a_3, $\ldots$. This should not cause a lot of confusion. It should be possible to figure out, from the context, when $\sum_{n=1}^{\infty} a_n$ denotes the sum of the numbers a_1, a_2, a_3, $\ldots$ or the series with terms a_1, a_2, a_3, $\ldots$.

Example 295 *Consider the constant sequence $\{\frac{3}{4}, \frac{3}{4}, \frac{3}{4}, \ldots\}$. Its partial sums are*

$$s_1 = \tfrac{3}{4}, \; s_2 = \tfrac{6}{4}, \; s_3 = \tfrac{9}{4}, \; \ldots, \; s_k = \tfrac{3k}{4}, \ldots$$

Therefore $\sum_{n=1}^{\infty} \frac{3}{4}$ is the divergent sequence $\{\frac{3}{4}, \frac{6}{4}, \frac{9}{4}, \ldots, \frac{3n}{4}, \ldots\}$.

Example 296 *Let $\{a_n\}_{n=1}^{\infty}$ be an A.P. with first term a_1 and common difference d. Thus $a_2 = a_1 + d$, $a_3 = a_1 + 2d$, in general, $a_n = a_1 + (n-1)d$. To get the series $\sum_{n=1}^{\infty} a_n$, we have to determine the partial sums*

$$s_k = a_1 + [a_1 + d] + [a_1 + 2d] + \cdots + [a_1 + (n-1)d], \; k = 1, 2, \ldots \quad (7.5)$$

A neat way of adding the k numbers $a_1, [a_1 + d], [a_1 + 2d], \cdots, [a_1 + (n-1)d]$, in the right hand side of (7.5) is to write out s_k twice, with the numbers in the second row written in the reverse order:

$$
\begin{array}{ccccccc}
s_k & = & a_1 & + & a_1 + d & + \cdots + & a_1 + (k-1)d \\
s_k & = & a_1 + (k-1)d & + & a_1 + (k-2)d & + \cdots + & a_1 \\
 & & \downarrow & & \downarrow & & \downarrow \\
 & & 2a_1 + (k-1)d & & 2a_1 + (k-1)d & \cdots & 2a_1 + (k-1)d
\end{array}
$$

When one adds the two numbers in any of the columns to the right one gets the same result $2a_1 + (k-1)d$, (the number below the arrow). Since there are k columns, it follows that

$$2s_k = k[2a_1 + (k-1)d].$$

This implies that $s_k = \frac{1}{2}k[2a_1 + (k-1)d]$. Clearly the sequence $\{s_k\}_{k=1}^{\infty}$ has no limit, (because it is not bounded), therefore the series $\sum_{n=1}^{\infty} a_n = \sum_{n=1}^{\infty}(a_1 + (n-1)d)$ diverges.

Example 297 *Let $\{a_n\} = \{a, ar, ar^2, \ldots\}_{n=1}^{\infty}$ be a G.P. with a non-zero first term a and a non-zero common ratio r. Thus $a_1 = a$, $a_2 = ar$, $a_3 = ar^2$, in general, $a_n = ar^{n-1}$. To get the series $\sum_{n=1}^{\infty} a_n$, we have to determine the partial sums*

$$s_k = a + ar + ar^2 + ar^3 + \cdots + ar^{k-1}, \; k = 1, 2, \ldots \quad (7.6)$$

One way of determining a simple expression for the sum of the terms $a, ar, ar^2, ar^3, \ldots, ar^{k-1}$ is to write down s_k and rs_k as below

$$
\begin{array}{ccccccccc}
s_k & = & a & + & ar & + & ar^2 & + \cdots + & ar^{k-1} \\
rs_k & = & & & ar & + & ar^2 & + \cdots + & ar^{k-1} & + & ar^k
\end{array}
$$

When we subtract the second row from the first row we get

$$s_k - rs_k = a - ar^k = a\left(1 - r^k\right)$$

Note that $s_k - rs_k$ factors as $(1-r)\,s_k$. Solving for s_k gives

$$s_k = \frac{a\left(1-r^k\right)}{(1-r)} = \frac{a}{1-r} - \frac{ar^k}{1-r}$$

*If $|r| < 1$ then the $\{r^k\}_{k=1}^{\infty}$ converges to 0, therefore $\{s_k\}_{k=1}^{\infty}$ converges to $\frac{a}{1-r}$. On the other hand $\{r^k\}_{k=1}^{\infty}$ diverges if $|r| > 1$. Therefore **the series** $\sum\limits_{n=1}^{\infty} ar^{n-1}$ **converges to** $\frac{a}{1-r}$ **if** $|r| < 1$ **and it diverges if** $|r| > 1$. Thus*

- $\sum\limits_{n=1}^{\infty} 3\left(\frac{1}{2}\right)^{n-1}$ *converges to* $\frac{3}{1-\frac{1}{2}} = 6$, *therefore we may write* $\sum\limits_{n=1}^{\infty} 3\left(\frac{1}{2}\right)^{n-1} = 6$.

- $\sum\limits_{n=1}^{\infty} 32000\left(\frac{104}{100}\right)^{n-1}$ *diverges because* $\left|\frac{104}{100}\right| > 1$.

Example 298 *Consider the series* $\sum\limits_{n=1}^{\infty} \frac{1}{\sqrt{n}}$. *The corresponding partial sums are*

$$s_1 = 1, \qquad s_2 = 1 + \frac{1}{\sqrt{2}} > \frac{1}{\sqrt{2}} + \frac{1}{\sqrt{2}} = \frac{2}{\sqrt{2}} = \sqrt{2}$$

$$s_3 = 1 + \frac{1}{\sqrt{2}} + \frac{1}{\sqrt{3}} > \frac{1}{\sqrt{3}} + \frac{1}{\sqrt{3}} + \frac{1}{\sqrt{3}} = \sqrt{3}$$

$$s_4 = 1 + \frac{1}{\sqrt{2}} + \frac{1}{\sqrt{3}} + \frac{1}{\sqrt{4}} > \frac{1}{\sqrt{4}} + \frac{1}{\sqrt{4}} + \frac{1}{\sqrt{4}} + \frac{1}{\sqrt{4}} = \sqrt{4}$$

$$\vdots$$

$$s_k = 1 + \frac{1}{\sqrt{2}} + \frac{1}{\sqrt{3}} + \cdots + \frac{1}{\sqrt{k}} > \frac{1}{\sqrt{k}} + \frac{1}{\sqrt{k}} + \frac{1}{\sqrt{k}} + \cdots + \frac{1}{\sqrt{k}} = \sqrt{k}$$

$$\vdots$$

The sequence of partial sums is not bounded, therefore $\sum\limits_{n=1}^{\infty} \frac{1}{\sqrt{n}}$ *diverges.*

In Example 295, we saw that the series $\sum\limits_{n=1}^{\infty} \left(\frac{3}{4}\right)$ whose terms are all equal to the constant $\frac{3}{4}$ diverges. This should be expected. The partial sums get larger without bound. There is nothing special about the number $\frac{3}{4}$. Any nonzero constant sequence $\{c, c, c, \dots\}$ gives a divergent series. In general, any sequence $\{a_k\}_{k=1}^{\infty}$ whose terms a_n do not shrink to 0 gives a divergent series $\sum\limits_{n=1}^{\infty} a_n$. This follows from the following theorem:

Theorem 299 *If a series* $\sum\limits_{n=1}^{\infty} a_n$ *converges then* $\lim\limits_{n\to\infty} a_n = 0$.

For a proof, suppose $\sum\limits_{n=1}^{\infty} a_n$ converges to s. This means that the sequence $\{s_k\}_{k=1}^{\infty}$ of partial sums $s_k = a_1 + a_2 + \cdots + a_k$ converges to s. Consider the sequence $\{s_{k-1}\}_{k=2}^{\infty}$. It is really $\{s_k\}_{k=1}^{\infty}$ in disguise, therefore it also converges to s. Note that

$$a_k = s_k - s_{k-1}$$

It follows from the properties of limits that $\lim\limits_{k\to\infty} a_k = \lim\limits_{k\to\infty} s_k - \lim\limits_{k\to\infty} s_{k-1} = s - s = 0$.

Summary 300 *According to Theorem 299, if a series $\sum\limits_{n=1}^{\infty} a_n$ converges then its terms must shrink to 0. It stands to reason that a series whose terms do not shrink to 0 cannot converge, so it must diverge. Among these, (as pointed out above), are the series $\sum\limits_{n=1}^{\infty} c$ where c is a non-zero constant. Another one whose terms are not constant is $\sum\limits_{n=1}^{\infty} \frac{n}{10n+15}$. The limit of $\frac{n}{10n+15}$ is $\frac{1}{10}$ which is not zero.*

You should not misinterpret the above conclusion. **It does not say** that if the terms of a series shrink to zero then the series converges. That is far from true. For example, the terms of $\sum\limits_{n=1}^{\infty} \frac{1}{\sqrt{n}}$ shrink to zero, (i.e. $\lim\limits_{n\to\infty} \frac{1}{\sqrt{n}} = 0$), but the series diverges.

A more famous one is $\sum\limits_{n=1}^{\infty} \frac{1}{n}$, called the **harmonic series**. We already know that $\lim\limits_{n\to\infty} \frac{1}{n} = 0$, but the series diverges. To prove it, group its terms, starting from the third, as follows:

Group 1 contains the 2 terms $\frac{1}{3}$, $\frac{1}{4}$.

Group 2 contains the $4 = 2^2$ terms $\frac{1}{5}$, $\frac{1}{6}$, $\frac{1}{7}$, $\frac{1}{8}$.

Group 3 contains the $8 = 2^3$ terms $\frac{1}{9}$, $\frac{1}{10}$, $\frac{1}{11}$, $\frac{1}{12}$, $\frac{1}{13}$, $\frac{1}{14}$, $\frac{1}{15}$, $\frac{1}{16}$.

In general, group k contains the 2^k terms $\frac{1}{1+2^k}$, $\frac{1}{2+2^k}$, $\frac{1}{3+2^k}$, $\cdots$, $\frac{1}{2^{k+1}}$.

Clearly,

The sum of the terms in group 1 is bigger than $\frac{1}{4} + \frac{1}{4} = \dfrac{1}{2}$.

The sum of the terms in group 2 is bigger than $\frac{1}{8} + \frac{1}{8} + \frac{1}{8} + \frac{1}{8} = \frac{1}{2}$.

Indeed the sum of the terms in each group is bigger than $\frac{1}{2}$, (prove this). Therefore we have the following picture

$1 + \frac{1}{2} +$	$\frac{1}{3} + \frac{1}{4} +$	$\frac{1}{5} + \frac{1}{6} + \frac{1}{7} + \frac{1}{8} +$	$\frac{1}{9} + \frac{1}{10} + \frac{1}{11} + \cdots + \frac{1}{15} + \frac{1}{16} +$	$\cdots$
$> \frac{1}{2}$	$> \frac{1}{2}$	$> \frac{1}{2}$	$> \frac{1}{2}$	$\cdots$

It follows that

$$s_{2^1} = s_2 > \tfrac{1}{2}$$

$$s_{2^2} = s_4 > \tfrac{1}{2} + \tfrac{1}{2} = 1$$

$$s_{2^3} = s_8 > \tfrac{1}{2} + \tfrac{1}{2} + \tfrac{1}{2} = \tfrac{3}{2}$$

In general $s_{2^k} > \frac{k}{2}$, therefore $\{s_n\}_{n=1}^{\infty}$ diverges because it is not bounded.

236

Exercise 301

1. Split $a_n = \frac{6}{n(n+2)}$ into partial fractions then show that the kth partial sum s_k of the sequence $\{a_n\}_{n=1}^{\infty}$ is $s_k = 3\left(\frac{3}{2} - \frac{1}{k+1} - \frac{1}{k+2}\right)$. Use this to determine $\sum_{n=1}^{\infty} \frac{6}{n(n+2)}$.

2. Split $\frac{12}{(3n-1)(3n+2)}$ into partial fractions then find $\sum_{n=1}^{\infty} \frac{12}{(3n-1)(3n+2)}$.

3. Show that the sum of the first n positive integers is

$$1 + 2 + 3 + \cdots + n = \sum_{i=1}^{n} i = \frac{n(n+1)}{2} \qquad (7.7)$$

4. Show that the series $\sum_{n=1}^{\infty} \frac{n^n}{n!}$ diverges.

5. Prove that $\sum_{n=1}^{\infty} \frac{n}{n+2}$ diverges.

6. What is the sum of the series $\sum_{n=1}^{\infty} \left(-\frac{2}{3}\right)^n$?

7. Find the sum of the series $\sum_{n=1}^{\infty} \frac{4}{(-5)^n}$.

8. The repeating decimal $0.121212121212\overline{12}$ may be written as

$$\frac{12}{100} + \frac{12}{10000} + \frac{12}{1000000} + \cdots = 12\left(\frac{1}{100}\right) + 12\left(\frac{1}{100}\right)^2 + 12\left(\frac{1}{100}\right)^3 + \cdots$$

   Find the sum of the series $\sum_{n=1}^{\infty} 12\left(\frac{1}{100}\right)^n$ and use it to write $0.121212121212\overline{12}$ as a quotient of two integers.

9. Show that $3.799999\overline{9} = 3.8$

10. Let $a \neq 0$ and r be constants, and consider the series $\sum_{n=1}^{\infty} ar^n$.

    (a) Determine the kth partial sum of the series when $r = 1$ and deduce that, in this case, the series diverges.

    (b) Does the series converge when $r = -1$? Defend your answer.

11. Fix a real number x. Then $\sum_{n=0}^{\infty} x^n = 1 + x + x^2 + x^3 + \cdots$ and $\sum_{n=0}^{\infty} (-x)^n = 1 - x + x^2 - x^3 + \cdots$ are geometric progressions. Show that if $|x| < 1$ then

$$\sum_{n=0}^{\infty} x^n = \frac{1}{1-x} \qquad and \qquad \sum_{n=0}^{\infty} (-x)^n = \frac{1}{1+x}$$

Deriving the Sum of Squares, Cubes, ... of natural numbers

Formula (7.7) may be used to determine the sum of the squares of the first n positive integers as follows:

Start with $\sum_{i=1}^{n} \left[(i+1)^3 - i^3 \right] = (n+1)^3 - 1$ which you can verify directly. On expanding $(i+1)^3 - i^3$, the result is $3i^2 + 3i + 1$. Therefore

$$\sum_{i=1}^{n} \left[3i^2 + 3i + 1 \right] = (n+1)^3 - 1 = n^3 + 3n^2 + 3n \qquad (7.8)$$

The left hand side of (7.8) expands into $3\sum_{i=1}^{n} i^2 + 3\sum_{i=1}^{n} i + n$. Now we use (7.7) to

obtain $3\sum_{i=1}^{n} i^2 + \dfrac{3n(n+1)}{2} + n = n^3 + 3n^2 + 3n$. We then solve for $\sum_{i=1}^{n} i^2$ to get

$$\sum_{i=1}^{n} i^2 = \frac{n(n+1)(2n+1)}{6}. \qquad (7.9)$$

Exercise 302

1. *Use the fact that* $\sum_{i=1}^{n} \left[(i+1)^4 - i^4 \right] = (n+1)^4 - 1$ *and the results (7.7) and*

   (7.9) above to show that $\sum_{i=1}^{n} i^3 = \left[\dfrac{n(n+1)}{2} \right]^2.$

2. *Determine a formula for* $\sum_{i=1}^{n} i^4.$

Some Properties of Series

Let $\sum\limits_{n=1}^{\infty} a_n$ and $\sum\limits_{n=1}^{\infty} b_n$ be given series and c be a constant. Then:

1. The series $\sum\limits_{n=1}^{\infty} (a_n + b_n)$ is called the sum of $\sum\limits_{n=1}^{\infty} a_n$ and $\sum\limits_{n=1}^{\infty} b_n$.

2. The series $\sum\limits_{n=1}^{\infty} ca_n$ is called the product of c and $\sum\limits_{n=1}^{\infty} a_n$.

The following theorem states what you would expect:

Theorem 303 *Suppose the series* $\sum\limits_{n=1}^{\infty} a_n$ *and* $\sum\limits_{n=1}^{\infty} b_n$ *converge with* $\sum\limits_{n=1}^{\infty} a_n = s$ *and*

$\sum\limits_{n=1}^{\infty} b_n = t.$ *Let c be a constant. Then*

1. $\sum_{n=1}^{\infty} (a_n + b_n)$ converges and $\sum_{n=1}^{\infty} (a_n + b_n) = s + t$.

2. $\sum_{n=1}^{\infty} ca_n$ converges and $\sum_{n=1}^{\infty} ca_n = cs$.

Example 304 $\sum_{n=1}^{\infty} \left(\frac{3}{4}\right)^n = \sum_{n=1}^{\infty} \frac{3}{4} \left(\frac{3}{4}\right)^{n-1}$ converges to $\frac{3/4}{1-\frac{3}{4}} = 3$ and $\sum_{n=1}^{\infty} \frac{1}{n(n+1)}$ converges to 1, therefore $\sum_{n=1}^{\infty} \left(\left(\frac{3}{4}\right)^n + \frac{1}{n(n+1)}\right)$ converges to $3 + 1 = 4$

Example 305 Since $\sum_{n=1}^{\infty} \frac{1}{n(n+1)}$ converges to 1, $\sum_{n=1}^{\infty} \frac{-7}{3n(n+1)}$ converges to $-\frac{7}{3}$.

Exercise 306

1. Show that if $\sum_{n=1}^{\infty} a_n$ converges and $\sum_{n=1}^{\infty} b_n$ diverges then $\sum_{n=1}^{\infty} (a_n + b_n)$ diverges. (Hint: If $\sum_{n=1}^{\infty} (a_n + b_n)$ converges, what about $\sum_{n=1}^{\infty} (a_n + b_n) - \sum_{n=1}^{\infty} a_n$?)

2. Show that if $\sum_{n=1}^{\infty} b_n$ diverges and c is a nonzero constant then $\sum_{n=1}^{\infty} cb_n$ also diverges.

The p Series

These are special series of the form $\sum_{n=1}^{\infty} \frac{1}{n^p}$ where p is a constant. Examples

(a) $\sum_{n=1}^{\infty} \frac{1}{n}$, $(p = 1)$ (b) $\sum_{n=1}^{\infty} \frac{1}{\sqrt{n}} = \sum_{n=1}^{\infty} \frac{1}{n^{1/2}}$, $(p = \frac{1}{2})$ (c) $\sum_{n=1}^{\infty} \frac{1}{n^4}$, $(p = 4)$

The main result of this section is the following theorem:

Theorem 307 The p-series $\sum_{n=1}^{\infty} \frac{1}{n^p}$ converges if $p > 1$ and diverges if $p \leq 1$.

The proof that $\sum_{n=1}^{\infty} \frac{1}{n^p}$ diverges if $p \leq 1$ is just an extension of the proof that the harmonic series $\sum_{n=1}^{\infty} \frac{1}{n}$ diverges. In fact if $p < 1$ then $n^p < n$ and so $\frac{1}{n} < \frac{1}{n^p}$ for all n. Let

$$t_n = \frac{1}{1^p} + \frac{1}{2^p} + \cdots + \frac{1}{n^p} \quad \text{and} \quad s_n = \frac{1}{1} + \frac{1}{2} + \cdots + \frac{1}{n}.$$

Then $t_n > s_n$ for all n. In the proof that the harmonic series diverges, we showed that $\{s_n\}_{n=1}^{\infty}$ is unbounded. It follows that $\{t_n\}_{n=1}^{\infty}$, which has bigger terms, is also unbounded, hence it diverges. This proves that $\sum_{n=1}^{\infty} \frac{1}{n^p}$ diverges.

The proof that $\sum\limits_{n=1}^{\infty} \frac{1}{n^p}$ converges if $p > 1$ is more involved. We again consider its sequence $\{t_n\}_{n=1}^{\infty}$ of partial sums but this time show that it is increasing and bounded. By Theorem 285 on page 225, that suffices to prove that it converges. To see that it is increasing, simply note that $t_{n+1} = t_n + \frac{1}{(n+1)^p} > t_n$, therefore $t_n < t_{n+1}$ for all n.

To prove that it is bounded, group its terms from the second onwards as below:

Group 1 contains the 2 terms $\frac{1}{2^p}$, $\frac{1}{3^p}$.

Group 2 contains the $4 = 2^2$ terms $\frac{1}{4^p}$, $\frac{1}{5^p}$, $\frac{1}{6^p}$, $\frac{1}{7^p}$.

Group 3 contains the $8 = 2^3$ terms $\frac{1}{8^p}$, $\frac{1}{9^p}$, $\frac{1}{10^p}$, $\frac{1}{11^p}$, $\frac{1}{12^p}$, $\frac{1}{13^p}$, $\frac{1}{14^p}$, $\frac{1}{15^p}$.

$\vdots$

Group k contains the 2^k terms $\frac{1}{(2^k)^p}$, $\frac{1}{(1+2^k)^p}$, $\frac{1}{(2+2^k)^p}$, $\cdots$, $\frac{1}{(2^{k+1}-1)^p}$.

To simplify notation, introduce the number $r = \frac{1}{2^{p-1}}$. Since $p > 1$, $0 < r < 1$

The sum of the terms in group 1 is less than

$$\frac{1}{2^p} + \frac{1}{2^p} = \frac{2}{2^p} = \frac{1}{2^{p-1}} = r.$$

The sum of the terms in group 2 is less than

$$\frac{1}{4^p} + \frac{1}{4^p} + \frac{1}{4^p} + \frac{1}{4^p} = \frac{4}{4^p} = \frac{2^2}{2^{2p}} = \left(\frac{2}{2^p}\right)^2 = r^2.$$

The sum of the terms in group 3 is less than

$$\frac{1}{8^p} + \frac{1}{8^p} + \frac{1}{8^p} + \frac{1}{8^p} + \frac{1}{8^p} + \frac{1}{8^p} + \frac{1}{8^p} + \frac{1}{8^p} = \frac{8}{8^p} = \left(\frac{2}{2^p}\right)^3 = r^3$$

In general, the sum of the terms in the kth group is less than r^k, (prove this). Therefore we have the following picture.

$1+$	$\frac{1}{2^p} + \frac{1}{3^p} +$	$\frac{1}{4^p} + \frac{1}{5^p} + \frac{1}{6^p} + \frac{1}{7^p} +$	$\frac{1}{8^p} + \frac{1}{9^p} + \cdots + \frac{1}{14^p} + \frac{1}{15^p} +$	$\cdots$
	$< r$	$< r^2$	$< r^3$	$\cdots$

It follows that: $\qquad t_{2^2-1} = t_3 < 1 + r = \frac{1-r^2}{1-r} < \frac{1}{1-r}$

$$t_{2^3-1} = t_7 < 1 + r + r^2 = \frac{1-r^3}{1-r} < \frac{1}{1-r}$$

$$t_{2^4-1} = t_{15} < 1 + r + r^2 + r^3 = \frac{1-r^4}{1-r} < \frac{1}{1-r}$$

In general; $\qquad t_{2^k-1} < 1 + r + r^2 + \cdots + r^{k-1} = \frac{1-r^k}{1-r} < \frac{1}{1-r}, \quad k = 2, 3, \ldots$

This proves that all the terms of $\{t_n\}_{n=1}^{\infty}$ are between 0 and $\frac{1}{1-r}$, therefore it is bounded.

Some More Convergence Tests

The statement "*The series $\sum_{n=1}^{\infty} ar^{n-1}$ converges to $\frac{a}{1-r}$ if $|r| < 1$ and it diverges if $|r| > 1$*" is an example of a test for convergence. It tells us when a geometric progression converges or diverges. Another one is Theorem 307 which asserts that "*The p-series $\sum_{n=1}^{\infty} \frac{1}{n^p}$ converges if $p > 1$ and diverges if $p \leq 1$*". This section introduces some more tests for convergence.

The Comparison Test for Convergence

The comparison test applies to sequences whose terms are non-negative. Say we want to show that such a series $\sum_{n=1}^{\infty} a_n$ converges. We produce a series $\sum_{n=1}^{\infty} b_n$ which we known converges and has **bigger** terms, then argue that if $\sum_{n=1}^{\infty} b_n$ converges, then $\sum_{n=1}^{\infty} a_n$ with smaller terms must also converge.

However, if we want to prove that a given series $\sum_{n=1}^{\infty} u_n$ with non-negative terms diverges, we produce a series $\sum_{n=1}^{\infty} v_n$ which we are sure diverges and has **smaller** terms than $\sum_{n=1}^{\infty} u_n$. It stands to reason that if $\sum_{n=1}^{\infty} v_n$ diverges, then $\sum_{n=1}^{\infty} u_n$ with bigger terms must also diverge.

The following is a more precise statement of this test. (Its proof is outlined in Exercise 5 on page 245).

Theorem 308 *Let $\sum_{n=1}^{\infty} a_n$ and $\sum_{n=1}^{\infty} b_n$ be given series. We assume that all but a finite number of their terms are non-negative. Suppose there is an integer M such that $0 \leq a_n \leq b_n$ for all $n \geq M$. If $\sum_{n=1}^{\infty} b_n$ converges then $\sum_{n=1}^{\infty} a_n$ also converges. If $\sum_{n=1}^{\infty} a_n$ diverges then $\sum_{n=1}^{\infty} b_n$ also diverges.*

Example 309 *Consider the series $\sum_{n=1}^{\infty} \frac{1}{n^2+2n+3}$ and $\sum_{n=1}^{\infty} \frac{1}{n^2}$. Since $n^2+2n+3 > n^2$*

for all n, it follows that $0 < \frac{1}{n^2+2n+3} < \frac{1}{n^2}$ for all n. But $\sum_{n=1}^{\infty} \frac{1}{n^2}$ converges, (because

it is a p-series with $p = 2$). It follows that $\sum_{n=1}^{\infty} \frac{1}{n^2+2n+3}$ also converges.

Example 310 *Consider the series $\sum_{n=1}^{\infty} \frac{1}{n}$ and $\sum_{n=1}^{\infty} \frac{1}{\ln n}$. Since $\ln n < n$ for all*

positive integers n, (see Exercise 9 on page 125), it follows that $\frac{1}{n} \leq \frac{1}{\ln n}$ for all

$n > 1$. We know that the harmonic series $\sum_{n=1}^{\infty} \frac{1}{n}$ diverges. It follows that $\sum_{n=1}^{\infty} \frac{1}{\ln n}$

also diverges.

Example 311 *Consider the series* $\sum_{n=1}^{\infty} \frac{1}{n^3 - 10n}$. *To get a p-series out of this, replace* $10n$ *with a suitable constant multiple of* n^3. *To this end, note that if* $n > 5$ *then* $\frac{1}{2}n^3 > 10n$, *therefore replace* $10n$ *with* $\frac{1}{2}n^3$. *This implies that* $n^3 - \frac{1}{2}n^3 < n^3 - 10n$, *(because you subtract more from* n^3 *in* $n^3 - \frac{1}{2}n^3$*). It follows that if* $n > 5$ *then*

$$\frac{1}{n^3 - 10n} < \frac{1}{n^3 - \frac{1}{2}n^3} = \frac{2}{n^3}$$

Since $\sum_{n=1}^{\infty} \frac{2}{n^3}$ *converges,* $\sum_{n=1}^{\infty} \frac{1}{n^3 - 10n}$, *with smaller terms, must also converge.*

The Limit Comparison Test

Let $\sum_{n=1}^{\infty} a_n$ and $\sum_{n=1}^{\infty} b_n$ be given series. Assume that all but a finite number of their terms are positive. We say that they have comparable terms if $\lim_{n \to \infty} \left(\frac{a_n}{b_n} \right)$ exists and is a nonzero number.

The limit comparison test simply states that if $\sum_{n=1}^{\infty} a_n$ and $\sum_{n=1}^{\infty} b_n$ have comparable terms, then when it comes to convergence/divergence, whatever one of the two does, so does the other. Thus, if one of them converges, so does the other; and if one of them diverges, the other one also diverges. (See exercise 6 on page 245 for an outline of a proof.)

Theorem 312 *Let* $\sum_{n=1}^{\infty} a_n$ *and* $\sum_{n=1}^{\infty} b_n$ *be given series. Assume that all but a finite number of their terms are positive. Suppose* $\lim_{n \to \infty} \left(\frac{a_n}{b_n} \right)$ *exists and is nonzero. If one of the series converges then the other one also converges. If one of them diverges then the other one also diverges.*

Example 313 *Consider the series* $\sum_{n=1}^{\infty} \frac{n^2 + 1}{3n^4 + 5}$ *and* $\sum_{n=1}^{\infty} \frac{1}{n^2}$. *The second one is a p-series with* $p > 1$, *therefore it converges. Since*

$$\lim_{n \to \infty} \left(\frac{n^2 + 1}{3n^4 + 5} \right) / \left(\frac{1}{n^2} \right) = \lim_{n \to \infty} \left(\frac{n^4 + n^2}{3n^4 + 5} \right) = \lim_{n \to \infty} \left(\frac{1 + \frac{1}{n^2}}{3 + \frac{5}{n^2}} \right) = \frac{1}{3} \neq 0,$$

it follows that $\sum_{n=1}^{\infty} \left(\frac{n^2 + 1}{3n^4 + 5} \right)$ *also converges.*

Example 314 *The series* $\sum_{n=1}^{\infty} \frac{\sqrt{n}}{5n + 12}$ *diverges. To prove it take the series with nth term* $\frac{\sqrt{n}}{5n + 0} = \frac{1}{5\sqrt{n}}$. *Thus take* $\sum_{n=1}^{\infty} \frac{1}{5\sqrt{n}}$. *We know that it diverges. Since*

$$\lim_{n \to \infty} \left(\frac{\sqrt{n}}{5n + 12} \right) / \left(\frac{1}{5\sqrt{n}} \right) = \lim_{n \to \infty} \frac{5}{5 + \frac{12}{5\sqrt{n}}} = 1,$$

the series $\sum_{n=1}^{\infty} \frac{\sqrt{n}}{5n + 12}$ *must diverge.*

If, in Theorem 312, $\lim\limits_{n\to\infty}\frac{a_n}{b_n}=0$ then the terms of $\sum\limits_{n=1}^{\infty}a_n$ are not comparable to those of $\sum\limits_{n=1}^{\infty}b_n$. In fact the a_n's must be much smaller than the b_n's, so it is possible for $\sum\limits_{n=1}^{\infty}a_n$ to converge while $\sum\limits_{n=1}^{\infty}b_n$ diverges. For an example, take $\sum\limits_{n=1}^{\infty}a_n=\sum\limits_{n=1}^{\infty}\frac{1}{n^3}$ and $\sum\limits_{n=1}^{\infty}b_n=\sum\limits_{n=1}^{\infty}\frac{1}{\sqrt{n}}$. Then $\lim\limits_{n\to\infty}\frac{a_n}{b_n}=0$, the first one converges but the second one diverges.

The Integral Test

We use the series $\sum\limits_{n=2}^{\infty}\frac{1}{n(\ln n)^2}$ and $\sum\limits_{n=2}^{\infty}\frac{1}{n\ln n}$ to illustrate what the integral test says. Introduce the function $f(x)=\frac{1}{x(\ln x)^2}$. Its values at the integer points 2, 3, 4, ..., are the terms of the series $\sum\limits_{n=2}^{\infty}\frac{1}{n(\ln n)^2}$. Since

$$\int_2^{\infty}\frac{1}{x(\ln x)^2}dx=\lim_{N\to\infty}\int_2^{N}\frac{1}{x(\ln x)^2}dx=\lim_{N\to\infty}\left[\frac{-1}{\ln x}\right]_2^{N}=\frac{1}{\ln 2}$$

the improper integral $\int_2^{\infty}\frac{1}{x(\ln x)^2}dx$ converges. We use this fact to deduce that the series $\sum\limits_{n=2}^{\infty}\frac{1}{n(\ln n)^2}$ converges. The strategy is to view $\int_2^{\infty}\frac{1}{x(\ln x)^2}dx$ as the area of the region R enclosed by the x-axis and the graph of f on the interval $[2,\infty)$. By definition, it is the limit, as N approaches infinity, of the area of the region $R(N)$ enclosed by the x-axis and the graph of f on an interval $[2, N]$. The left figure below shows such a region. The right figure shows rectangles of width 1 and heights $f(3)$, $f(4)$, ..., $f(N)$ inside $R(N)$.

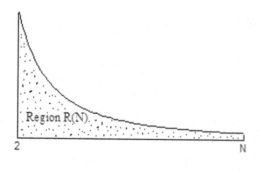
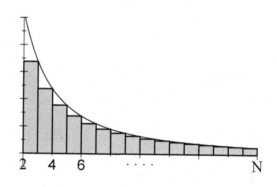

The total area of these rectangles is $f(3)+f(4)+\cdots+f(N)$. Clearly, their total area is less than the area of $R(N)$, therefore

$$f(3)+f(4)+\cdots+f(N)<\int_2^{N}\frac{1}{x(\ln x)^2}dx.$$

Since this is true for all positive integers $N > 2$, it follows that

$$\lim_{N\to\infty}[f(3)+f(4)+\cdots+f(N)] \le \lim_{N\to\infty}\int_2^N \frac{1}{x(\ln x)^2}dx = \int_2^\infty \frac{1}{x(\ln x)^2}dx = \frac{1}{\ln 2}.$$

This proves that the partial sums for the series $\sum_{n=3}^\infty f(n)$ are bounded. Since they are increasing, the series converges.

Turning to $\sum_{n=2}^\infty \frac{1}{n\ln n}$, its terms are the values of the function $f(x) = \frac{1}{x\ln x}$ at the integer points $2, 3, \ldots$. This time the improper integral $\int_2^\infty \frac{1}{x\ln x}dx$ diverges since

$$\int_2^N \frac{1}{x\ln x}dx = [\ln(\ln x)]_2^N = \ln(\ln N) - \ln(\ln 2)$$

and $\lim_{N\to\infty}\ln(\ln x) = \infty$. This is used to deduce that $\sum_{n=2}^\infty \frac{1}{n\ln n}$ diverges. The strategy is the same: regard $\int_2^\infty \frac{1}{x\ln x}dx$ as the area of the region enclosed by the x-axis and the graph of g on the interval $[2, \infty)$. It is also the limit of the area of the region $W(N)$ enclosed by the x-axis and the graph of g on the interval $[2, N]$. The left figure below shows such a region and the right figure shows the

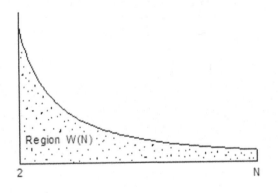
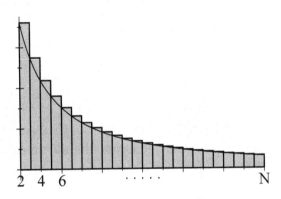

rectangles with width 1 and heights $f(2), f(3), \ldots, f(N-1)$ respectively. These rectangles enclose $W(N)$ and their total area is $f(2)+f(3)+\cdots+f(N-1)$, which is bigger than the area of $W(N)$, therefore

$$\int_2^N \frac{1}{x\ln x}dx < f(2)+f(3)+\cdots+f(N-1).$$

Since $\lim_{N\to\infty}\int_2^N \frac{1}{x\ln x}dx = \infty$, it follows that $\lim_{N\to\infty}\sum_{n=2}^{N-1}\frac{1}{n\ln n} = \infty$ and so the series $\sum_{n=2}^\infty \frac{1}{n\ln n}$ diverges.

The integral test is a generalization of the above arguments. It may be stated as follows:

Theorem 315 *Let f be a positive strictly decreasing function whose values $f(1)$, $f(2)$, $f(3)$, ... are the terms a_1, a_2, a_3, ... of a given series $\sum_{n=1}^{\infty} a_n$. If the improper integral $\int_1^{\infty} f(x)dx$ converges then the series converges. If $\int_1^{\infty} f(x)dx$ diverges then the series diverges.*

The proof is a generalization of the above steps. One considers the region $R(N)$ enclosed by the x-axis and the graph of f on the interval $[1, N]$. Because f is strictly decreasing, the rectangles with width 1 and heights $f(2)$, $f(3)$, ..., $f(N)$ are inside the region $R(N)$. On the other hand, the rectangles with width 1 and length $f(1)$, $f(2)$, ..., $f(N-1)$ enclose $R(N)$. It follows that

$$f(2) + f(3) + \cdots + f(N) < \int_1^N f(x)dx < f(1) + f(2) + \cdots + f(N-1)$$

If $\int_1^{\infty} f(x)dx$ converges then $\sum_{n=2}^{\infty} f(n)$ must be finite, hence $\sum_{n=1}^{\infty} f(n)$ converges. On the other hand, if $\int_1^{\infty} f(x)dx$ diverges then $\sum_{n=1}^{\infty} f(n)$ must be infinite, therefore the series diverges.

The Alternating Series Test

An **alternating series** is a series whose terms have strictly alternating signs. It has one of the two forms

$$\sum_{n=1}^{\infty} (-1)^n a_n \quad \text{or} \quad \sum_{n=1}^{\infty} (-1)^{n+1} a_n.$$

The numbers a_1, a_2, a_3, ... themselves are all positive. It is the factors $(-1)^n$ or $(-1)^{n+1}$ that generate the alternating sign. The alternating series test may be states as follows:

Theorem 316 *Let $\sum_{n=1}^{\infty} (-1)^n a_n$ be an alternating series. If $a_1 > a_2 > a_3 > \cdots > a_n > a_{n+1} > \cdots$, (i.e. the numbers a_1, a_2, a_3, ... strictly decrease), and $\lim_{n\to\infty} a_n = 0$, (i.e. the terms shrink to 0), then the series converges.*

Example 317 *Although the harmonic series $\sum_{n=1}^{\infty} \dfrac{1}{n}$ diverges, the alternating series $\sum_{n=1}^{\infty} \dfrac{(-1)^n}{n}$ converges since $1 > \dfrac{1}{2} > \dfrac{1}{3} > \dfrac{1}{4} > \cdots$ and $\lim_{n\to\infty} \dfrac{1}{n} = 0$.*

Example 318 $\sum_{n=2}^{\infty} \dfrac{(-1)^n}{\ln n}$ *converges because* $\dfrac{1}{\ln 2} > \dfrac{1}{\ln 3} > \dfrac{1}{\ln 4} > \cdots$ *and* $\lim_{n\to\infty} \dfrac{1}{\ln n} = 0.$

Exercise 319

1. Use an appropriate test to determine whether or not the given series converges:

(a) $\sum_{n=1}^{\infty} \frac{2n}{3n^3+4}$ (b) $\sum_{n=1}^{\infty} \frac{(-1)^n 2^n}{2^n+17}$ (c) $\sum_{n=1}^{\infty} \frac{1}{2n-1}$ (d) $\sum_{n=1}^{\infty} \frac{3}{(2n+1)^2}$

(e) $\sum_{n=1}^{\infty} \frac{1}{n+2^n}$ (f) $\sum_{n=1}^{\infty} \frac{n^2}{3n^3+4}$ (g) $\sum_{n=1}^{\infty} \frac{(-1)^n}{2n-1}$ (h) $\sum_{n=1}^{\infty} \sin n$

(i) $\sum_{n=1}^{\infty} \frac{\cos n\pi}{5n}$ (j) $\sum_{n=1}^{\infty} \frac{\sqrt{n}}{n^2+2n+5}$ (k) $\sum_{n=1}^{\infty} \frac{2}{\sqrt[3]{n}}$ (l) $\sum_{n=1}^{\infty} \frac{2}{n^2-12}$

(m) $\sum_{n=1}^{\infty} \frac{\cos^2 n}{n^2}$ (n) $\sum_{n=1}^{\infty} 3^{-n}$ (o) $\sum_{n=1}^{\infty} \frac{2^{-n}}{n}$ (p) $\sum_{n=1}^{\infty} \frac{n^3}{(n+4)^4}$

2. Give an example of divergent series $\sum_{n=1}^{\infty} a_n$ and $\sum_{n=1}^{\infty} b_n$ such that $\lim_{n \to \infty} \frac{a_n}{b_n} = 0$.

3. Use the integral test to show that the p-series $\sum_{n=1}^{\infty} \frac{1}{n^p}$ converges if $p > 1$ and diverges if $p \leq 1$.

4. We showed that $\sum_{n=2}^{\infty} \frac{1}{n \ln n}$ diverges. Deduce that $\sum_{n=2}^{\infty} \frac{1}{\sqrt{n} \ln n}$ also diverges.

5. Fill in the necessary details in the following proof of the comparison test, (i.e. Theorem 308 on page 240)

   To simplify the proof, assume that the terms of $\sum_{n=1}^{\infty} a_n$ and $\sum_{n=1}^{\infty} b_n$ are all positive and that $a_n \leq b_n$ for all positive integers n. Let $s_n = a_1 + a_2 + \cdots + a_n$ and $t_n = b_1 + b_2 + \cdots + b_n$. Show that:

   (a) $s_n \leq t_n$ for all positive integers n.
   (b) $\{s_n\}$ and $\{t_n\}$ are increasing.
   (c) If $\{t_n\}$ converges then it must be bounded. Use this to deduce that $\{s_n\}$ must converge.
   (d) If $\{s_n\}$ diverges then it must be unbounded. Use this to deduce that $\{t_n\}$ must diverge.

6. Let $\sum_{n=1}^{\infty} a_n$ and $\sum_{n=1}^{\infty} b_n$ have positive comparable terms. Thus $\lim_{n \to \infty} \frac{a_n}{b_n}$ exists and is non-zero. Show that there are constants K and M such that for every positive integers n, $Kb_n \leq a_n \leq Mb_n$. Use this to show that if one of the series converges the so does the other. Also show that if one of them diverges then the other one also diverges.

Tests for Absolute Convergence

We say that a series $\sum_{n=1}^{\infty} a_n$ **converges absolutely** if the series $\sum_{n=1}^{\infty} |a_n|$, obtained by taking the absolute value of each of its terms, converges.

Example 320 *The series* $\sum_{n=1}^{\infty} \frac{(-1)^n}{2^n}$ *converges absolutely because* $\sum_{n=1}^{\infty} \left| \frac{(-1)^n}{2^n} \right| = \sum_{n=1}^{\infty} \frac{1}{2^n}$ *which is a G.P. with* $r = \dfrac{1}{2}$.

If a series $\sum_{n=1}^{\infty} a_n$ converges but the series $\sum_{n=1}^{\infty} |a_n|$, diverges then we say that $\sum_{n=1}^{\infty} a_n$ **converges conditionally**.

Example 321 *The series* $\sum_{n=1}^{\infty} \frac{(-1)^n}{\sqrt{n}}$ *converges by the alternating series test. However,* $\sum_{n=1}^{\infty} \frac{1}{\sqrt{n}}$ *diverges. Therefore* $\sum_{n=1}^{\infty} \frac{(-1)^n}{\sqrt{n}}$ *converges conditionally.*

It turns out that if $\sum_{n=1}^{\infty} |a_n|$ converges then $\sum_{n=1}^{\infty} a_n$ itself converges. Therefore verifying that a given series converges absolutely automatically verifies that it converges. This is a useful fact to know because there are several tests for absolute convergence. We mention two of them here:

The Ratio Test for Absolute Convergence

The ratio test states that:

Theorem 322 *Let* $\sum_{n=1}^{\infty} a_n$ *be a series of real numbers. Suppose* $\lim_{n \to \infty} \frac{|a_{n+1}|}{|a_n|}$ *exists and is a number* L. *If* $L < 1$ *then the series converges absolutely. If* $L > 1$ *then the series diverges. If* $L = 1$, *no meaningful conclusion can be arrived at because there are convergent series for which* $L = 1$ *and there are also divergent series for which* L *is 1.*

Example 323 *Consider the series* $\sum_{n=1}^{\infty} \frac{n}{(-2)^n}$. *In this case* $a_n = \frac{n}{(-2)^n}$ *and* $a_{n+1} = \frac{n+1}{(-2)^{n+1}}$. *It follows that*

$$\lim_{n \to \infty} \frac{|a_{n+1}|}{|a_n|} = \lim_{n \to \infty} \frac{n+1}{2^{n+1}} \cdot \frac{2^n}{n} = \lim_{n \to \infty} \frac{1}{2} \left(\frac{n+1}{n} \right) = \frac{1}{2}.$$

By the ratio test, the series converges absolutely. Since an absolutely convergent series also converges, $\sum_{n=1}^{\infty} \frac{n}{(-2)^n}$ *converges.*

Example 324 *Consider the series* $\sum_{n=1}^{\infty} \frac{6^n}{n!}$. *In this case* $a_n = \frac{6^n}{n!}$ *and* $a_{n+1} = \frac{6^{n+1}}{(n+1)!}$. *It follows that*

$$\lim_{n \to \infty} \frac{|a_{n+1}|}{|a_n|} = \lim_{n \to \infty} \frac{6^{n+1}}{(n+1)!} \cdot \frac{n!}{6^n} = \lim_{n \to \infty} \frac{6}{(n+1)} = 0.$$

By the ratio test, the series converges absolutely.

Example 325 *Consider the series* $\sum\limits_{n=1}^{\infty} \frac{2^n}{n^4}$. *In this case* $a_n = \frac{2^n}{n^4}$ *and* $a_{n+1} = \frac{2^{n+1}}{(n+1)^4}$.
It follows that

$$\lim_{n\to\infty} \frac{|a_{n+1}|}{|a_n|} = \lim_{n\to\infty} \frac{2^{n+1}}{(n+1)^4} \cdot \frac{n^4}{2^n} = \lim_{n\to\infty} 2\left(\frac{n}{n+1}\right)^4 = 2.$$

Therefore $\sum\limits_{n=1}^{\infty} \frac{2^n}{n^4}$ *diverges.*

Example 326 *Consider the series* $\sum\limits_{n=1}^{\infty} \frac{n}{n^3+1}$. *In this case*

$$\lim_{n\to\infty} \frac{|a_{n+1}|}{|a_n|} = \lim_{n\to\infty} \frac{n+1}{(n+1)^3+1} \cdot \frac{n^3+1}{n} = 1.$$

The ratio test fails. One has to use another test. The limit comparison test reveals that it converges, (its terms are comparable to the terms of $\sum\limits_{n=1}^{\infty} \frac{1}{n^2}$ *which converges).*

Example 327 *Let* $a_n = \frac{n!}{n^n}$. *Then*

$$\lim_{n\to\infty} \frac{|a_{n+1}|}{|a_n|} = \lim_{n\to\infty} \frac{(n+1)!}{(n+1)^{n+1}} \cdot \frac{n^n}{n!} = \lim_{n\to\infty} \frac{(n+1)n!}{(n+1)(n+1)^n} \cdot \frac{n^n}{n!} = \lim_{n\to\infty} \frac{n^n}{(n+1)^n}$$

$$= \lim_{n\to\infty} \left(\frac{n}{n+1}\right)^n = \lim_{n\to\infty} \left(\left(1+\frac{1}{n}\right)^n\right)^{-1} = e^{-1} < 1$$

Therefore $\sum\limits_{n=1}^{\infty} \frac{n!}{n^n}$ *converges.*

The Root Test for Absolute Convergence This may be stated as follows:

Theorem 328 *Let* $\sum\limits_{n=1}^{\infty} a_n$ *be a series of real numbers. Suppose* $\lim\limits_{n\to\infty} |a_n|^{1/n}$ *exists and is equal to* L. *If* $L < 1$ *then the series converges absolutely. If* $L > 1$ *then the series diverges. If* $L = 1$, *no meaningful conclusion can be arrived at because there are convergent series with* $L = 1$ *and there are also divergent series with* $L = 1$.

Example 329 *Consider* $\sum\limits_{n=1}^{\infty} \frac{n^4}{3^n}$. *In this case* $|a_n|^{1/n} = \frac{n^{4/n}}{3} = \frac{1}{3}\left(n^{1/n}\right)^4$. *Since* $\lim\limits_{n\to\infty} n^{1/n} = 1$, *(see problem 2 on page 230), it follows that* $\lim\limits_{n\to\infty} |a_n|^{1/n} = \frac{1}{3}$, *therefore* $\sum\limits_{n=1}^{\infty} \frac{n^4}{3^n}$ *converges absolutely.*

Example 330 *Consider* $\sum\limits_{n=1}^{\infty} \frac{2^{n^2}}{(2n)^n}$. *We find that* $\left|\frac{2^{n^2}}{(2n)^n}\right|^{1/n} = \frac{2^n}{2n}$. *One may follow the steps of problem 4 on page 231 to show that* $\lim\limits_{n\to\infty} \frac{2^n}{2n} = \infty$. *Therefore the series diverges.*

Exercise 331

1. Determine $\lim\limits_{n\to\infty}\left(\frac{1}{n^2}\right)^{1/n}$. Does $\sum\limits_{n=1}^{\infty}\left(\frac{1}{n^2}\right)^{1/n}$ converge? Defend your answer.

2. Let a and k be positive numbers. Use the ratio test to establish whether or not the series $\sum\limits_{n=1}^{\infty}\frac{n^k}{2^{an}}$ converges.

3. Determine $\lim\limits_{n\to\infty}\frac{\sin\left(\frac{1}{n^2}\right)}{\frac{1}{n^2}}$ then deduce that $\sum\limits_{n=1}^{\infty}\sin\left(\frac{1}{n^2}\right)$ converges absolutely.

4. Use the root test to determine whether or not the series $\sum\limits_{n=2}^{\infty}\frac{n}{(\ln n)^n}$ converges.

5. Let $a_n = \frac{3^n n^n}{(n!)^2}$. Use the ratio test to prove that $\sum\limits_{n=1}^{\infty} a_n$ converges.

6. Let y be a constant. Use the ratio test to show that the series $\sum\limits_{n=1}^{\infty}\frac{y^n}{n!}$ converges, then use Theorem 299 to deduce that $\lim\limits_{n\to\infty}\left(\frac{y^n}{n!}\right) = 0$ for every real number y. (We use this result in the next section.)

7. Give an example of a series $\sum\limits_{n=1}^{\infty} a_n$ with the property that $\sum\limits_{n=1}^{\infty} a_n$ converges but $\sum\limits_{n=1}^{\infty} a_n^2$ diverges.

8. Test the given series for convergence. Any test that works is acceptable.

(a) $\sum\limits_{n=1}^{\infty}\frac{3^n}{n^3}$ (b) $\sum\limits_{n=1}^{\infty}\frac{(-1)^n}{\sqrt{n+5}}$ (c) $\sum\limits_{n=1}^{\infty}\frac{1}{n^2-7}$ (d) $\sum\limits_{n=1}^{\infty}\frac{(-1)^n 3^n}{n!}$

(e) $\sum\limits_{n=1}^{\infty}\frac{1}{\sqrt{n^2+n}}$ (f) $\sum\limits_{n=1}^{\infty}\frac{5^n}{n!}$ (g) $\sum\limits_{n=1}^{\infty}\left(\frac{3}{5}\right)^n$ (h) $\sum\limits_{n=1}^{\infty} n\left(\frac{3}{4}\right)^n$

(i) $\sum\limits_{n=1}^{\infty}\frac{(-1)^n n}{2n+1}$ (j) $\sum\limits_{n=1}^{\infty}\frac{2^n}{n^n}$ (k) $\sum\limits_{n=1}^{\infty}\frac{\ln n}{e^n}$ (l) $\sum\limits_{n=1}^{\infty}\frac{n^3}{n+2^n}$

9. Let $\sum\limits_{n=1}^{\infty} a_n$ be a series and $L = \lim\limits_{n\to\infty}\left|\frac{a_{n+1}}{a_n}\right|$ be less than 1. Show that there is a number $0 < K < 1$ and an integer N such that $\left|\frac{a_{n+1}}{a_n}\right| < K$ for all $n \geq N$. Since $\left|\frac{a_{N+1}}{a_N}\right| < K$, it follows that $|a_{N+1}| < K|a_N|$. Also $\left|\frac{a_{N+2}}{a_{N+1}}\right| < K$, therefore $|a_{N+2}| < K|a_{N+1}| < K^2|a_N|$. In general, $|a_{N+i}| < K^i|a_N|$. Use the fact that $\sum\limits_{i=1}^{\infty}|a_N|K^i$ converges to deduce that $\sum\limits_{n=1}^{\infty} a_n$ converges absolutely.

Taylor Series and Maclaurin Series

We start with a derivation of Taylor's theorem. To this end, let f be a function that has derivatives of all orders on an interval $[a, b]$. Fix c and x in the interval $[a, b]$. We know that

$$f(x) - f(c) = \int_c^x f'(t)dt \qquad (7.10)$$

To go further than this, we integrate $\int_c^x f'(t)dt$ by parts. We use the integration by parts formula in the form

$$\int_c^x v(t)\frac{du}{dt} = \left[v(t)u(t)\right]_c^x - \int_c^x u(t)\frac{dv}{dt}$$

Take $\dfrac{du}{dt} = 1$ and $v(t) = f'(t)$. Then

$$\int_c^x f'(t)dt = [tf'(t)]_c^x - \int_c^x tf''(t)dt = xf'(x) - cf'(c) - \int_c^x tf''(t)dt.$$

Now (7.10) may be written as

$$f(x) = f(c) + xf'(x) - cf'(c) - \int_c^x tf''(t)dt.$$

To get rid of the term $xf'(x)$, we use the fact that

$$f'(x) - f'(c) = \int_c^x f''(t)dt$$

It implies that $xf'(x) = xf'(c) + x\int_c^x f''(t)dt = xf'(c) + \int_c^x xf''(t)dt$, (since x is a constant). Therefore

$$f(x) = f(c) + (x - c)\,f'(c) + \int_c^x (x - t)f''(t)dt.$$

The next step is to integrate $\int_c^x (x - t)f''(t)dt$ by parts. You are invited to do so. You should get

$$\int_c^x (x - t)f''(t)dt = \tfrac{1}{2}(x - c)^2\, f''(c) + \tfrac{1}{2}\int_c^x (x - t)^2 f^{(3)}(t)dt$$

Therefore

$$f(x) = f(c) + (x - c)\,f'(c) + \tfrac{1}{2}(x - c)^2\, f''(c) + \tfrac{1}{2}\int_c^x (x - t)^2 f^{(3)}(t)dt$$

Another integration by parts yields

$$f(x) = f(c)+(x - c)\,f'(c)+\tfrac{1}{2}(x - c)^2\, f''(c)+\tfrac{1}{3!}(x - c)^3\, f^{(3)}(c)+\tfrac{1}{3!}\int_c^x (x-t)^3 f^{(4)}(t)dt$$

After n such integrations, you should have

$$f(x) = f(c) + (x - c) f'(c) + \tfrac{1}{2} (x - c)^2 f''(c) + \cdots + \tfrac{1}{n!} (x - c)^n f^{(n)}(c)$$
$$+ \tfrac{1}{n!} \int_c^x (x - t)^n f^{(n+1)}(t) dt$$

By the second Mean Value Theorem for integrals, there is a number θ_n between c and x such that

$$\tfrac{1}{n!} \int_c^x (x - t)^n f^{(n+1)}(t) dt = \tfrac{1}{n!} f^{(n+1)}(\theta_n) \int_c^x (x - t)^n \, dt$$

Therefore

$$\tfrac{1}{n!} \int_c^x (x - t)^n f^{(n+1)}(t) dt = \tfrac{f^{(n+1)}(\theta_n)}{(n+1)!} (x - c)^{n+1}$$

Now we can state Taylor's theorem.

Theorem 332 *If f has derivatives of all orders on an interval $[a, b]$, n is a positive integer and c, x are numbers in $[a, b]$ then there is a number θ_n between c and x such that*

$$f(x) = f(c) + (x - c) f'(c) + \cdots + \tfrac{1}{n!} (x - c)^n f^{(n)}(c) + \tfrac{f^{(n+1)}(\theta_n)}{(n+1)!} (x - c)^{n+1}$$

On page **??**, we called

$$p_n(x) = f(c) + (x - c) f'(c) + \tfrac{1}{2} (x - c)^2 f''(c) + \cdots + \tfrac{1}{n!} (x - c)^n f^{(n)}(c)$$

the nth order Taylor polynomial for f at c. Now we know the error in approximating $f(x)$ with $p_n(x)$. It is

$$R_n(x) = \tfrac{f^{(n+1)}(\theta_n)}{(n+1)!} (x - c)^{n+1}$$

In many cases, θ_n is not known, therefore we are forced to settle for an an upper bound of $|R_n(x)|$. To get one, determine the largest value of $\left| f^{(n+1)}(x) \right|$ on the interval with end-points c and x. (For many familiar functions, this can be done.) Let it be M. Then

$$|R_n(x)| \leq \tfrac{M \left| (x-c)^{n+1} \right|}{(n+1)!}$$

This bound was mentioned in Remark 154 on page 110.

For some functions f and suitable numbers x, $\displaystyle\lim_{n \to \infty} \tfrac{f^{(n+1)}(\theta_n)}{(n+1)!} (x - c)^{n+1} = 0$. This implies that if the series $f(c) + \displaystyle\sum_{n=1}^{\infty} \tfrac{1}{n!} f^{(n)}(c) (x - c)^n$ converges then its sum is $f(x)$. In other words,

$$f(x) = f(c) + \sum_{n=1}^{\infty} \tfrac{1}{n!} f^{(n)}(c) (x - c)^n \tag{7.11}$$

It is convenient to define $f^{(0)}(c) = f(c)$ and there is no inconsistency if we take $0!$ to be 1. Then (7.11) may be written as

$$f(x) = \sum_{n=0}^{\infty} \tfrac{1}{n!} f^{(n)}(c)(x-c)^n$$

The series $\sum_{n=0}^{\infty} \tfrac{1}{n!} f^{(n)}(c)(x-c)^n$ is called the **Taylor series** for f at c. When $c = 0$, we get the special series $\sum_{n=0}^{\infty} \tfrac{1}{n!} f^{(n)}(0)x^n$, called the **Maclaurin series** for f.

The Maclaurin Series for Some Elementary Functions

1. Consider $f(x) = e^x$ defined on an interval $[-M, M]$. To write down its Maclaurin series, one must have the numbers $f'(0)$, $f'(0)$, $f''(0)$, ... These are easy to get: $f'(0) = e^0 = 1$, $f''(0) = e^0 = 1$, in general, $f^{(n)}(0) = 1$ for every positive integer n. Take any number x in $[-M, M]$ and fix a positive integer n. By Theorem 332, there is a number θ_n between 0 and x such that

$$e^x = 1 + x + \tfrac{x^2}{2!} + \tfrac{x^3}{3!} + \cdots + \tfrac{x^n}{n!} + \tfrac{e^{\theta_n}x^{n+1}}{(n+1)!}$$

The error term R_n satisfies the inequality

$$|R_n| = \left| \tfrac{e^{\theta_n}x^{n+1}}{(n+1)!} \right| \leq \tfrac{e^M M^{n+1}}{(n+1)!}$$

By Exercise 6 on page 248, $\lim_{n\to\infty} \tfrac{M^{n+1}}{(n+1)!} = 0$, therefore

$$e^x = 1 + x + \tfrac{x^2}{2!} + \tfrac{x^3}{3!} + \cdots + \tfrac{x^n}{n!} + \cdots = \sum_{n=0}^{\infty} \tfrac{x^n}{n!}$$

This is the Maclaurin series for $f(x) = e^x$.

2. Let $g(x) = \cos x$, $-M \leq x \leq M$. Then $g'(0) = 0$, $g''(0) = -1$, $g'''(x) = 0$, $g^{(4)}(0) = 1$, in general

$$g^{(n)}(0) = \begin{cases} 0 & \text{if } n \text{ is odd} \\ (-1)^{\frac{1}{2}n} & \text{if } n \text{ is even} \end{cases}$$

By Theorem 332, there is a number θ_{2n+1} between 0 and x such that

$$\cos x = 1 - \tfrac{x^2}{2!} + \tfrac{x^4}{4!} - \tfrac{x^6}{6!} + \cdots + \tfrac{(-1)^n x^{2n}}{(2n)!} + \tfrac{(-1)^{n+1}\sin(\theta_{2n+1})x^{2n+1}}{(2n+1)!}$$

The error term R_{2n+1} satisfies

$$|R_{2n+1}| = \left| \tfrac{(-1)^{n+1}\sin(\theta_{2n+1})x^{2n+1}}{(2n+1)!} \right| \leq \tfrac{M^{2n+1}}{(2n+1)!}$$

thus it shrinks to 0 as $n \to \infty$. Therefore $\cos x = \sum_{n=0}^{\infty} \tfrac{(-1)^n x^{2n}}{(2n)!}$ is the Maclaurin series for $f(x) = \cos x$

3. In the exercises, you are asked to show that the Maclaurin series for $h(x) = \sin x$ is

$$\sin x = \sum_{n=0}^{\infty} \frac{(-1)^{n+1} x^{2n-1}}{(2n-1)!}$$

4. Let $f(x) = \ln(1+x)$, $x > -1$. Then $f(0) = 0$, $f'(x) = \dfrac{1}{1+x}$ so that $f'(0) = 1$, $f''(x) = -\frac{1}{(1+x)^2}$ which implies that $f''(0) = -1$. You should go on to confirm that $f'''(0) = 2!$, $f^{(4)}(0) = -3!$, in general, $f^{(n)}(0) = (-1)^{n-1}(n-1)!$. Let $x > -1$ and n be a given integer. By Theorem 332, there is a number θ_n between 0 and x such that

$$\ln(1+x) = x - \frac{x^2}{2!} + \frac{(2!)x^3}{3!} - \cdots + \frac{(-1)^{n-1}(n-1)! x^n}{n!} + \frac{(-1)^n n! (\theta_n)^{n+1}}{(n+1)!}$$

This simplifies to

$$\ln(1+x) = x - \frac{x^2}{2} + \frac{x^3}{3} - \frac{x^4}{4} + \cdots + \frac{(-1)^{n-1} x^n}{n} + \frac{(-1)^n n! (\theta_n)^{n+1}}{(n+1)!}$$

The error term R_n satisfies $|R_n| = \frac{|\theta_n|^{n+1}}{n+1} \le \frac{|x|^{n+1}}{n+1}$. If $|x| > 1$ then

$$\lim_{n \to \infty} \frac{|x|^{n+1}}{n+1} = \infty,$$

therefore the error term does not shrink to 0. If $|x| \le 1$

$$\text{then} \lim_{n \to \infty} \frac{|x|^{n+1}}{n+1} = 0$$

so that the error term shrinks to 0, therefore we may write

$$\ln(1+x) = x - \frac{x^2}{2} + \frac{x^3}{3} - \frac{x^4}{4} + \cdots + \frac{(-1)^{n-1} x^n}{n} + \cdots = \sum_{n=1}^{\infty} \frac{(-1)^{n-1} x^n}{n}$$

if $-1 < x \le 1$. (The series does not converge when $x = -1$ even though the error term shrinks to 0 when $x = -1$.)

Exercise 333

1. Determine the Taylor series for:

   (a) $f(x) = \ln x$ at $c = 1$ (b) $g(x) = \cos x$ at $c = \frac{\pi}{2}$.

2. Determine the Maclaurin series for $f(x) = \sin x$.

3. Determine the Maclaurin series for $f(x) = \dfrac{1}{(1-x)}$ and $g(x) = \dfrac{1}{(1+x)}$, $|x| < 1$ and compare to the result of Exercise 11 on page 236.

4. It can be proved that if a series $\sum_{n=0}^{\infty} a_n (x-c)^n$ converges on some interval I, and if we define a function f on I by

$$f(x) = \sum_{n=0}^{\infty} a_n (x-c)^n$$

then f can be differentiated on I and its derivative is given by

$$f'(x) = \sum_{n=1}^{\infty} n a_n (x - c)^{n-1}$$

Furthermore, the series for $(x - c) f(x)$ is given by

$$(x - c) f(x) = \sum_{n=0}^{\infty} a_n (x - c)(x - c)^n = \sum_{n=0}^{\infty} a_n (x - c)^{n+1}$$

You showed that if $|x| < 1$ then $\sum_{n=0}^{\infty} x^n = \dfrac{1}{(1 - x)}$.

(a) By differentiating both sides of $\dfrac{1}{(1 - x)} = \sum_{n=0}^{\infty} x^n$, show that

$$\frac{1}{(1 - x)^2} = \sum_{n=1}^{\infty} n x^{n-1}$$

(b) Multiply both sides of $\dfrac{1}{(1 - x)^2} = \sum_{n=1}^{\infty} n x^{n-1}$ by x and deduce that

$$\frac{x}{(1 - x)^2} = \sum_{n=1}^{\infty} n x^n$$

(c) Use the result in (b) to determine the sum of the series $\sum_{n=1}^{\infty} n^2 x^n$ when $|x| < 1$.

5. Use the Maclaurin series for $g(x) = \dfrac{1}{(1 + x)}$ to deduce the Maclaurin series for $h(y) = \dfrac{1}{(1 + y^2)}$. (Hint: substitute $x = y^2$ in the Maclaurin series for $\dfrac{1}{(1 + x)}$.

6. It can be proved that if a series $\sum_{n=0}^{\infty} a_n (x - c)^n$ converges on some interval I, and if we define a function f on I by

$$f(x) = \sum_{n=0}^{\infty} a_n (x - c)^n$$

then f can be integrated on I and its integral is given by

$$\int_a^b f(x) dx = \sum_{n=0}^{\infty} a_n \int_a^b (x - c)^n dx$$

Consider the Maclaurin series for $\dfrac{1}{(1 + y^2)}$. Integrate on $[0, x]$ to deduce the Maclaurin series for $\arctan x$.

254

7. *Integrate the Maclaurin series for* $\dfrac{1}{(1+y)}$ *on* $[0, x]$ *to deduce the Maclaurin series for* $\ln(1+x)$ *and compare your result to the series we derived directly above.*

8. *Let* $f(x) = \dfrac{1}{\sqrt{1-x}} = (1-x)^{-1/2}$. *Show that* $f'(0) = \dfrac{1}{2}$, $f''(0) = \dfrac{(1)(3)}{2^2}$, $f'''(0) = \dfrac{(1)(3)(5)}{2^3}$, $\ldots$, $f^{(n)}(0) = \dfrac{(1)(3)(5)\cdots(2n-1)}{2^n}$

9. *Use the results of the above exercise to show that the Maclaurin series for* $f(x) = \dfrac{1}{\sqrt{1-x}}$ *is* $1 + \displaystyle\sum_{n=1}^{\infty} \dfrac{(1)(3)(5)\cdots(2n-1)}{(2^n)(n!)} x^n$

10. *Use the Maclaurin series for* $f(x) = \dfrac{1}{\sqrt{1-x}}$ *to deduce the Maclaurin series for* $g(x) = \dfrac{1}{\sqrt{1-x^2}}$ *and then deduce the Maclaurin series for* $h(x) = \arcsin x$.

Chapter 8   APPENDIX

Functions and Function Notation

Statements that one variable quantity is a *function* of another variable quantity are quite common. Here are some:

1. The gas mileage you get out of your car, on a highway, is a function of the speed at which you drive. Roughly, very high speeds or fairly low speeds translate into poor mileages.

2. The number of daylight hours, at a place like Chicago far from the equator, is a function of the day of the year. (Summer days are the most generous, and winter the least.)

3. The grade one gets in course is a function of the time and effort one invests in the course.

These are rather imprecise. In mathematics, when we say that one variable quantity is a function of another variable quantity we mean that their "values" are somehow related in some precise way. Here are two examples:

Example 334 *A child's dosage for a cough syrup is a function of the weight of the child. By this, we mean that every weight of a child may be matched with a precise dosage. The table below, pulled from a bottle of an off-the-counter children's cough syrup, gives such a matching for children weighing between 24 and 95 pounds.*

Weight of child, to the nearest pound	24-35	36-47	48-59	60-71	72-95
Dosage in teaspoons	1	$1\frac{1}{2}$	2	$2\frac{1}{2}$	3

Example 335 *The letter grade one gets in most mathematics courses is a function of the numeric grade, which is a whole number between 0 and 100, one scores in the course. The table below shows how letter grades are typically related to the numeric grades.*

Numerical grade	90 - 100	79 - 89	68 - 78	57 - 67	Below 57
Letter Grade	A	B	C	D	F

In general, whenever one variable is a function of another variable in a precise way, we can construct a table with two rows showing precisely how they are related. For this reason, we define a function as follows:

Definition 336 *A function is a table with two rows, and with the property that every element in the top row is matched with exactly one element in the bottom row.*

The quantity whose values are listed in the first row is called the independent variable. The other quantity is called the dependent variable. Thus in Example 334, the independent variable is the weight of the children and the dependent variable is the dosage.

Note that two or more elements in the top row of a function may be matched with the same element in the bottom row. In Example 334, for instance, the twelve weights 36, 37, ..., 46, 47 in the top row are all matched with the same number $1\frac{1}{2}$ in the bottom. But you cannot have one element, in the top row of a function, matched with two or more elements in bottom row. For example, it would be absurd to prescribe 2 teaspoon to a 49-pound child and also prescribe 3 teaspoons to the same child. Therefore the following table is not a function:

Integers from 0 to 4	0	1	2	3	4
Square root of integer	0	1 or -1	$\sqrt{2}$ or $-\sqrt{2}$	$\sqrt{3}$ or $-\sqrt{3}$	2 or -2

Some New Terms   *Given a function, (i.e. a two-row table that qualifies to be a function), the set of all the elements that may be put in the top row is called the **domain** of the function. The **range** of the function consists of every element that may go in the bottom and can be matched with at least one element of the top row.*

- The domain of the function in Example 334 is the set of all the integers from 24 to 95. Its range is the set of numbers 1, $1\frac{1}{2}$, 2, $2\frac{1}{2}$, 3.

- The domain of the function in Example 335 is the set of integers from 0 to 100. Its range is the set of letters A, B, C, D, F

It is convenient to denote a function by a letter of the alphabet. The common choices are f, g, h, among many.

Example 337 *Denote the function in Example 334 by D, (for dosage). Let x be any number in the domain of D. The number in the bottom row that is matched with x is called the **image** of x under D, or the value of D at x, and it is denoted by $D(x)$, pronounced "D of x". For example, 40 is matched with $1\frac{1}{2}$, therefore $1\frac{1}{2}$ is the image of 40 under D, and we may write*

$$D(40) = 1\tfrac{1}{2}$$

It is a mistake to think of $D(x)$ as the product of D and x because there is no product involved here. Think of $D(x)$ as an abbreviation for the phrase "the dosage for an x pound child". More examples,

$$D(31) = 1, \quad D(27) = 1, \quad D(88) = 3, \quad D(70) = 2\tfrac{1}{2}, \quad D(91) = 3, \quad D(60) = 1\tfrac{1}{2}$$

Example 338 *Denote the function in Example 335 by g, (for grade). A number like 85 is matched with the letter B, therefore we may write $g(85) = B$, which you should consider to be an abbreviation for the phrase "the grade assigned to 85 points is B". The following are more images of g:*

$$g(79) = B, \quad g(88) = B, \quad g(52) = F, \quad g(65) = D, \quad g(92) = A, \quad g(68) = C$$

In the two examples above, we were able to list all the numbers in the domain of each function. This is rarely the case. Here is one example in which we cannot:

Example 339 *A truck rental company charges $40.00 per day plus mileage at the rate of 10 cents per mile. Suppose you rent a truck from this company for one day. The amount of money you will be charged is a function of the number of miles you clock. If, for example, you clock 125.6 miles then you will be charged*

$$40 + \tfrac{10}{100} \times 125.6 = 40 + 12.56 = 52.56 \ dollars$$

It is clearly impossible to give a table listing all the possible mileages you may clock and the corresponding charges because they are infinitely many. One way around this problem is to use an algebraic expression to describe the possible pairs in the table. To this end, note that if you clock x miles (x can be any positive number) then you pay

$$40 + \tfrac{10}{100} \times x = 40 + 0.1x \ dollars$$

In other words, an arbitrary mileage x corresponds to a charge of $40+0.1x$ dollars. Now, to display this function as a table, we list a few pairs with concrete mileages plus a pair with a general mileage x as shown below

Mileage	10	59	125	200	x
Charge in $	41	45.90	52.50	60	$40 + 0.1x$

Formula for a function

Denote the function in Example 339 above, by C (for cost). We pointed out that the cost of x miles is $40 + 0.1x$ dollars, therefore we may write $C(x) = 40 + 0.1x$. Once again, think of $C(x)$ as the cost of x miles, not as C times x. In particular, $C(10) = 41$, and $C(200) = 60$. The equation $C(x) = 40 + 0.1x$ enables us to determine the image of any number x in the domain of C by simply substituting the value of x into the equation. For this reason, we call it the "**formula for C**".

In general, if there is an algebraic equation which we may use to determine the image of any element in the domain of a given function f then we call it a formula for f.

Because a table is so cumbersome to draw, whereas a formula is generally easy to write down, it is customary to define a function by giving its formula. We simply say that "let f be the function with formula $f(x) = \dots$". It is then understood that its domain is the set of all the numbers that can be substituted into the formula.

Example 340 *Let f be the function with formula $f(x) = 6x + 1$. As a table, (which you do not have to give), this function has the following sample pairs*

x	-1	0	2	2.1	x
$f(x)$	-5	1	13	7.6	$6x + 1$

Its domain is the set of all numbers (because any number x can be substituted into the expression $6x + 1$).

Example 341 *Let g be the function with formula $g(x) = \frac{x+1}{x-2}$. As a table, this function has the following sample pairs.*

x	-1	0	1	$3,5$	x
$g(x)$	0	$-\frac{1}{2}$	-2	3	$\frac{x+1}{x-2}$

Note that with the exception of the number 2, all the numbers can be substituted into the expression $\frac{x+1}{x-2}$. Therefore the domain of this function is the set of all the numbers except 2. Using interval notation, the domain may be given as $(-\infty, 2) \cup (2, \infty)$.

Example 342 *Let h be the function with formula $h(x) = \sqrt{x + 5}$. As a table, this function has the following sample pairs.*

x	-4	-3	0	3.1	x
$h(x)$	1	$\sqrt{2}$	$\sqrt{5}$	$\sqrt{8.1}$	$\sqrt{x+5}$

All the numbers that are bigger than or equal to -5 may be substituted into the expression $\sqrt{x+5}$. The numbers smaller than -5 cannot be substituted into $\sqrt{x+5}$ because doing so requires us to find square roots of negative numbers. Therefore the domain of h is the set of all the numbers bigger than or equal to -5, denoted by $[-5, \infty)$.

Example 343 *Let w be the function with formula*

$$w(x) = \frac{\sqrt{4-x}}{x+2}.$$

Clearly, x cannot be -2. Secondly, x cannot be bigger than 4. Therefore the domain of w is the set $(-\infty, -2) \cup (-2, 4]$.

Compositions of functions

We use examples to describe the concept of a composition of functions. For our first example, consider the function of Example 334 on page 255. We denoted it by g. Its domain is the set of integers from 0 to 100. If n is such an integer then $g(n)$ is a letter grade. It is usual for a letter grade to be assigned one of the integers 0, 1, 2, 3, or 4 for the purpose of computing a grade point average. The table below shows how these numbers are typically assigned.

Letter grade	A	B	C	D	F
Points assigned	4	3	2	1	0

(8.1)

Clearly, table (8.1) is a function. Its domain is the set of letters A, B, C, D, F, and its range is the set of integers 0, 1, 2, 3, 4. Denote it by p, (for points). Then $p(A) = 4$, $p(B) = 3$, and so on. Let n be any one of the integers from 0 to 100. Then $g(n)$ is one of the letters A, B, C, D, F, and it may be matched, using table (8.1), with one of the numbers 0, 1, 2, 3, 4. The number matched with $g(n)$ is denoted by $p(g(n))$ or $p \circ g(n)$. The process of determining $g(n)$ followed by $p(g(n))$ is called evaluating the composition $p \circ g$ at n. For example:

- When we determine $g(83)$ to get B then follow it by determining $p(B)$ to get 3 then we have evaluated the composition $p \circ g$ at 83 and we write $p \circ g(83) = 3$.

- When we determine $g(51)$ to get F then follow it by determining $p(F)$ to get 0 then we have evaluated the composition $p \circ g$ at 51 and we write $p \circ g(51) = 0$.

The function whose domain consists of the integers n from 0 to 100 and has values $p(g(n))$ is called the composition of p and g, (in that order). It is denoted by $p \circ g$ or $p \circ g(n)$.

More examples of compositions:

Example 344 *Let f have formula $f(x) = x^2 + 4$ and g have formula $g(x) = 2x + 1$. Then, (the numbers x in the domain are chosen randomly):*

- $g(1) = 2 + 1 = 3$ *and* $f(3) = 9 + 4 = 13$, *therefore* $f \circ g(1) = 13$.

- $g(-2.3) = -4.6 + 1 = -3.6$ *and* $f(-3.6) = 12.96 + 4 = 16.96$, *therefore* $f \circ g(-3.6) = 16.96$.

- $g(5) = 10 + 1 = 11$ *and* $f(11) = 121 + 4 = 125$, *therefore* $f \circ g(5) = 125$.

- *In general, if x is any number then $g(x) = 2x + 1$ and $f(2x + 1) = (2x + 1)^2 + 4 = 4x^2 + 4x + 5$, therefore $f \circ g(x) = 4x^2 + 4x + 5$. This is the formula for $f \circ g$*

Example 345 *Let $h(x) = \frac{2}{x}$ and $f(x) = \frac{3}{x}$, with $x \neq 0$. Thus the domain of these functions is the set of non-zero numbers. We take a sample of values picked randomly and evaluate $f(g(x))$:*

- $f(3) = 1$ *and* $h(1) = 2$, *therefore* $h \circ f(3) = 2$.

- $f(-4) = -\frac{3}{4}$ *and* $h(-\frac{3}{4}) = 2 \cdot (-\frac{4}{3}) = -\frac{8}{3}$, *therefore* $h \circ f(-4) = -\frac{8}{3}$.

- *In general, if $x \neq 0$ then $f(x) = \frac{3}{x}$ and $h(\frac{3}{x}) = 2 \cdot \frac{x}{3} = \frac{2x}{3}$, therefore*

$$h \circ f(x) = \frac{2x}{3}$$

This is the formula for $h \circ f$.

To summarize what we have investigated so far; if f and g are given functions and $g(x)$ is in the domain of f for all possible values of x then the composition of f and g, (in that order), is the function whose value at x is $f(g(x))$. It is denoted and we evaluate a value $f \circ g(x)$ as follows: determine the value of g at x, then determine the value of f at $g(x)$.

For practice, do the following exercise:

Exercise 346

1. *Consider a credit card company that rewards you with 2 points for every dollar you charge to your credit card account. To simplify matters, we assume that your purchases are rounded off to the nearest dollar. You may redeem your points for cash as follows:*

Number of points	Below 2500	2500 − 9999	10000 − 24999	Over 24999
Cash value in dollars	0	0.5% of points	0.8% of points	1% of points

Let $p(x)$ be the number of points gained for purchases totaling x dollars and c be the function represented by the above table.

(a) *Determine the following:*

(i) $c \circ p(14000)$,   (ii) $c \circ p(24000)$,   (iii) $c \circ p(1000)$,   (iv) $c \circ p(35000)$.

(b) *Let x be a positive integer. Complete the following formula for the composition $c \circ p$ by replacing every question mark ? with an appropriate number or expression.*

$$
c \circ p(x) = \begin{cases}
0 & \text{if} & x < 1,250 \\
? & \text{if} & ? \leq x \leq ? \\
\frac{1.6}{100}x & \text{if} & 5,000 \leq x < 12500 \\
? & \text{if} & ? \leq x
\end{cases}
$$

2. *Let $f(x) = x^2 + 3$ and $g(x) = \sqrt{4 - x}$, $x \leq 4$. Determine*

(i) $f \circ g(0)$,   (ii) $f \circ g(-9)$ *and* (iii) $f \circ g(x)$ *when $x \leq 4$.*

Why must $x \leq 4$ in (iii)?

3. *Write each given function as a composition of two simpler functions of your choice. For example, given $f(x) = \frac{1}{\sqrt{3x+5}}$, one may write it as a composition $f(x) = g \circ h(x)$ where $h(x) = 3x + 5$ and $g(x) = \frac{1}{\sqrt{x}}$.*

(a) $f(x) = \sqrt{x^2 - 9} + 7$ (b) $h(x) = 2(x + 1)^3 - 1$

Inverse of a function

Before introducing the inverse of a function, we need to introduce the idea of a one-to-one function. It is a function f that matches distinct elements u and v in its domain with different images $f(u)$ and $f(v)$. Stated differently, f is one-to-one if the only way images $f(u)$ and $f(v)$ can be the same is if u and v are the same.

Example 347 *Consider the function f with formula $f(x) = 3x+1$. If $f(u) = f(v)$ then $3u + 1 = 3v + 1$. But this implies that $3u = 3v$, or $u = v$. Therefore f is one-to-one because the only way $f(u)$ can be equal to $f(v)$ is if $u = v$.*

Example 348 *The function g with formula $g(x) = x^2+1$ is not one-to-one because we can find different numbers in its domain that have the same image. For example, -2 and 2 have the same image 5.*

Turning to inverses, only a one-to-one function can have an inverse. The inverse of such a function f is the table you get when you swap the two rows in the table for f. It is denoted by f^{-1}. Thus the domain of f^{-1} is the range of f and its range is the domain of f.

Example 349 *Consider f with formula $f(x) = 3x + 1$. It is displayed as a table below.*

x	1	2	3	4.5	x
$f(x)$	4	7	10	14.5	$3x + 1$

Its inverse f^{-1} is displayed as a table below.

y	4	7	10	14.5	y
$f^{-1}(y)$	1	2	3	4.5	$\frac{1}{3}(y - 1)$

Its formula is $f^{-1}(y) = \frac{1}{3}(y - 1)$. We chose to use y, instead of x, in the formula for f^{-1} purely for convenience.

Evaluate $f^{-1}(f(2))$ and $f^{-1}(f(4.5))$. Why must $f^{-1}(f(x)) = x$ for all x in the domain of f?

Example 350

x	3	4	5.2	7	x
$g(x)$	5	7	9.4	13	$2x - 1$

has inverse

y	5	7	9.4	13	y
$g^{-1}(y)$	3	4	5.2	7	$\frac{1}{2}(y + 1)$

Evaluate $g(g^{-1}(9.4))$ and $g(g^{-1}(5))$. Why must $g(g^{-1}(y)) = y$ for all y in the range of g?

One way of determining the formula for f^{-1} given the formula for f is to write down the equation $f(x) = y$ then solve for x in terms of y. For example, let f have formula $f(x) = 3x + 1$. Write $3x + 1 = y$ and solve for x. You should get $x = \frac{1}{3}(y - 1)$, therefore

$$f^{-1}(y) = \tfrac{1}{3}(y - 1)$$

Remark 351 *If f has inverse f^{-1} then $f^{-1}(f(x)) = x$ for all x in the domain of f and $f(f^{-1}(y)) = y$ for all y in the range of f.*

Exponential Functions

You must have heard of the living organisms called amoebas. They "reproduce" by splitting into two. In other words, an amoeba divides into two parts and each part becomes an independent living amoeba. Suppose, for the sake of an illustration, each amoeba divides into two every day. If you start with one amoeba then 1 day later there will be 2 of them. Two days later, there will be 4. Three days later there will be 8; four days later there will be 16, and so on. Assume that there is an inexhaustible supply of nutrients for the amoebas and that there are no other creatures to eat them. Then their numbers will change as shown in the following table.

Number of days later	0	1	2	3	4	5	x
Number of amoebas	1	2	4	8	16	32	2^x

Denote the above table by f. Its formula is $f(x) = 2^x$. This time the variable x is an exponent. For this reason, f is called an exponential function. Since we get the images of f by raising 2 to different exponents, 2 is called the base of the exponential function, and we say that f is an exponential function to base 2. The number of amoebas increases with time x. This kind of increase is called an "exponential growth" and we say that the number of amoebas grows exponentially.

Some quantities decrease exponentially with time. An example is the resale value of motor vehicles. For some cars, the value decreases by about $\frac{1}{8}$ every year. This means that at the end of every year, its value drops by $\frac{1}{8}$ of the value it had at the beginning of the year. For instance, if you buy it for $18,000 then one year later, its resale value in dollars, is down to

$$18,000 - 18,000 \times \tfrac{1}{8} = 18,000 \left(1 - \tfrac{1}{8}\right) = 18,000 \left(\tfrac{7}{8}\right) \text{ dollars}$$

Two years later its dollar value will be

$$18,000 \left(\tfrac{7}{8}\right) - 18,000 \left(\tfrac{7}{8}\right) \times \tfrac{1}{8} = 18,000 \left(\tfrac{7}{8}\right) \left(1 - \tfrac{1}{8}\right) = 18,000 \left(\tfrac{7}{8}\right)^2$$

Three years later its dollar value will be

$$8,000 \left(\tfrac{7}{8}\right)^2 - 18,000 \left(\tfrac{7}{8}\right)^2 \times \tfrac{1}{8} = 18,000 \left(\tfrac{7}{8}\right)^2 \left(1 - \tfrac{1}{8}\right) = 18,000 \left(\tfrac{7}{8}\right)^3$$

Following the pattern, the value 4 years later will be $18,000 \left(\frac{7}{8}\right)^4$ dollars. In general, the value x years later is $18,000 \left(\frac{7}{8}\right)^x$ dollars. This information is summarized in the table below.

Number of years later	0	1	2	3	x
Value of car in $	$18,000$	$18,000 \left(\frac{7}{8}\right)$	$18,000 \left(\frac{7}{8}\right)$	$18,000 \left(\frac{7}{8}\right)^3$	$18,000 \left(\frac{7}{8}\right)^x$

Denote this table by g. Its formula is $g(x) = 18000 \left(\frac{7}{8}\right)^x$, and it is an exponential function to base $\left(\frac{7}{8}\right)$. The numbers $18,000 \left(\frac{7}{8}\right)^x$ get smaller as x increases, and we say that the resale value of the car decreases exponentially.

Exponential functions are common in financial mathematics. If, for example, you deposit $1000 in an account that pays 6% annual interest compounded monthly then the amount of money in the account after t months is

$$A(t) = 1000 \left(1 + \frac{0.06}{12}\right)^t = 1000 \left(1.005\right)^t.$$

This is an exponential function to base (1.005).

In general, an exponential function is characterized by having a formula in which the independent variable appears at least once as an exponent of some constant, called the base of the function. The most common ones, (which are on most scientific calculators) are the exponential to base 10 with formula

$$f(x) = 10^x$$

and the exponential to base e, with formula

$$g(x) = e^x.$$

Exercise 352 *A biology experiment begins with 5 thousand cells. The number of cells $N(t)$ left after t minutes is given by the formula*

$$N(t) = 5000e^{-0.598t}$$

Find the number of cells remaining after $t = 5$, 10, 15, and 20 minutes then draw your results on a graph.

Logarithm functions

We need logarithms to solve problems like the following:

The resale value of a car drops by $\frac{1}{8}$ every year. How many years does it take the value of an $18,000 car to drop below $3500?

As we pointed out earlier on, its value after x years is $18000 \left(\frac{7}{8}\right)^x$ dollars. Therefore we must find the smallest integer x such that

$$18000 \left(\frac{7}{8}\right)^x < 3500$$

One way to do this is to find the number t such that $18000 \left(\frac{7}{8}\right)^t = 3500$ then take the smallest integer bigger than t. When we simplify, we find that t must satisfy the equation

$$\left(\tfrac{7}{8}\right)^t = \tfrac{3500}{18000} = \tfrac{7}{36}$$

Therefore we must answer the question: What exponent of $\frac{7}{8}$ equals $\frac{7}{36}$? That exponent is called the logarithm of $\frac{7}{36}$ to base $\frac{7}{8}$.

With this in mind, consider the exponential function f with formula $f(x) = 2^x$. The following is a table of its sample values.

x	-2	-1	0	0.5	1	1.5	2	2.5	x
$f(x)$	0.25	0.5	1	$\sqrt{2}$	2	2.82	4	5.64	2^x

The following is a table of samples of values for its inverse, denoted by $\log_2 x$ and called the logarithm function to base 2. It is obtained by simply swapping the rows for f.

x	0.25	0.5	1	$\sqrt{2}$	2	2.82	4	5.64	x
$f^{-1}(x)$	-2	-1	0	0.5	1	1.5	2	2.5	$\log_2 x$

Thus $\log_2 x$ is the exponent of 2 which equals x. For example, $\log_2 64 = 6$ because $2^6 = 64$, $\log_2 8 = 3$ because $2^3 = 8$, $\log_2 128 = 7$ because $2^7 = 128$, $\log_2 \left(\frac{1}{8}\right) = -3$ because $2^{-3} = \frac{1}{8}$ and $\log_2 \left(\frac{1}{16}\right) = -4$ because $2^{-4} = \frac{1}{16}$

In general, given a positive base b different from 1, the logarithm function to base b is the inverse of the exponential function $f(x) = b^x$. It is denoted by $\log_b x$. Clearly, if x is a positive number then $\log_b x$ is the exponent of b which equals x. Use your knowledge of exponents to complete the table below.

$\log_2 32 =$ $\qquad$ $\log_3 9 =$ $\qquad$ $\log_5 25 =$

$\log_2 \left(\frac{1}{4}\right) =$ $\qquad$ $\log_3 \left(\frac{1}{81}\right) =$ $\qquad$ $\log_5 \left(\frac{1}{5}\right) =$

$\log_2 256 =$ $\qquad$ $\log_3 243 =$ $\qquad$ $\log_5 625 =$

$\log_2 2 =$ $\qquad$ $\log_3 3 =$ $\qquad$ $\log_5 5 =$

$\log_2 1 =$ $\qquad$ $\log_3 1 =$ $\qquad$ $\log_5 1 =$

$\log_2 4 =$ $\qquad$ $\log_3 27 =$ $\qquad$ $\log_5 125 =$

$\log_2 \left(\frac{1}{64}\right) =$ $\qquad$ $\log_3 \left(\frac{1}{3}\right) =$ $\qquad$ $\log_5 \left(\frac{1}{25}\right) =$

The two most common logarithm functions are $\log_{10} x$, (the inverse of 10^x), and $\log_e x$ (the inverse of e^x). Their values may be obtained from most scientific calculators. The symbols $\log_{10}$ and $\log_e x$ are abbreviated to $\log x$ and $\ln x$ respectively. For example, $\log 25$ is an abbreviation for $\log_{10} 25$ and $\ln 100$ is an abbreviation for $\log_e 100$.

Some Properties of Logarithms

Take any positive base $b \neq 1$. Let x and y be positive numbers. There is an exponent u such that $b^u = x$. By definition, u is what we called the logarithm of x to base b and we write $u = \log_b x$. Likewise, there is an exponent v such that $b^v = y$. By definition, $v = \log_b y$. Using rules of exponents yields

$$xy = b^u b^v = b^{u+v}$$

Therefore, the exponent of b that equals xy is $u + v$. In the language of logarithms,

$$\log_b xy = u + v = \log_b x + \log_b y$$

When we divide x by y and use rules of indices we get

$$\frac{x}{y} = \frac{b^u}{b^v} = b^{u-v}$$

Thus the exponent of b that equals $\frac{x}{y}$ is $u - v$. In the language of logarithms,

$$\log_b \left(\frac{x}{y} \right) = u - v = \log_b x - \log_b y$$

Finally, note that

$$\log_b x^2 = \log_b xx = \log_b x + \log_b x = 2 \log_b x$$

$$\log_b x^3 = \log_b xxx = \log_b x + \log_b x + \log_b x = 3 \log_b x$$

$$\log_b x^4 = \log_b xxxx = \log_b x + \log_b x + \log_b x + \log_b x = 4 \log_b x$$

In general, if n is any real number then $\log_b x^n = n \log_b x$

We have derived three useful properties that are used in solving problems involving exponents and logarithms. We state them formally:

Let $b \neq 1$ be a positive number, x and y be any positive numbers, and n be any number. Then

$$\log_b xy = \log_b x + \log_b y \tag{8.2}$$

$$\log_b \left(\frac{x}{y} \right) = \log_b x - \log_b y \tag{8.3}$$

$$\log_b x^n = n \log_b x \tag{8.4}$$

To see property (8.4) in use, consider the problem of determining t such that

$$\left(\tfrac{7}{8} \right)^t = \tfrac{7}{36}$$

The complication here is that the variable is an exponent. We can bring it down by taking logarithms of both sides. We may take logarithms to base 10 or e (because they are on common calculators). Let us take logarithms to base e. The result is

$$\ln\left(\tfrac{7}{8}\right)^t = \ln\tfrac{7}{36}$$

Property (8.4) of logarithms enables us to simplify this to

$$t\ln\left(\tfrac{7}{8}\right) = \ln\left(\tfrac{7}{36}\right) \tag{8.5}$$

Since $\ln\left(\tfrac{7}{8}\right) = -0.134$ and $\ln\left(\tfrac{7}{36}\right) = -1.638$, Equation (8.5) may be written as

$$-0.1335t = -1.638$$

which we may solve to get $t = 1.638 \div 0.1335 = 12.27$. Therefore the answer to the question we posed on page 264 is that the value of the car will fall below \$3500 after the 12th year.

We have to mention the **change of base formula for logarithms**. To this end, suppose a, b, and x are positive numbers and $y = \log_a x$. Thus y is the exponent of a that equals x. We may write this as

$$x = a^y$$

It follows that $\log_b x = \log_b a^y$, and if we apply property (8.4) to the right hand side, we get

$$\log_b x = y \log_b a$$

Therefore $y = (\log_b x) / (\log_b a)$. In other words

$$\log_a x = \frac{\log_b x}{\log_b a}.$$

This is called the change of base formula.

Trigonometric Functions

Trigonometric functions are the right tools to use when describing periodic observations; i.e. observations that keep on recurring at regular time intervals. These are abundant in nature and here are some examples:

- The average day temperatures at places like Chicago that are far from the equator change periodically. They are very low during the winter period. As summer approaches, they rise and reach extreme highs in the middle of summer. They then start going down as summer fades away, and plummet into freezing as winter approaches. The cycle then starts all over.

- Chances are, you have seen a pendulum clock. Its pendulum swings, (at regular time intervals), from one side to the other, then back to the first side, and so on.

- If you pluck a taut string, it moves from side to side very rapidly. This movement is also periodic, although more complicated than the swing of a pendulum.

A quick way of defining the trigonometric sine, cosine and tangent of an angle x is as follows:

Draw a Cartesian coordinate system and denote the origin by O. Now draw a line segment OR making an angle x with the horizontal axis.

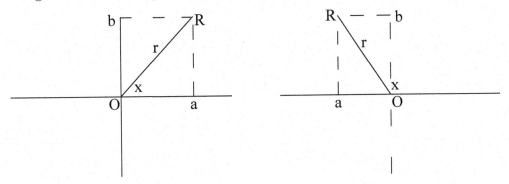

An angle x between 0 and 90°   A different angle x between 90° and 180°

Let R have coordinates (a, b). For convenience, we denote the length of the line segment OR by r, which is a positive number. Then the sine, cosine and tangent of x are denoted by $\sin x$, $\cos x$, and $\tan x$ respectively, and are defined by

$$\sin x = \tfrac{b}{r}, \quad \cos x = \tfrac{a}{r}, \quad \tan x = \tfrac{b}{a} = \tfrac{\sin x}{\cos x}$$

Since $a^2 + b^2 = r^2$, it follows that

$$\tfrac{a^2}{r^2} + \tfrac{b^2}{r^2} = 1.$$

In other words $(\cos x)^2 + (\sin x)^2 = 1$. We write $(\cos x)^2$ and $(\sin x)^2$ more briefly as $\cos^2 x$ and $\sin^2 x$ respectively. Therefore, for any angle x, it is true that

$$\cos^2 x + \sin^2 x = 1$$

This is a very useful identity. You should take steps to remember it.

For some angles x, we may use the geometry of the figure corresponding to x to evaluate $\sin x$, $\cos x$, and $\tan x$. For example:

- If $x = 45°$ and OR has length r then, (because $a = b$), $a^2 + a^2 = r^2$. This implies that $a = \tfrac{r}{\sqrt{2}} = b$. It follows that

$$\sin x = \tfrac{b}{r} = \tfrac{1}{\sqrt{2}} = \tfrac{\sqrt{2}}{2}, \quad \cos x = \tfrac{a}{r} = \tfrac{1}{\sqrt{2}} = \tfrac{\sqrt{2}}{2} \text{ and } \tan x = \tfrac{b}{a} = 1.$$

- If $x = 60°$ and OR has length r then $a = \tfrac{r}{2}$, and $b = \tfrac{\sqrt{3}r}{2}$.

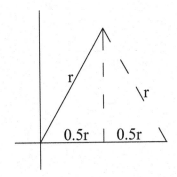

Therefore

$$\sin x = \frac{\sqrt{3}r/2}{r} = \frac{\sqrt{3}}{2}, \quad \cos x = \frac{r/2}{r} = \frac{1}{2} \text{ and } \tan x = \frac{\sqrt{3}r/2}{r/2} = \sqrt{3}.$$

Complete the following table in a similar way

x	$0°$	$30°$	$45°$	$60°$	$90°$	$120°$	$135°$	$150°$	$180°$
$\sin x$			$\frac{\sqrt{2}}{2}$	$\frac{\sqrt{3}}{2}$					
$\cos x$			$\frac{\sqrt{2}}{2}$	$\frac{1}{2}$					
$\tan x$			1	$\sqrt{3}$	Undefined				
x	$210°$	$225°$	$240°$	$270°$	$300°$	$315°$	$330°$	$360°$	$390°$
$\sin x$									
$\cos x$									
$\tan x$				Undefined					

The graphs of $f(x) = \sin x$, $g(x) = \cos x$ and $h(x) = \tan x$ are given below

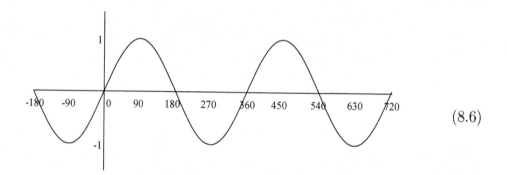

(8.6)

Graph of $f(x) = \sin x$, for $-180° \le x \le 720°$

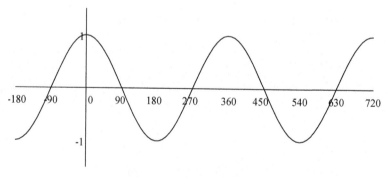

Graph of $g(x) = \cos x$, for $-180° \le x \le 720°$

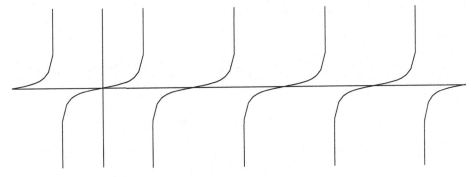

Graph of $h(x) = \tan x$, for $-180° \le x \le 720°$

Clearly, $\sin(x + 360°) = \sin x$ and $\cos(x + 360°) = \cos x$ for all numbers x. We say that the sine and cosine functions are periodic with period $360°$. In the case of $\tan x$, we find that $\tan(x + 180°) = \tan x$ for all numbers x, therefore the tangent function is periodic with period $180°$.

Inverse Trigonometric Functions We wish to define inverses of $\sin x$, $\cos x$, and $\tan x$; but, as their graphs show, they are not one-to-one functions. We have to restrict their domains to achieve this. For $f(x) = \sin x$ and $g(x) = \tan x$, it is standard to restrict their domains to the interval $[-\frac{\pi}{2}, \frac{\pi}{2}]$. Actually, $\tan x$ is not defined at $-\frac{\pi}{2}$ or $\frac{\pi}{2}$ therefore we restrict its domain to $(-\frac{\pi}{2}, \frac{\pi}{2})$. For $h(x) = \cos x$, the standard domain is $[0, \pi]$. Their graphs on these restricted domains are shown below.

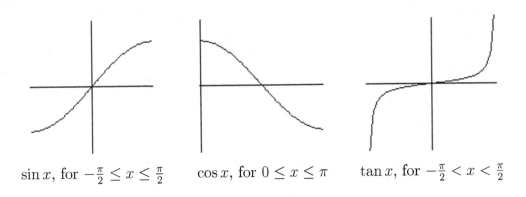

$\sin x$, for $-\frac{\pi}{2} \le x \le \frac{\pi}{2}$ $\qquad$ $\cos x$, for $0 \le x \le \pi$ $\qquad$ $\tan x$, for $-\frac{\pi}{2} < x < \frac{\pi}{2}$

Their sample values on these restricted domains are given below, as are sample values of their inverses. The inverse of $f(x) = \sin x$ is denoted by $\arcsin x$. You may view this as an abbreviation for the phrase "the angle between $-\frac{\pi}{2}$ and $\frac{\pi}{2}$ radians whose sine is x". The inverse of $g(x) = \cos x$ is denoted by $\arccos x$ which you may read as "the angle between $0°$ and π radians whose cosine is x". The inverse of $h(x) = \tan x$ is denoted by $\arctan x$ which you may read as "the angle between $-\frac{\pi}{2}$ and $\frac{\pi}{2}$ radians whose tangent is x".

The following are sample values for $\sin x$:

x	$-\frac{\pi}{2}$	$-\frac{\pi}{3}$	$-\frac{\pi}{4}$	$-\frac{\pi}{6}$	0	$\frac{\pi}{6}$	$\frac{\pi}{4}$	$\frac{\pi}{2}$	$\frac{\pi}{2}$
$\sin x$	-1	$-\frac{\sqrt{3}}{2}$	$-\frac{\sqrt{2}}{2}$	$-\frac{1}{2}$	0	$\frac{1}{2}$	$\frac{\sqrt{2}}{2}$	$\frac{\sqrt{3}}{2}$	1

The following are sample values for arcsin x:

x	-1	$-\frac{\sqrt{3}}{2}$	$-\frac{\sqrt{2}}{2}$	$-\frac{1}{2}$	0	$\frac{1}{2}$	$\frac{\sqrt{2}}{2}$	$\frac{\sqrt{3}}{2}$	1
arcsin x	$-\frac{\pi}{2}$	$-\frac{\pi}{3}$	$-\frac{\pi}{4}$	$-\frac{\pi}{6}$	0	$\frac{\pi}{6}$	$\frac{\pi}{4}$	$\frac{\pi}{2}$	$\frac{\pi}{2}$

A graph of arcsin x is given below, with the dependent variable in radians.

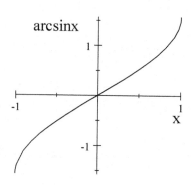

Graph of arcsin x

Sample values for $\cos x$ and its inverse arccos x, plus its graph are given below.

x	0	$\frac{\pi}{6}$	$\frac{\pi}{4}$	$\frac{\pi}{3}$	$\frac{\pi}{2}$	$\frac{2\pi}{3}$	$\frac{3\pi}{4}$	$\frac{5\pi}{6}$	π
$\cos x$	1	$\frac{\sqrt{3}}{2}$	$\frac{\sqrt{2}}{2}$	$\frac{1}{2}$	0	$-\frac{1}{2}$	$-\frac{\sqrt{2}}{2}$	$-\frac{\sqrt{3}}{2}$	-1

x	-1	$-\frac{\sqrt{3}}{2}$	$-\frac{\sqrt{2}}{2}$	$-\frac{1}{2}$	0	$\frac{1}{2}$	$\frac{\sqrt{2}}{2}$	$\frac{\sqrt{3}}{2}$	1
arccos x	π	$\frac{5\pi}{6}$	$\frac{3\pi}{4}$	$\frac{2\pi}{3}$	$\frac{\pi}{2}$	$\frac{\pi}{3}$	$\frac{\pi}{4}$	$\frac{\pi}{6}$	0

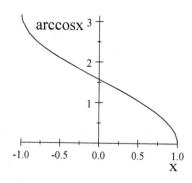

Graph of arccos x

Sample values for $\tan x$ and its inverse are given below.

x	-1.3	$-\frac{\pi}{3}$	$-\frac{\pi}{4}$	$-\frac{\pi}{6}$	0	$\frac{\pi}{6}$	$\frac{\pi}{4}$	$\frac{\pi}{3}$	1.3
$\tan x$	-3.7	$-\sqrt{3}$	-1	$-\frac{\sqrt{3}}{3}$	0	$\frac{\sqrt{3}}{3}$	1	$\sqrt{3}$	3.7

x	-3.7	$-\sqrt{3}$	-1	$-\frac{\sqrt{3}}{3}$	0	$\frac{\sqrt{3}}{3}$	1	$\sqrt{3}$	3.7
arctan x	-1.3	$-\frac{\pi}{3}$	$-\frac{\pi}{4}$	$-\frac{\pi}{6}$	0	$\frac{\pi}{6}$	$\frac{\pi}{4}$	$\frac{\pi}{3}$	1.3

The graph of arctan x, with the dependent variable in radians, is given.

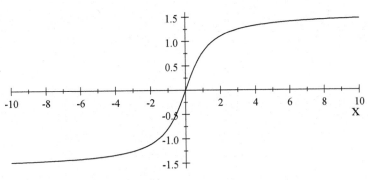

Graph of arctan x

Some Useful Properties of the Sine and Cosine Functions

- *Let x be any angle. Then $\sin(-x) = -\sin x$ and $\cos(-x) = \cos x$.*

There are 4 different possibilities: (i) $0 \le x < 90°$, (ii) $90° \le x < 180°$, (iii) $180° \le x < 270°$, and (iv) $270° \le x < 360°$. We verify the identities for cases (i) and (ii) and leave (iii) and (iv) for the exercises.

(i) $0 \le x < 90°$. The diagram below shows x and $-x$. We assume that the length of OR is 1 unit.

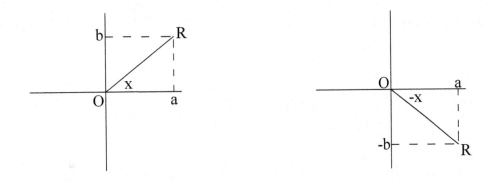

It follows from the definition that $\sin(-x) = -b = -\sin x$ and $\cos(-x) = a = \cos x$.

(ii) $90° \le x < 180°$. The diagram below shows x and $-x$. As before, we have assumed that the length of OR is 1 unit.

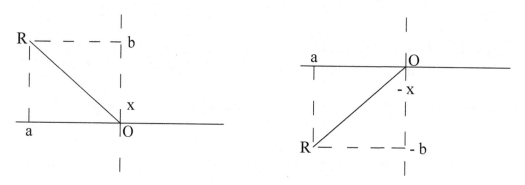

It follows from the definition that $\sin(-x) = -b = -\sin x$ and $\cos(-x) = a = \cos x$.

- *Let x be any angle. Then $\cos(90° - x) = \sin x$.*

Again there are 4 different possibilities; and we verify it for cases (i) and (ii) and leave (iii) and (iv) to you.

(i) $0 \leq x < 90°$. The diagram below shows x and $90° - x$. Again, we have assumed that OR has length 1 unit.

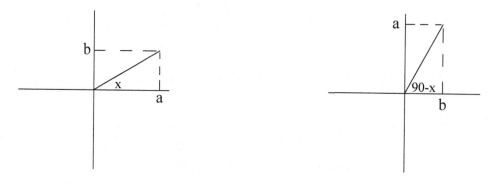

From the definition, $\cos(90° - x) = b = \sin x$.

(ii) $90° \leq x < 180°$. Then $x = 90° + y$ where $0 < y < 90°$, as shown in the left diagram. $90° - x$ is shown in the right diagram. Once again, we have assumed that OR has length 1 unit.

$$90 - x = -y$$

As expected, $\cos(90° - x) = b = \sin x$.

- *Let x and y be any angles. Then*

$$\cos(x+y) = \cos x \cos y - \sin x \sin y \qquad (8.7)$$

To prove this, consider the figures below showing the following angles: angle AOB which we have labelled x, angle AOC which we have labelled y, angle AOD, which happens to be the sum of x and y, and angle FOC which is obtained by giving the figure to its left a clockwise rotation of x units. Consequently, angle AOF must be $-x$ degrees. Each circle has a radius of 1 unit.

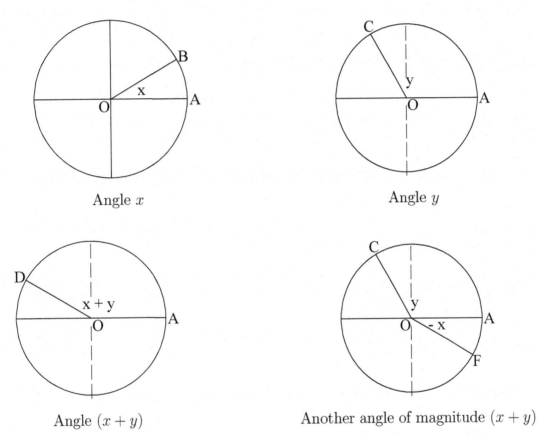

Angle x	Angle y
Angle $(x+y)$	Another angle of magnitude $(x+y)$

The line segments DA and CF have equal lengths because they are facing angles of equal magnitude. The coordinates of D are $(\cos(x+y), \sin(x+y))$ while A has coordinates $(1,0)$. Denote the length of DA by $|DA|$. Then

$$|DA|^2 = [\cos(x+y) - 1]^2 + [\sin(x+y)]^2 = 2 - 2\cos(x+y)$$

(You should work this out in detail.)

The coordinates as C are $(\cos y, \sin y)$, and F has coordinates $(\cos x, -\sin x)$. It follows that

$$|CF|^2 = [\cos y - \cos x]^2 + [\sin y + \sin x]^2 = 2 - 2\cos x \cos y + 2\sin x \sin y$$

(You should also work this out in detail.) Since $|DA|^2 = |CF|^2$, we equate the two expressions:

$$2 - 2\cos(x+y) = 2 - 2\cos x \cos y + 2\sin x \sin y$$

The required identity now follows.

- $\cos(x - y) = \cos x \cos y + \sin x \sin y$

This may be deduced from (8.7) by writing $\cos(x - y)$ as $\cos(x + (-y))$ then use the fact that $\sin(-y) = -\sin y$ and $\cos(-y) = \cos y$. Thus

$$\cos(x - y) = \cos(x + (-y)) = \cos x \cos(-y) - \sin x \sin(-y) = \cos x \cos y + \sin x \sin y$$

- $\sin(x + y) = \sin x \cos y + \cos x \sin y$

This may be deduced from the fact that $\sin A = \cos(90 - A)$ and $\cos(A - B) = \cos A \cos B + \sin A \sin B$ for any angles A and B. In particular,

$$\sin(x + y) = \cos(90 - (x + y)) = \cos((90 - x) - y).$$

If we substitute $A = (90 - x)$ and $B = y$ in $\cos(A - B)$ we get

$$\begin{aligned}\sin(x + y) &= \cos((90 - x) - y) = \cos(90 - x)\cos y + \sin(90 - x)\sin y \\ &= \sin x \cos y + \cos x \sin y\end{aligned}$$

- $\sin(x - y) = \sin x \cos y - \cos x \sin y$

We may deduce this from $\sin(x + y) = \sin x \cos y + \cos x \sin y$ by writing $\sin(x - y)$ as $\sin(x + (-y))$. The result is

$$\sin(x - y) = \sin(x + (-y)) = \sin x \cos(-y) + \cos x \sin(-y) = \sin x \cos y - \cos x \sin y$$

Three more trigonometric functions are defined as follows:

$$\sec x = \frac{1}{\cos x}, \qquad \csc x = \frac{1}{\sin x}, \qquad \cot x = \frac{1}{\tan x} = \frac{\cos x}{\sin x}$$

Exercise 353

1. Verify the identities $\sin(-x) = -\sin x$ and $\cos(-x) = \cos x$ when $180 \leq x < 270$, and when $270 \leq x < 360$.

2. Verify the identity $\cos(90 - x) = \sin x$ when $180 \leq x < 270$, and when $270 \leq x < 360$.

3. Use the identity $\cos^2 x + \sin^2 x = 1$ to show that $1 + \tan^2 x = \sec^2 x$. (Hint: divide both sides of $\cos^2 x + \sin^2 x = 1$ by $\cos^2 x$.)

4. Use the identity $\cos^2 x + \sin^2 x = 1$ to show that $1 + \cot^2 x = \csc^2 x$.

Solutions to some of the problems

Exercise 5 on page 9

Question 1: (a) (i) Since $f'(x) = -\frac{1}{x^2}$, slope of tangent at $(1,1)$ is -1. Its equation is $y = 1 - (x-1) = -x + 2$. (ii) Slope of tangent at $\left(3, \frac{1}{3}\right)$ is $-\frac{1}{9}$. Its equation is $y = \frac{1}{3} - \frac{1}{9}(x-3) = \frac{2}{3} - \frac{1}{9}x$.

(b) (ii) Since $f'(x) = -\frac{1}{2}x^{-\frac{3}{2}}$, slope of tangent at $\left(4, \frac{1}{2}\right)$ is $-\frac{1}{16}$. Its equation is $y = \frac{1}{2} - \frac{1}{16}(x-4) = -\frac{1}{16}x + \frac{3}{4}$.

(c) (iii) Slope of tangent at $\left(a, a^{3/2}\right)$ is $\frac{3}{2}a^{1/2}$. Its equation is $y = \frac{3}{2}a^{1/2}x - \frac{1}{2}a^{3/2}$.

Exercise 13 on page 10

Question 1: $v'(-2) \simeq 3.2$, $v'(-1) = 0$, and $v'(1.5) \simeq 1.2$

Question 5: $\frac{g(x+h)-g(x)}{h} = \frac{-1}{\left(\sqrt{x(x+h)}\right)\left(\sqrt{x+h}+\sqrt{x}\right)}$. $g'(x) = -\frac{1}{2x^{3/2}}$

Question 7: (c) $f(x) = x^{-5}$, $f'(x) = -5x^{-6}$, (e) $f'(x) = \frac{2}{3}x^{-1/3}$, (f) $f'(x) = \frac{7}{4}x^{3/4}$, (g) $f(x) = x^{-4/5}$, $f'(x) = -\frac{4}{5}x^{-9/5}$, (j) $f'(x) = -\frac{7}{2}x^{-9/2}$, (l) $f'(x) = \sqrt{3}x^{\sqrt{3}-1}$

Exercise 19 on page 13

Question 1: (a) $f'(x) = 6x$, (c) $h'(x) = 18x^2 + 5$, (e) $f'(x) = x^2 - 8x$, (h) $u'(x) = -\frac{1}{4}x^{-2} = -\frac{1}{4x^2}$, (j) $v'(x) = \frac{10x^4}{3}$, (l) $u'(x) = \frac{1}{4}x^{-1/2} + \frac{1}{3} - \frac{3}{8}x^{1/2}$, (n) $g'(x) = -\frac{2}{x^2} - \frac{6}{x^3} + \frac{12}{5x^4}$, (q) $h'(x) = \sqrt{x} - \frac{1}{3}x^{-4/5}$, (u) $h'(x) = \frac{9}{7}x^2 + \frac{20}{9}x^4$

Question 2: (b) $\frac{1}{4}x^4$, (d) $\frac{1}{12}x^6$, (f) $-\frac{1}{3}x^{-3}$.

Question 4: Line has slope $\frac{3}{4}$. $f'(x) = 3x^2$. We need x such that $3x^2 = \frac{3}{4}$. Clearly, $x = \pm\frac{1}{2}$. Points are $\left(\frac{1}{2}, \frac{1}{8}\right)$ and $\left(-\frac{1}{2}, -\frac{1}{8}\right)$.

Question 6: $\frac{v(x+h)-v(x)}{h} = \frac{5}{\sqrt{5x+5h+3}+\sqrt{5x+3}}$. $v'(x) = \frac{5}{2\sqrt{5x+3}}$

Question 7: $\frac{u(x+h)-u(x)}{h} = \frac{-3}{\sqrt{3x+1}\sqrt{3x+3h+1}\left(\sqrt{3x+3h+1}+\sqrt{3x+1}\right)}$, $u'(x) = \frac{-3}{2(3x+1)^{3/2}}$.

Slope of tangent at $\left(1, \frac{1}{2}\right)$ is $-\frac{3}{16}$. Its equation is $y = -\frac{3}{16}x + \frac{11}{16}$.

Test your skill - 1 on page 17

Question 1 $f(x+h) = 6 - 4x - 4h + 3x^2 + 6xh + 3h^2$. Slope of secant line is $\frac{f(x+h)-f(x)}{h} = -4 + 6x + 3h$. Slope of tangent at $(x, f(x))$ is $f'(x) = \lim_{h \to 0}(-4 + 6x + 3h) = -4 + 6x$. Slope of tangent at $(1, 5)$ is $f'(1) = 2$, and its equation $y = 2x + 3$

Question 2: (a) $-12x^{11}$, (b) $12 - 48x^{-13}$, (c) $30x^4 - 24x^5$, (d) $3x^3 - \frac{5}{3x^2}$, (e) $16x^{1/3} + \frac{4}{x^{3/2}} - \frac{2}{x^2}$, (f) $-4\pi - 6x^{-1/4}$, (g) $\frac{6}{x^2} - \frac{1}{6}$, (h) $18x^{1/2} + \frac{5}{12x^2}$

Exercise 26 on page 21

Question 1:

c) $f(x) = x^2 + 2x + 8$, $n = \frac{1}{2}$, derivative of $(f(x))^{1/2}$ is
$\frac{1}{2}(x^2 + 2x + 8)^{-\frac{1}{2}}(2x + 2) = (x+1)(x^2 + 2x + 8)^{-\frac{1}{2}}$

e) $f(x) = 3 + \sqrt{x}$, $n = \frac{1}{2}$, derivative of $(f(x))^{1/2}$ is
$\frac{1}{2}(3 + \sqrt{x})^{-1/2}\left(\frac{1}{2}x^{-1/2}\right) = \frac{1}{4}[x(3 + \sqrt{x})]^{-1/2}$

f) $f(x) = x^4 + 2x^2 + 5$, $n = -3$, derivative of $(f(x))^{-3}$ is
$-3(x^4 + 2x^2 + 5)^{-4}(4x^3 + 4x)$

i) $f(x) = \sqrt{2x+1} + x^3$, $n = -2$, derivative of $(f(x))^{-2}$ is

$$-2\left(\sqrt{2x+1}+x^3\right)^{-3}\left((2x+1)^{-1/2}+3x^2\right)$$

Question 2: (a) $f'(x) = 15\left(5x+2\right)^2$, (c) $h'(x) = \frac{3}{4}\left[\frac{1}{2}\left(x^2+x\right)^{-\frac{1}{2}}\right]\left(2x+1\right) = \frac{3(2x+1)}{8\sqrt{x^2+x}}$, (e) $v'(s) = 24\left(2s-3\right)^3$, (g) $f'(y) = \frac{1}{2\sqrt{y}} - \frac{1}{\sqrt{2y+1}}$, (i) $h'(t) = t + \frac{2t}{(2t^2+3)^{3/2}}$, (k) $w'(y) = 6\left(3y-2\right) - 6\left(2y+3\right)^2$

Question 8: $g(x) = \frac{3}{x+1} - \frac{2}{x+3} = 3\left(x+1\right)^{-1} - 2\left(x+3\right)^{-1}$, $g'(x) = \frac{-3}{(x+1)^2} + \frac{2}{(x+3)^2}$

Exercise 27 on page 23.

Question 1: (a) $2x^3 + 25x + c$, (c) $-\frac{1}{4}x^{-4} + c = -\frac{1}{4x^4} + c$, (e) $\frac{3}{4}\pi x^4 + \frac{2}{x^2} + c$, (g) $\frac{3}{4}x^{4/3} + 5x + c$, (i) $-\frac{2}{3}x^{-3/2} + c$.

Question 2: (a) $U(x) = \frac{3}{4}x^{4/3} + 5x + \frac{5}{4}$, (c) $H(x) = -\frac{2}{x} + \frac{2}{x^2} + 5$, (e) $V(x) = \frac{1}{2}x^2 + \frac{1}{2x} + \frac{5}{4}$

Exercise 31 on page 25.

Question 1: (a) $2\left(x+2\right)^{\frac{1}{2}} + c$. (c) $\frac{1}{7}\left(4x-3\right)^{\frac{7}{4}} + c$, (e) $\frac{1}{3}\left(x^2+1\right)^{\frac{3}{2}} + c$.

Question 3: $F(x) = \frac{1}{3}\left(x^2+4\right)^{\frac{3}{2}} + c$. Since its graph passes through $(0,3)$, $3 = \frac{1}{3}\left(4\right)^{\frac{3}{2}} + c = \frac{8}{3} + c$. It follows that $c = \frac{1}{3}$ and $F(x) = \frac{1}{3}\left(x^2+4\right)^{\frac{3}{2}} + \frac{1}{3}$.

Exercise 34 on page 29.

Question 3: If the length of the rectangle is x then its width is $\frac{1}{2}\left(2-x\right)$ and its area is $A(x) = \frac{1}{2}x\left(2-x\right) = x - \frac{1}{2}x^2$. $A'(x) = 1 - x$ and $a'(x) = 0$ when $x = 1$. Maximum area when length is 1 and width is $\frac{1}{2}$.

Question 4: If slope of PQ is m then its equation is $y = mx - 2m + 4$, and the coordinates of P and Q are $(0, 4-2m)$ and $\left(2 - \frac{4}{m}, 0\right)$ respectively. The triangle has area $A(m) = \frac{1}{2}\left(4-2m\right)\left(2-\frac{4}{m}\right) = 8 - \frac{8}{m} - 2m$. $A'(m) = \frac{8}{m^2} - 2$ which is zero when $m = \pm 2$. But slope of PQ must be negative. Minimum area when $m = -2$ and the minimum value is 16.

Test your skills - 2,   (on page 31)
Question 1:

(a) $8 + \frac{6}{5}x^{-3}$ (b) $40x(x^2-1)^3$ (c) $27x\left(3x^2+4\right)^{-1/2}$

(d) $x\left(x^2+2\right)^{-1/2}$ (e) $\frac{2}{3}x\left(x^2+2\right)^{-2/3}$ (f) $1 + 6\left(3x+2\right)^{-2}$

(g) $-2x\left(x^2+1\right)^{-2}$ h) $\frac{2}{5}\left(3x - \frac{2}{3x}\right)^{-3/5}\left(3 + \frac{2}{3x^2}\right)$ (i) $-9(6x+1)^{-3/2}$.

Question 2: $f'(x) = (2x+1)^{-1/2}$, and $f'(4) = \frac{1}{3}$. Normal has slope -3 and equation $y = 15 - 3x$.

Question 3: $2x - \frac{2}{3}\left(3x-1\right)^{3/2}\left(\frac{1}{3}\right) + c = 2x - \frac{2}{9}\left(3x-1\right)^{3/2} + c$.

Question 4: Find the smallest value of $g(x) = x^2 - 8x + 26$. Since $g'(x) = 2x - 8$, slope of tangent to the graph of g is zero when $x = 4$. $g(4) = 10$, therefore minimum distance is $\sqrt{g(4)} = \sqrt{10}$.

Exercise 50 on page 38.

Question 1a: Approximate values are -2, (a point of rel. max.) and 1.5, (a point of rel. min.). The exact values are the roots of the equation $3x^2 + 2x - 8 = (3x-4)(x+2) = 0$, and they are -2 and $\frac{4}{3}$.

Question 2(b): $g'(x) = 3x^2 - 2x - 8 = (3x + 4)(x - 2)$. The critical points are $x_1 = -\frac{4}{3}$ and $x_2 = 2$. Since $g'(-2) = 8$ which is positive, and $g'(1) = -7$ which is negative, x_1 is a point of relative maximum. Since $g'(1)$ is negative but $g'(3)$ is positive, x_2 is a point of relative minimum.

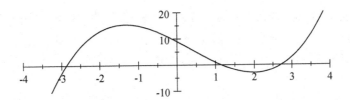

Question 2(d):

$$v'(x) = 5\left(x^3 - 9x\right)^{2/3}\left(x^2 - 3\right) = 5x^{2/3}\left(x - 3\right)^{2/3}\left(x + 3\right)^{2/3}\left(x - \sqrt{3}\right)\left(x + \sqrt{3}\right).$$

The critical points are $x_1 = -3$ (neither a point of relative maximum nor a minimum), $x_2 = -\sqrt{3}$, (a point of relative maximum), $x_3 = 0$ (neither a maximum nor a minimum), $x_4 = \sqrt{3}$ (a point of relative minimum), and $x_5 = 3$ (neither a point of relative maximum nor a minimum). Sketch the graph.

Question 4: In order for the figure drawn inside the parabola to be a rectangle, x and z must satisfy the equation $f(x) = f(z)$. Solve to get $z = a + b - x$. The length and height of the rectangle are $z - x = (a + b - 2x)$ and $f(x) = -(x - a)(x - b)$ respectively. Complete the solution.

Exercise 52 on page 41.

Question 1: $f'(27) = \frac{1}{3}(27)^{-2/3} = \frac{1}{27}$. The equation of the tangent at $(27, 3)$ is $T(x) = 3 + \frac{1}{27}(x - 27)$. $\sqrt[3]{25.2} = f(25.2) \simeq T(25.2) = 3 + \frac{1}{27}(-1.8) \simeq 2.93$. Error is approximately $2.9318 - 2.93 = 0.0018$.

Exercise 55 on page 44.

Question 1(b): Since $g'(x) = \sqrt{5 - x^2} - \frac{x^2}{\sqrt{5 - x^2}}$, $g'(2) = -3$, hence dg_2 has formula $dg_2(h) = -3h$.

Question 3: The volume of a cube with edges of length x is $V(x) = x^3$. The error in measuring x is ± 0.2, (thus the volume of the cube could be as big as $V(6.2)$ and as small as $V(5.8)$). The estimated error in the volume is $\pm V'(6) \cdot 0.2 = \pm 21.6$ and the estimated percentage error is $\pm \frac{21.6 \times 100}{V(6)}\% = \pm 10\%$.

Question 5: $V(r) = 4\pi r^3 / 3$. If the error in r is $\triangle r$ then the estimated error in V is $4\pi r^2 \triangle r$ and the estimated percentage error is $\frac{4\pi r^2 \triangle r \times 100}{\frac{4}{3}\pi r^3}\% = \frac{300 \triangle r}{r}\%$. If this is to be smaller than 1%, then $|\triangle r| < \frac{r}{300}$. This implies that the estimated percentage error in r must be less than $\frac{1}{3}\%$.

Test your skill - 3 on page 45

Question 1: $f'(x) = -8 + 6x - x^2$. Therefore $f'(x) = 0$ when $x^2 - 6x + 8 = 0$. Critical points are 2, (which is a point of relative minimum) and 4, (a point of relative maximum).

Question 2: (a) $g'(x) = (3x + 5)^3$, (b) $h'(x) = \frac{1}{2}(x^2 - x + 5)^{-1/2}(2x - 1)$, (c) $w'(x) = \frac{1}{2}(4 + \sqrt{x})^{-1/2}\left(\frac{1}{2}x^{-1/2}\right) = \frac{1}{4}\left(4x + x\sqrt{x}\right)^{-1/2}$

Question 3: $\frac{2}{3}\left(3+4x^2\right)^{3/2}\left(\frac{1}{8}\right)+c=\frac{1}{12}\left(3+4x^2\right)^{3/2}+c.$

Question 4: Take $c=2$ because $f(2)$ is known exactly and 1.8 is close to 2. Clearly, $f'(2)=\frac{5}{6}$, therefore the equation of tangent at $(2,3)$ is $T(x)=\frac{5}{6}x+\frac{4}{3}$ and $f(1.8)\simeq T(1.8)=\frac{8.5}{3}\simeq 2.83$

Exercise 58 on page 47 Question 1: $x_1=2-\frac{4-2\sqrt{2}}{1+2\sqrt{2}}=1.694.$ To solve the equation, set $\sqrt{x}=y$ to get the quadratic equation $y^2+y-3=0$ with positive solution $y=1.303$, hence $x=1.303^2=1.698.$

Exercise 60 on page 49

Question 1: $x_1=1-\frac{f(1)}{f'(1)}=1-\frac{(-1)}{2.5}=1.4,\ x_2=1.356$

Question 5: $\quad 1.21$

Exercise 68 on page 59

Question 1(a) Slope of chord joining $(-3,-27)$ and $(0,0)$ is 9. We need θ between -3 and 0 such that $3\theta^2=9$. Clearly, $\theta=-\sqrt{3}.$

(c) $\theta=-\frac{3}{4}.$

Question 2: Let $f(x)=x\left(x-1\right)\left(x-2\right)\left(x-3\right)$ and $[a,b]=[0,3]$. There is a number θ_1 between 0 and 1 such that $f'\left(\theta_1\right)=0$. There is a number θ_2 between 1 and 2 such that $f'\left(\theta_2\right)=0$. Finally, there is a number θ_3 between 2 and 3 such that $f'(\theta_3)=0$. These numbers have the required properties.

Test your skills - 4, on page 61

Question 1: (a) $f'(x)=x^3+\frac{7}{2}x^{-3/2}$, (b) $g'(x)=\frac{2}{9}x+2\left(x+1\right)^{-2}$, (c) $h'(x)=\frac{1}{2}x\left(x^2+1\right)^{-3/4}.$

Question 2: $F(x)=\frac{1}{2}\left(2x^2+1\right)^{1/2}$

Question 3: Let $f(x)=x^3-3x+1$. With $x_0=-2$ as an approximate solution, a better approximation, (to 3 decimal places) is $x_1=x_0-\frac{f(x_0)}{f'(x_0)}=-2+\frac{1}{9}=-\frac{17}{9}=-1.889.$ A still better approximation is $x_2=x_1-\frac{f(x_1)}{f'(x_1)}=1.879.$

Question 4:

Length of square	0	1.5	2	3	4.5	x
Volume of box	0	101.25	112	108	60.75	$2x^3-30x^2+108x$

The volume is 0 when $x=0$. It increases as x increases from 0, reaches a maximum value when x is some number c between 2 and 3, then starts decreasing as x increases, and it is back to 0 when x is 6. The number c is a solution of $V'(x)=0$. Thus solve $6x^2-60x+108=0$, or $x^2-10x+18=0$. The solution between 0 and 6 is $x=\dfrac{10-\sqrt{100-4\times 18}}{2}=2.35$ (to 2 dec. places) and $V(2.35)=114.1$ (to 1 decimal place). Therefore, the maximum volume is approximately 114.1 units.

Question 5: When its radius and height are r and h respectively then $2\pi r^2+2\pi rh=400$, therefore $h=\frac{400-2\pi r^2}{2\pi r}$. This implies that its volume is $V(r)=\dfrac{\pi r^2\left(400-2\pi r^2\right)}{2\pi r}=200r-\pi r^3$. Therefore $V'(r)=200-3\pi r^2$. The critical points are $\pm\left(\frac{200}{3\pi}\right)^{1/2}$. Since the radius must be positive and $\left(\frac{200}{3\pi}\right)^{1/2}$ is a point of relative maximum, the maximum volume is $V\left(\left(\frac{200}{3\pi}\right)^{1/2}\right) = 614.2$ cubic units, (to 1 decimal place).

Exercise 86 on page 71

Question 1(a) $\lim_{x\to-2} 2x^3 = -16$, $\lim_{x\to-2} 3x^2 = 12$ and $\lim_{x\to-2} 9 = 9$, therefore $\lim_{x\to-2}(2x^3 + 3x^2 + 9) = 5$.

(c) $\frac{x^2-25}{x^2-6x+5} = \frac{(x-5)(x+5)}{(x-1)(x-5)}$. If $x \neq 5$ then $\frac{x^2-25}{x^2-6x+5} = \frac{(x+5)}{(x-1)}$, therefore $\lim_{x\to5}\frac{x^2-25}{x^2-6x+5} = \lim_{x\to5}\frac{(x+5)}{(x-1)} = \frac{5}{2}$.

(e) Rationalizing the denominator gives $\frac{x+2}{\sqrt{x+6}-2} = \frac{(x+2)\left(\sqrt{x+6}+2\right)}{(x+2)}$. If $x \neq -2$ then $\frac{x+2}{\sqrt{x+6}-2} = \sqrt{x+6} + 2$, therefore $\lim_{x\to-2}\frac{x+2}{\sqrt{x+6}-2} = \sqrt{-2+6} + 2 = 4$

Question 5(b) $\lim_{\theta\to0}\frac{\sin 4\theta}{\theta} = \lim_{\theta\to0} 4\left(\frac{\sin 4\theta}{4\theta}\right) = 4(1) = 4$.

(d) $\lim_{\theta\to0}\frac{\theta - \sin\frac{1}{3}\theta}{2\theta} = \lim_{\theta\to0}\left(\frac{\theta}{2\theta} - \frac{\sin\frac{1}{3}\theta}{2\theta}\right) = \lim_{\theta\to0}\left(\frac{1}{2} - \frac{1}{6}\left(\frac{\sin\frac{1}{3}\theta}{\frac{1}{3}\theta}\right)\right) = \frac{1}{2} - \frac{1}{6} = \frac{1}{3}$

Exercise 103 on page 78

Question 1(a): (i) $\lim_{x\to2^+} f(x) = 2$, (ii) $\lim_{x\to2^-} f(x) = 1$, (iii) $\lim_{x\to-5^+} f(x) = -4$, (iv) $\lim_{x\to-5^-} f(x) = -5$

Question 1(b): For $n \geq 0$, limits are n and $n - 1$ and for $n < 0$ they are $n + 1$ and n respectively.

Question 4(d): $\lim_{x\to n^+} f(x) = n$ and $\lim_{x\to n^-} f(x) = n - 1$

Question 7: $\lim_{x\to0^-} g(x) = \lim_{x\to0^+} g(x) = \infty$

Test your skills - 5, on page 83

Question 1: (a) 51, (b) 2, (c) 126;   Question 2: (a) 2, (b) 4, (c) $-\frac{1}{4}$;   Question 3: 23;   Question 4: Since $\lim_{x\to0}\frac{f(x)}{x}$ exists and is 6 and we know that $\lim_{x\to0} x$ exists and is 0, it follows from the product rule that $\lim_{x\to0}\frac{f(x)}{x} \cdot x$ exists and is $6 \cdot 0 = 0$.

Question 5: $|f(x) - 3| = |x^2 + 2x - 3| = |x + 3|\,|x - 1|$. Consider only numbers x such that $0 < x < 2$. Then $|x + 3| < 5$, hence it suffices to find x such that $5|x - 1| < \varepsilon$. If $|x - 1| < \frac{\varepsilon}{5}$ then $5|x - 1| < \varepsilon$, therefore any $\delta < \min\left\{1, \frac{\varepsilon}{5}\right\}$ will do.

Question 6: $\lim_{x\to-3^+} f(x) = 0$, $\lim_{x\to-3^-} f(x) = 4$, $\lim_{x\to-3} f(x)$ does not exist because the limit from the left is different from the limit to the right. $\lim_{x\to-4} f(x) = 2$.

Question 7: $f(x) = \frac{x-6}{(x-10)(x-6)}$. The function is not continuous at $c = 6$ and $c = 10$. It is not defined at $c = 6$ and $\lim_{x\to10} f(x)$ does not exist.

Question 8: (a) $\lim_{x\to0}\left(\frac{1-\cos x}{x^2}\right)\left(\frac{\sin x}{x}\right) = \frac{1}{2} \cdot 1 = \frac{1}{2}$.

(b) $\lim_{x\to4}\left(\frac{4-x}{\sqrt{x^2+9}-5}\right) = \lim_{x\to4}\left(\frac{4-x}{\sqrt{x^2+9}-5} \cdot \frac{\sqrt{x^2+9}+5}{\sqrt{x^2+9}+5}\right) = \lim_{x\to4}\frac{-\sqrt{x^2+9}+5}{x+4} = -\frac{10}{8} = -\frac{5}{4}$

Exercise 106 on page 86

Question 1 (b) $g'(x) = 4\cos x - 5\sin x$, (d) $u'(x) = -\frac{2}{3}\sin x - \frac{4}{5}\cos x$, (h) $q'(x) = -3e^x$, (k) $f'(x) = 12(2 + 3\sin x)^3 \cos x$, (m) $h'(x) = 3x(4 + 3x^2)^{-1/2}$, (n) $v(x) = -(3x + 2)^{-4}$, (o) $\frac{3}{2}(4 - 3\cos x)^{-1/2}\sin x$, (q) $-8(4x + 3)^{-2}$, (r) $e^x(2e^x + 3)^{-1/2}$

Question 2   (a) $4x - 3e^x + c$ (c) $\frac{1}{4}(x + 1)^4 + c$ (f) $\frac{1}{18a}(ax + 1)^{18} + c$

Question 4:

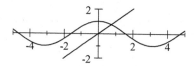

Approximate solution is $x_0 = 1$. First iterate given by Newton's method is $x_1 = x_0 - \dfrac{f(x_0)}{f'(x_0)}$ where $f(x) = x - \cos x$. A calculator gives $x_1 = 0.999850307$. Second iterate is $x_2 = x_1 - \frac{f(x_1)}{f'(x_1)} = 0.999847784$ or 0.9998 to four decimal places.

Exercise 117 on page 90

Question 1 (a) $u'(x) = 8x\cos x - 4x^2\sin x$, (c) $w'(x) = -3x^{-4}e^x + x^{-3}e^x$, (d) $f'(x) = 10x - 3\sin x - 3x\cos x$, (e) $u'(x) = \dfrac{2}{(2 + x)^2}$, (f) $h'(x) = \dfrac{4e^x}{(4 + e^x)^2}$, (n) $s'(x) = 4\csc x - 4x\csc x\cot x - 9x^2$, (o) $v'(x) = -2\cos x\sin x - 2x$, (p) $f'(x) = 4\sec^2 x\tan x - 15x^2$

Question 8: $f(x) = \dfrac{1}{x^n}$ thus $f'(x) = \dfrac{0 \cdot x^n - 1 \cdot nx^{n-1}}{(x^n)^2} = -\dfrac{nx^{n-1}}{x^{2n}} = -nx^{-n-1}$.

Question 9: (a) $F(x) = -4\cos x + 3\sin x - \pi x + c$, (c) $H(x) = 3x - \frac{4}{5}\tan x + c$, (e) $V(x) = -5\csc x - \frac{1}{3}x^3 + c$, (g) $H(x) = e^x + e^{-x} + c$, (h) $F(x) = 2x^3 - 5\sec x + c$, (i) $G(x) = 2e^x - 3e^{-x} + c$.

Exercise 125 on page 94 Question 1

(b) $w'(x) = 6\cos 3x + 6\sin 2x$, (d) $f'(x) = 4x - 2\csc\frac{1}{2}x\cot\frac{1}{2}x$, (f) $h'(x) = \frac{2\pi^2}{3}\sec^2\pi x$, (h) $u'(x) = -2(\csc^2 x)e^{\cot x}$, (j) $g'(x) = 6x\tan\sqrt{x} + \frac{3}{2}x^{3/2}\sec^2\sqrt{x}$, (l) $u'(x) = 6\sec^2 3x\tan 3x$, (n) $w'(x) = \dfrac{6\sin 3x}{(\cos 3x + 2)^{3/2}}$, (p) $g'(x) = -6\cot^2 2x\csc^2 2x + 3$

Exercise 132 on page 96: (i) $42xe^{3x^2}$, (ii) $-\dfrac{2\sec^2 x}{(\tan x)^{3/2}}$, (iii) $3e^{3x}(\tan 3x + \sec^2 3x)$

Exercise 136 on page 98

Question 1 (b) $g'(x) = \frac{1}{x^2}\sec^2\left(\frac{\pi}{3} - \frac{1}{x}\right)$, (c) $h'(x) = \cos 2x - 2x\sin 2x$, (f) $w'(x) = \frac{9}{2}\sec^3 2x\tan 2x$, (g) $f'(x) = 4x^3\cot\left(\frac{1}{x^3}\right) + 3\csc^2\left(\frac{1}{x^3}\right)$, (h) $g'(x) = 2e^{-x^2}\cos 2x - 2e^{-x^2}x\sin 2x$, (j) $\frac{3}{5}\cos\left(\frac{3x-1}{5}\right) - \frac{1}{2\pi}$

Question 4

(a) $F(x) = \frac{2}{5}\sin 2x + c$, (c) $H(x) = \frac{1}{3}e^{3x} + 5x + c$, (e) $W(x) = 3x + \frac{2}{5}\cos 5x + c$, (f) $\frac{2}{3}(x^2 + 4)^{3/2} + c$, (h) $W(x) = \frac{3}{4}e^{4x} - \frac{4}{3}x^3 + c$, (k) $H(x) = (x + 5)^{1/2} + c$

Question 6: $RQ = RC\tan 2x = \tan 2x$. If you join P and C with a line segment then the geometry of the figure implies that angle CPR is x radians. Use triangle CPR to calculate PR. Use $PQ = PR + RQ$ to get the required length then complete the solution.

Exercise 143 on page 103

Question 3(b): $f'(x) = 4x^3 - 8x = 4x\left(x^2 - 2\right) = 4x\left(x - \sqrt{2}\right)\left(x + \sqrt{2}\right)$. Critical points are $x_1 = 0$, $x_2 = -\sqrt{2}$ and $x_3 = \sqrt{2}$. $f''(x) = 12x^2 - 8$. Since $f''(0) = -8 < 0$, x_1 is a point of relative maximum. However, $f''(\sqrt{2}) = f''(-\sqrt{2}) = 16$ which is positive. Therefore x_2 and x_3 are points of relative minimum.

Question 7: $f'(x) = 4x^3 - 32$. Critical point is $x = 2$ and it is a point of relative minimum because $f''(2) = 48$ which is positive.

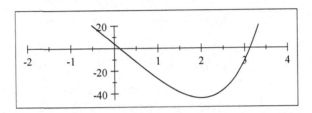

There is a root of the equation between $x = 0$ and $x = 2$. Since $f(0)$, (which is 4), is nearer 0 than $f(2)$, (which is -44), $x_0 = 0$ is a better approximate root. By Newton's method, a better approximate root is $x_1 = x_0 - \frac{f(0)}{f'(0)} = \frac{1}{8} = 0.125$. An even better approximation is $x_2 = \frac{1}{8} - \frac{f\left(\frac{1}{8}\right)}{f'\left(\frac{1}{8}\right)} = 0.125008$. To 3 decimal places, the smallest root is $x = 0.125$.

Question 8: (a) $f^{(n)}(x) = \begin{cases} (-1)^{\frac{1}{2}n} \sin x & \text{if } n \text{ is even} \\ (-1)^{\frac{1}{2}(n-1)} \cos x & \text{if } n \text{ is odd} \end{cases}$,

(b) $g^{(n)}(x) = a^n e^{ax}$, (c) $u^{(n)}(x) = e^x (n + x)$, (d) $h^{(n)}(x) = \frac{(-1)^n n!}{x^{n+1}}$

Question 10 : The third derivative is $f^{(3)}(x)g(x) + 3f^{(2)}(x)g^{(1)}(x) + 3f^{(1)}(x)g^{(2)}(x) + f(x)g^{(3)}(x)$. Note the coefficients 1 3 3 1.

Exercise 147 on page 107: (a) $\frac{3}{5}$,   (e) $\frac{1}{3}$

Exercise 151 on page 108: (a) 0,   (b) 0,   (c) 0.

Exercise 155 on page 111

Question 1: (a) $p_3(x) = 2 + \frac{3}{4}(x - 1) - \frac{9}{64}(x - 1)^2 + \frac{27}{512}(x - 1)^3$. (d) $p_5(x) = x + \frac{1}{3}x^3 + \frac{2}{15}x^5$. (e) $p_n(x) = 1 + x + \frac{1}{2}x^2 + \frac{1}{3!}x^3 + \cdots + \frac{1}{n!}x^n$.

Question 5: $f(0) = a^k$, $f'(x) = k(a + bx)^{k-1}(b)$, hence $f'(0) = ka^{k-1}b$. $f''(x) = k(k - 1)(a + bx)^{k-2}b^2$ and $f''(0) = k(k - 1)a^{k-2}b^2$. In general, if $n \le k$ then
$$f^{(n)}(x) = k(k - 1)\cdots(k + 1 - n)(a + bx)^{k-n}b^n, \text{ and}$$
$$f^{(n)}(0) = k(k - 1)\cdots(k + 1 - n)a^{k-n}b^n$$

Lastly, $f^{(k)}(x) = b^k$ and $f^{(m)}(x) = 0$ for all $m > k$. Now it is easy to write down the Taylor polynomials for f. The kth one is

$$p_k(x) = a^k + ka^{k-1}bx + \frac{1}{2}k(k - 1)a^{k-2}b^2x^2 + \cdots + b^k x^k.$$

Since $f(x) = p_k(x)$, substitute $x = 1$ to get (3.10).

Exercise 161 on page 115
Question 1

(c) $\frac{dy}{dx} = -\sqrt{\frac{y}{x}}$ (d) $\frac{dy}{dx} = \frac{1 - 3x^2 - 2xy}{x^2 + 2y}$ (f) $\frac{dy}{dx} = \frac{-\sin y}{x \cos y + 2}$

Question 3: $\sec^2 y \frac{dy}{dx} - 1 = 0$. Therefore $\frac{dy}{dx} = \frac{1}{\sec^2 y}$. Since $\sec^2 y = 1 + \tan^2 y = 1 + x^2$, we get $\frac{dy}{dx} = \frac{1}{1+x^2}$.

Question 4 (a) $(1)(2) + (1)^2(2)^2 = 6$ hence point is on curve. $\frac{dy}{dx} = \frac{-y-2xy^2}{x+2yx^2}$. Slope of tangent at $(1,2)$ is $\frac{-10}{5} = -2$. Equation of tangent is $y - 2 = -2(x-1)$. (f) $y = -(x-1) = 1 - x$. (h) $\frac{dy}{dx} = \frac{e^y}{3-xe^y}$. Slope of tangent is $\frac{1}{2}$ and its equation is $y = \frac{1}{2}(x-1)$

Question 5: $\frac{dy}{dx} = \frac{2x+2}{3-2y}$. Slope is zero when $2x + 2 = 0$, which implies that $x = -1$. Substituting in $x^2 + 2x + y^2 - 3y = 0$ gives $y^2 - 3y - 1 = 0$ which has solutions $\frac{3\pm\sqrt{3}}{2}$. Tangents are horizontal at $\left(-1, \frac{3+\sqrt{3}}{2}\right)$ and $\left(-1, \frac{3-\sqrt{3}}{2}\right)$. Slope is infinite when $3 - 2y = 0$. Solve for y then determine the corresponding x's.

Question 7 (b): Take derivatives implicitly to get $2y\left(\frac{dy}{dx}\right)x^2 + 2xy^2 + 3 - 4\frac{dy}{dx} = 0$. Take derivatives again:

$$2\left(\frac{dy}{dx}\right)\left(\frac{dy}{dx}\right)x^2 + \left(\frac{d^2y}{dx^2}\right)2yx^2 + 2x\left(2y\left(\frac{dy}{dx}\right)\right) + 2y^2 + 2x\left(2y\left(\frac{dy}{dx}\right)\right) - 4\frac{d^2y}{dx^2} = 0$$

Now solve for $\frac{d^2y}{dx^2}$. The result should be $\left[2x^2\left(\frac{dy}{dx}\right)^2 + 8xy\left(\frac{dy}{dx}\right) + 2y^2\right] \div (4 - 2yx^2)$.

Question 9: Graph crosses x-axis when $y = 0$. This implies that $x = \pm\sqrt{7}$, hence it crosses the axis at $\left(\sqrt{7}, 0\right)$ and $\left(-\sqrt{7}, 0\right)$. Differentiate implicitly and solve for $\frac{dy}{dx}$. The result should be $\frac{dy}{dx} = \frac{-(2x+y)}{x+2y}$. Slope of tangent at $\left(\sqrt{7}, 0\right)$ and $\left(-\sqrt{7}, 0\right)$ is -2, therefore the two lines are parallel. Their equations are $y = -2\left(x - \sqrt{7}\right)$ and $y = -2\left(x + \sqrt{7}\right)$ respectively.

Exercise 166 on page 118

Question 1: Let h be the height and x be the angle of elevation at time t. Then $h = 6\tan x$ and

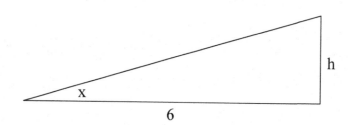

$\frac{dh}{dt} = (6\sec^2 x)\frac{dx}{dt}$. The observation was made when $x = \frac{\pi}{6}$ radians and $\frac{dx}{dt} = r$. Therefore $\frac{dh}{dt} = 6\left(\frac{4}{3}\right)r = 8r$

Question 5: Let height of water in tank at time t be h. Then volume of water in tank at time t is $V = \pi\left(6h^2 - \frac{1}{3}h^3\right)$, and $\frac{dV}{dt} = \pi(12h - h^2)\frac{dh}{dt}$. Therefore $\frac{dh}{dt} = \frac{1}{\pi(12h-h^2)}\frac{dV}{dt}$. We are given that $\frac{dV}{dt} = \frac{3}{100}$ cubic meters per minute. It follows that when $h = 4$, then $\frac{dh}{dt} = \frac{3}{100\pi(48-16)} = \frac{3}{3200\pi}$ meters per minute.

Test your skills - 6, on page 119

Question 1: (a) $16x^3 + 6x + 5$, (b) $4e^x + x^{-1/2} - 4$, (c) $-4x^{-3} + 6$, (d) $3\cos x + 2\sin x + 4x^{-2}$, (e) $5xe^{3x}(2 + 3x)$, (f) $\frac{1}{(\cos x - 2\sin x)^2}$, (g) $6e^{4x}(2\tan 2x + \sec^2 2x)$, (h)

$3\sin x+3x\cos x-8xe^x-4x^2e^x$, (i) $2e^x\left(x-\frac{1}{x}\right)^3\left(x+4-\frac{1}{x}+\frac{4}{x^2}\right)$, (j) $-20\csc 4x\cot 4x-\frac{3}{2}\csc^2 2x$, (k) $(2x+3x^2\cos 3x)\,e^{\sin 3x}$, (l) $8x\cos\left(\frac{2}{x}\right)+8\sin\left(\frac{2}{x}\right)$.

Question 2 (a) $F(x)=\frac{4}{3}\tan 3x+x+c$, (b) $G(x)=-\frac{3}{2}\sin\left(\frac{8-2x}{3}\right)-\frac{9}{2}x^2+c$, (c) $H(x)=\frac{1}{6}\left(1+x^4\right)^{3/2}+c$, (d) $U(x)=-\frac{1}{27}\left(1-3x\right)^9+c$, (e) $W(x)=\frac{1}{3}\left(1+\tan x\right)^3+c$, (f) $V(x)=\frac{2}{3}\left(\sin x\right)^{3/2}+c$.

Question 3: (a) $f'(x)=4x^2-9x^{-4}$, $f''(x)=8x+36x^{-5}$, (b) $v'(x)=-3\sin x-3x\cos x$, $v''(x)=-6\cos x+3x\sin x$, (c) $w'(x)=3e^{2x}\left(2\sin 3x+3\cos 3x\right)$, $w''(x)=e^{2x}\left(-15\sin 3x+36\cos 3x\right)$.

Question 4: Let the length of the rectangle be x units. Then its width is $2\left(\sqrt{1-\frac{1}{4}x^2}\right)=\sqrt{4-x^2}$ units. Its area is $A(x)=x\sqrt{4-x^2}$. Clearly, $0\le x\le 2$. When $x=0$, the area is 0. As x increases, the area increases, but must start decreasing at some point, because it is back to 0 when $x=2$. The number that gives the maximum area is a critical point of $A(x)$. Since $A'(x)=\frac{4-2x^2}{\sqrt{4-x^2}}$, its critical points are the solutions of $4-2x^2=0$. The critical point in the interval $[0,2]$ is $x=\sqrt{2}$. Maximum area is $\sqrt{2}\cdot\sqrt{2}=2$ square units

Question 5: Differentiating implicitly gives $2x-2y-2x\frac{dy}{dx}+3y^2\frac{dy}{dx}=0$. Solving gives $\frac{dy}{dx}=\frac{2(y-x)}{3y^2-2x}$. At $(1,-1)$, $\frac{dy}{dx}=\frac{-4}{1}=-4$, which is the slope of the tangent at the point. Its equation is $y=3-4x$.

Question 6: $g'(x)=3x^2+12x+9=3\left(x^2+4x+3\right)=3\left(x+1\right)\left(x+3\right)$. The critical points are -1 and -3. $g''(x)=6x+12$ and $g''(-1)=6$, which is positive, hence -1 is a point of relative minimum; and $g''(-3)=-6$ which is negative, therefore -3 is a point of relative maximum. Sketch the graph. If we use $x_0=0$ as an approximate solution then, by Newton's method, a better approximate solution is $x_1=x_0-\frac{g(0)}{g'(0)}=-\frac{1}{9}$.

Question 7: The speed of the stone at time t is $s'(t)=128-32t$ per second. It reaches the maximum height at the moment the speed is zero. The time is given by $128-32t=0$, hence $t=4$. Maximum height is $s(4)=256$ feet.

Exercise 167 on page 123

Question 1: (a) $f'(x)=2x\ln x+x$, (c) $u'(x)=\frac{e^x}{x}+(\ln x)\,e^x$, (d) $h'(x)=\frac{-4}{\sqrt{1-\frac{x^2}{9}}}\cdot\frac{1}{3}=\frac{-4}{\sqrt{9-x^2}}$, (g) $f'(x)=\frac{4}{\sqrt{1-36x^2}}$, (j) $v'(x)=\frac{1}{1+\frac{1}{x^2}}\left(-\frac{1}{x^2}\right)=\frac{-1}{1+x^2}$, (l) $u'(x)=\frac{2x}{1+x^4}$, (n) $h'(x)=\cot x$, (p) $w'(x)=\frac{b}{1+bx}$, (r) $z'(x)=\frac{1}{x}$, (t) $f'(x)=\frac{ab}{b^2+a^2x^2}$, (v) $f'(x)=\frac{a}{\sqrt{1-(ax+b)^2}}$, (x) $z'(x)=\frac{a}{\sqrt{b^2-a^2x^2}}$

Question 4: $h'(x)=-\csc^2\left(\arcsin x\right)\frac{1}{\sqrt{1-x^2}}=\frac{-1}{\sin^2\left(\arcsin x\right)}\cdot\frac{1}{\sqrt{1-x^2}}=\frac{-1}{\sqrt{x^4-x^6}}$

Question 9: By the Mean Value Theorem, there is a number c between 1 and x such that $f(x)-f(1)=(x-1)f'(c)$. Since $f'(c)=\frac{nc^n-1}{c}$ which is positive, $(x-1)f'(c)$ must be positive. ...

Exercise 171 on page 126

Question 2: (b) $v'(x)=\frac{4x}{(\ln 10)(3+2x^2)}$, (d) $f'(x)=2\left(\ln x+1\right)x^{2x}$, (g) $g'(x)=x\left(\ln x+\frac{1}{2}\right)\left(\sqrt{x}\right)^{x^2}$

Question 9: (a) $F(x)=3\ln x+c$ (c) $H(x)=\frac{3}{2}\arcsin x+c$ (e) $U(x)=\frac{5}{2}\arctan 2x+c$

Exercise 172 on page 129

Question 7: (a) $f'(x) = 12\cosh 3x$, (c) $h'(x) = (3 + \cosh 2x)^{-1/2}\sinh 2x$, (f) $w'(x) = \frac{2\operatorname{sech}^2\sqrt{x}}{\sqrt{x}}$

Question 8: (a) $F(x) = \frac{3}{4}\cosh 4x + \frac{4}{3}\sinh 3x + c$, (d) $U(x) = \frac{5}{2}e^{2x} - 2\coth\frac{1}{2}x + c$

Test your skills - 7, on page 132

Question 1: (a) $\frac{1}{9}x^{-2/3} + \frac{1}{16}x^{-5/4}$, (b) $10xe^{3x} + 15x^2e^{3x}$, (c) $\frac{45x}{\sqrt{1-9x^2}}$, (d) $-\operatorname{sech}^2\frac{3x}{2}$, (e) $\frac{4e^{2x}+e^{3x}}{(2+e^x)^2}$ (f) $2(10^{2x})[(\ln 10)\cosh 2x + \sinh 2x]$, (g) $\frac{12}{1+9x^2} - \pi$, (h) $-\frac{3}{(3x+1)\ln 5}$, (i) $10e^{2x}\sin x\cosh 3x + 5e^{2x}\cos x\cosh 3x + 15e^{2x}\sin x\sinh 3x$, (j) $20\sinh 6x\cosh 6x - 4x$.

Question 2: $f'(x) = \frac{\cos^2 x + 4\cos x\sin x + 4\sin^2 x + \sin^2 x - 4\sin x\cos x + 4\cos^2 x}{(\cos x + 2\sin x)^2} = \frac{5(\sin^2 x + \cos^2 x)}{(\cos x + 2\sin x)^2} = \frac{5}{(\cos x + 2\sin x)^2}$

Question 3: $f'(x) = 3a\sec 3x\tan 3x$, $f''(x) = 9a\sec 3x(\tan^2 3x + \sec^2 3x)$.

Question 5: $g'(x) = e^x + xe^x = e^x(1+x)$. Since e^x cannot be zero, $g'(x) = 0$ when $x + 1 = 0$, i.e. when $x = -1$. $g''(x) = e^x(2+x)$ which is positive when $x = -1$, therefore the critical point is a point of relative minimum. Sketch the graph.

Question 6: (a) $\lim_{x\to 0}\frac{x+2\sin x}{x\cos x} = \lim_{x\to 0}\frac{1+2\cos x}{\cos x - x\sin x} = \frac{3}{1} = 3$, (b) Consider $\ln\left[(1+x)^{1/x}\right] = \frac{\ln(1+x)}{x}$. By L'Hopital's rule $\lim_{x\to 0}\frac{\ln(1+x)}{x} = \lim_{x\to 0}\frac{1}{1+x} = 1$. It follows that $\lim_{x\to 0}(1+x)^{1/x} = e^1 = e$, (c) $\lim_{x\to\pi/4}\frac{1-\tan x}{\sin x - \cos x} = \lim_{x\to\pi/4}\frac{-\sec^2 x}{\cos x + \sin x} = -\frac{1}{\sqrt{2}}$.

Question 7: Differentiating implicitly gives $2x + y^2 + 2xy\frac{dy}{dx} - \frac{dy}{dx} = 0$. Therefore $\frac{dy}{dx} = \frac{2x+y^2}{1-2xy}$. At $(2,1)$, slope of tangent is $\frac{2(2)+1^2}{1-2(2)(1)} = -\frac{5}{3}$. Therefore equation of tangent is $y = 1 - \frac{5}{3}(x-2) = -\frac{5}{3}x + \frac{13}{3}$.

Question 8: Taking derivatives successively gives $f'(x) = (1+2x)^{-1/2}$, $f''(x) = -(1+2x)^{-3/2}$ and $f'''(x) = 3(1+2x)^{-5/2}$. Therefore $f(4) = 3$, $f'(4) = \frac{1}{3}$, $f''(4) = -\frac{1}{27}$, and $f'''(4) = \frac{1}{81}$. The third order polynomial is $p_3(x) = 3 + \frac{1}{3}(x-4) - \frac{1}{54}(x-4)^2 + \frac{1}{486}(x-4)^3$. Since $1+2x = 7.8$ when $x = 3.4$, it follows that $\sqrt{7.8} = f(3.4) \simeq p_3(3.4) = 3 + \frac{1}{3}(-0.6) - \frac{1}{54}(-0.6)^2 + \frac{1}{486}(-0.6)^3 = 2.79289$

Question 9: $f'(x) = \frac{1}{x+\sqrt{a^2+x^2}}\left[1 + \frac{x}{\sqrt{a^2+x^2}}\right] = \frac{1}{x+\sqrt{a^2+x^2}} \cdot \frac{x+\sqrt{a^2+x^2}}{\sqrt{a^2+x^2}} = \frac{1}{\sqrt{a^2+x^2}}$

Question 10: $F(x) = \frac{1}{2\cos^2 x} + c$

Exercise 179 on page 138

Question 1: $F(x) = 2x + \frac{1}{4}x^4$. Area $= 2 + \frac{1}{4} - 0 = \frac{9}{4}$

Question 2: (a) $F(x) = \frac{x^4}{4} + 2x$, Area $= 4 + 4 - (\frac{1}{4} - 2) = \frac{39}{4}$, (c) $F(x) = -\cos x$, Area $= 1 + 1 = 2$, (g) $F(x) = \frac{1}{2}x^2 + \frac{1}{x}$, Area $= \frac{25}{2} + \frac{1}{5} - (\frac{1}{2} + 1) = \frac{55}{2}$, (i) $F(x) = \frac{x^3}{3} + x^2 + 5x$, Area $= 33 - \frac{19}{5} - 4 = \frac{68}{3}$.

Exercise 185 on page 145. Question 1

1. $\int\sqrt{x}\,dx = \int x^{1/2}\,dx = \frac{2}{3}x^{3/2} + c$ 3. $\int\frac{1}{x^4}\,dx = \int x^{-4}\,dx = \frac{1}{-3}x^{-3} + c$

5. If $n \neq -1$ then $\int x^n\,dx = \frac{x^{n+1}}{n+1} + c$ 7. $\int e^{4x}\,dx = \frac{1}{4}e^{4x} + c$

9. $\int e^{-x}\,dx = -e^{-x} + c$ 11. $\int\left(\pi^2 + \frac{8}{x}\right)dx = \pi^2 x + 8\ln x + c$

13. $\int \frac{3}{x^2+1}dx = 3\arctan x + c$ 15. $\int \left(\sqrt{x} - \sqrt{2}\right) dx = \frac{2}{3}x^{3/2} - \sqrt{2}x + c$

18. $\int \left(1 + x^{3/2}\right) dx = x + \frac{2}{5}x^{5/2} + c$ 19. $\int \left(\pi x - \sec^2 x\right) dx = \frac{\pi}{2}x^2 - \tan x + c$

22. $\int \left(\frac{2}{\pi x^2} - 3\right) dx = -\frac{2}{\pi}x^{-1} - 3x + c$ 23. $\int \left(2x + 1\right)^2 dx = \frac{1}{6}\left(2x + 1\right)^3 + c$

25. $\frac{1}{60}\left(4x^3 + 1\right)^5 + c$ 27. $\frac{1}{12}\left(8x + 1\right)^{3/2} + c$

29. $\frac{2}{9}\left(x^3 + 1\right)^{3/2} + c$ 31. $\frac{1}{8}\left(8x^2 + 1\right)^{1/2} + c$

33. $\frac{1}{18}\left(2x - 3\right)^9 + c$ 35. $\frac{2}{27}\left(\frac{x^3}{2} - 3\right)^9 + c$

37. $\int \sin 5x \, dx = -\frac{1}{5}\cos 5x + c$ 39. $\int \sin(\frac{2}{3}x + 1)dx = -\frac{3}{2}\cos\left(\frac{2}{3}x + 1\right) + c$

41. $\int x \sec^2\left(x^2\right) dx = \frac{1}{2}\tan\left(x^2\right) + c$ 43. $\int x^3 \sec^2\left(x^4\right) dx = \frac{1}{4}\tan\left(x^4\right) + c$

45. $\int x e^{x^2} dx = \frac{1}{2}e^{x^2} + c$ 47) $\int x^3 e^{x^4} dx = \frac{1}{4}e^{x^4} + c$

Exercise 200 on page 152

Question 1: a) $\int_{-1}^{2} \frac{x}{\sqrt{x+2}}dx = \int_{1}^{4} \left(u^{1/2} - 2u^{-1/2}\right) du = \left[\frac{2}{3}u^{3/2} - 4u^{1/2}\right]_{1}^{4} = \frac{2}{3}$

(c) $\int_{5}^{10} \frac{x+1}{\sqrt{x-1}}dx = \int_{4}^{9} \left(u^{1/2} + 2u^{-1/2}\right) = \left[\frac{2}{3}u^{3/2} + 4u^{1/2}\right]_{4}^{9} = 16\frac{2}{3}$

(e) $\int_{0}^{\pi/3} (\tan x)(\sec x)^{3/2} dx = \int_{1}^{2} \sqrt{u}\,du = \frac{2}{3}\left(2^{3/2} - 1\right)$

(g) $\int_{0}^{1} \frac{1-\sqrt{x}}{1+\sqrt{x}}dx = 2\int_{1}^{2} \frac{(2-u)(u-1)}{u}du = 2\int_{1}^{2} \left(3 - u - \frac{2}{u}\right) du = 3 - 4\ln 2$

(i) $\int \frac{1}{\sqrt{1-3x^2}}dx = \frac{1}{\sqrt{3}} \int \frac{1}{\sqrt{1-u^2}}du = \frac{1}{\sqrt{3}}\arcsin u + c = \frac{1}{\sqrt{3}}\arcsin \sqrt{3}x + c$

(o) $\int \frac{1}{1+5x^2}dx = \frac{1}{\sqrt{5}}\arctan\left(\sqrt{5}x\right) + c$

Exercise 206 on page 156

Question 1: (b) Write as $-\int \frac{\cos x - \sin x}{\sin x + \cos x}dx = -\ln |\sin x + \cos x| + c$

(f) $\int \frac{2x+3}{x^2+2x+2}dx = \int \frac{(2x+2)}{x^2+2x+2}dx + \int \frac{1}{1+(x+1)^2}dx = \ln |x^2 + 2x + 2| + \arctan\left(x + 1\right) + c$

Question 3:

(a) $\int \frac{\cos x}{\sin x + 2\cos x}dx = \int \frac{2}{5}dx + \frac{1}{5}\int \frac{\cos x - 2\sin x}{\sin x + 2\cos x}dx = \frac{2}{5}x + \frac{1}{5}\ln |\sin x + 2\cos x| + c$

(f) $\int \frac{e^x + 3e^{-x}}{e^x + e^{-x}}dx = \int 2dx - \int \frac{e^x - e^{-x}}{e^x + e^{-x}}dx = 2x - \ln |e^x + e^{-x}| + c$

Exercise 210 on page 160

Question 1: $\int x \cos x \, dx = x \sin x - \int \sin x \, dx = x \sin x + \cos x + c$

Question 7: Take $g(x) = \arctan x$ and $\frac{df}{dx} = x$. Then

$$\int x \arctan x \, dx = \frac{1}{2}x^2 \arctan x - \frac{1}{2}\int \frac{x^2}{1+x^2}dx = \frac{1}{2}x^2 \arctan x - \frac{1}{2}\int \frac{1+x^2-1}{1+x^2}dx$$

$$= \frac{1}{2}x^2 \arctan x - \frac{1}{2}\int \left(1 - \frac{1}{1+x^2}\right) dx = \frac{1}{2}\left(x^2 + 1\right) \arctan x - \frac{1}{2}x + c$$

Question 9: Let $g(x) = x^n$ and $\frac{df}{dx} = e^{bx}$. Then $f(x) = \frac{1}{b}e^{bx}$ and $\frac{dg}{dx} = nx^{n-1}$, therefore

$$\int x^n e^{bx} = \frac{x^n e^{bx}}{b} - \frac{n}{b}\int x^{n-1}e^{bx}dx.$$

Exercise 219 on page 166:

Question 1: (a) Write $\cos^3 x$ as $(1 - \sin^2 x)\cos x$ then use the substitution $u = \sin x$ to get

$$\int \sin^3 x \cos^3 x dx = \int (u^3 - u^5)\, du = \frac{u^4}{4} - \frac{u^6}{6} + c = \frac{\sin^4 x}{4} - \frac{\sin^6 x}{6} + c.$$

(b) $\sin^2 x \cos^2 x = \frac{1}{4}(1 - \cos 2x)(1 + \cos 2x) = \frac{1}{4}(1 - \cos^2 2x) = \frac{1}{8}(1 - \cos 4x)$, therefore $\int \sin^2 x \cos^2 x dx = \frac{1}{8}x - \frac{1}{32}\sin 4x + c$

(f) $\int \sec^4 x \tan x dx = \int (\sec^3 x)\sec x \tan x dx$. Let $u = \sec x$. Then

$$\int \sec^4 x \tan x dx = \int u^3 du = \frac{1}{4}u^4 + c = \frac{1}{4}\sec^4 x + c.$$

(i) Write $\sec^4 x \tan^4 x$ as $(\sec^2 x \tan^4 x)\sec^2 x$. Then

$$\int \sec^4 x \tan^4 x dx = \int (\tan^4 x + \tan^6 x)\sec^2 x dx = \frac{1}{5}\tan^5 x + \frac{1}{7}\tan^7 x + c.$$

(k) $\int \sec^6 x dx = \int \sec^4 x \sec^2 x dx = \int (1 + \tan^2 x)^2 \sec^2 x dx$. Let $u = \tan x$. Then

$$\int (1 + \tan^2 x)^2 \sec^2 x dx = \int (1 + 2u^2 + u^4)\, du = \tan x + \frac{2}{3}\tan^3 x + \frac{1}{5}\tan^5 x + c$$

Alternatively, use a reduction formula.

(n) Let $x = \sin u$. Then

$$\int \frac{x^2}{\sqrt{1-x^2}}dx = \int \sin^2 u du = \frac{1}{2}u - \frac{1}{4}\sin 2u = \frac{1}{2}\left(\arcsin x - x\sqrt{1 - x^2}\right) + c.$$

(q) Let $x = \tan u$. Then

$$\int \frac{1}{(1+x^2)^2}dx = \int \frac{\sec^2 u}{\sec^4 u}du = \int \cos^2 u du = \frac{1}{2}\left(\arctan x + \frac{x}{1+x^2}\right) + c$$

Exercise 223 on page 169

Question 1: (a) $\frac{x}{(x-1)(x+2)} = \frac{1}{3(x-1)} + \frac{2}{3(x+2)}$. Therefore

$$\int \frac{x dx}{(x-1)(x+2)} = \frac{1}{3}\int \frac{1}{x-1}dx + \frac{2}{3}\int \frac{1}{x+2}dx = \frac{1}{3}\ln (x - 1)(x + 2)^2 + c$$

(c) $\int \frac{dx}{x^2(x+1)} = \int \left(\frac{1}{x^2} + \frac{1}{x+1} - \frac{1}{x}\right)dx = -\frac{1}{x} + \ln \left|\frac{x+1}{x}\right| + c$

(e) $\int \frac{(x-2)dx}{x^2(x-1)^2} = \int \left[\frac{-3}{x} - \frac{2}{x^2} + \frac{3}{x-1} - \frac{1}{(x-1)^2}\right]dx$

$$= -3\ln x + \frac{2}{x} + 3\ln |x - 1| + \frac{1}{x-1} + c$$

(g) The substitution $u = \sqrt{x}$ gives $\int \frac{dx}{x + \sqrt{x} - 2} = \int \frac{2u\,du}{u^2 + u - 2}$. Splitting the integrand into partial fractions gives

$$\int \frac{2u\,du}{(u-1)(u+2)} = \frac{2}{3} \int \frac{1}{u-1}\,du + \frac{4}{3} \int \frac{1}{u+2}\,du = \frac{2}{3} \ln \left(\sqrt{x} - 1 \right) \left(\sqrt{x} + 2 \right)^2 + c$$

Exercise 227 on page 171,
Question 1: (a) Let $x = \sinh u$ Then $dx = \cosh u\,du$ and the integral becomes

$$\int \frac{\sinh^2 u}{\cosh u} \cdot \cosh u\,du = \frac{1}{2} \int (\cosh 2u - 1)\,du = \frac{1}{4} \sinh 2u - \frac{1}{2} u + c$$

Since $\frac{1}{4} \sinh 2u - \frac{1}{2} u + c = \frac{1}{2} \sinh u \cosh u - \frac{1}{2} u + c$, use 3.21, 3.22 and 3.23 to get

$$\int \frac{\sinh^2 u}{\cosh u} \cdot \cosh u\,du = \frac{1}{2} x \sqrt{x^2 + 1} - \frac{1}{2} \ln \left(x + \sqrt{x^2 + 1} \right) + c.$$

(c) Let $x = \cosh u$. Then $dx = \sinh u\,du$ and the integral becomes

$$\int \frac{\cosh^3 u}{\sinh u} \cdot \sinh u\,du \;=\; \int \cosh^3 u\,du = \int \left(\sinh^2 u + 1 \right) \cosh u\,du$$
$$\;=\; \frac{\sinh^3 u}{3} + \sinh u + c = \frac{1}{3} \left(x^2 - 1 \right)^{3/2} + \left(x^2 - 1 \right)^{1/2} + c.$$

Test your skills - 8, on page 171
Question 1: (a) $-\frac{3}{4} \cos x - 2 \cot x + x + c$, (b) $\frac{3}{2} x^2 - \frac{3}{5} \ln x + \frac{2}{3x} + c$, (c) $\frac{10}{3} \sqrt{x} + 2 e^x - 3 \cos x + c$

Question 2: (a) $\left[3x - x^2 + \frac{5}{4} x^4 \right]_0^1 = \frac{13}{4}$, (b) $[3x + \tan x]_0^{\pi/4} = \frac{3\pi}{4} + 1$, (c) $\left[\frac{1}{3} \ln (x^3 + 3) \right]_0^2 = \frac{1}{3} \ln \left(\frac{11}{3} \right)$, (d) $\left[\frac{3}{7} \ln x + \frac{2}{x} \right]_1^2 = \frac{3}{7} \ln 2 - 1$

Question 3: Graphs intersect when $x^2 - 2x - 2 = 2x + 3$. Solve quadratic to get $x = -1$ or $x = 5$. Enclosed area is $\int_{-1}^{5} (-x^2 + 4x + 5)\,dx = 36$.

Question 4: Let $u = 3x + 1$. Then $dx = \frac{1}{3} du$. When $x = 0$, $u = 1$ and when $x = 5$, $u = 16$. The integral becomes $\int_1^{16} \frac{\frac{1}{3}(u-1)}{\sqrt{u}} \left(\frac{1}{3} du \right) = \frac{1}{9} \int_1^{16} \left(u^{1/2} - u^{-1/2} \right) du = \frac{1}{9} \left[\frac{2}{3} u^{3/2} - 2 u^{1/2} \right]_1^{16} = 4$

Question 5: Let $g(x) = 2x$ and $\frac{df}{dx} = \cos 3x$. Then $\frac{dg}{dx} = 2$ and $f(x) = \frac{1}{3} \sin 3x$. Therefore $\int 2x \cos 3x\,dx = \frac{2}{3} x \sin 3x - \frac{2}{3} \int \sin 3x\,dx = \frac{2}{3} x \sin 3x + \frac{2}{9} \cos 3x + c$.

Question 6: $\int \frac{x+5}{(x+3)(x+4)}\,dx = \int \frac{2}{x+3}\,dx - \int \frac{1}{x+4}\,dx = \ln \left| \frac{(x+3)^2}{(x+4)} \right| + c$

Question 7: Let $x = 3 \sin \theta$. Then $dx = 3 \cos \theta\,d\theta$. When $x = 0$, $\theta = 0$ and when $x = 3$, $\theta = \frac{\pi}{2}$. The integral becomes $\int_0^{\pi/2} \frac{9 \sin^2 \theta}{3 \cos \theta} \cdot 3 \cos \theta\,d\theta = 9 \int_0^{\pi/2} \frac{1 - \cos 2\theta}{2}\,d\theta = \frac{9}{2} \left[\theta - \sin 2\theta \right]_0^{\pi/2} = \frac{9\pi}{4}.$

Question 8: Let $u = 9 + x^2$. Then $dx = \frac{du}{2x}$. When $x = 0$, $u = 9$ and when $x = 4$, $u = 25$. The integral becomes $\int_9^{25} x^3 u^{1/2} \frac{du}{2x} = \frac{1}{2} \int_9^{25} x^2 u^{1/2} du = \frac{1}{2} \int_9^{29} (u - 9) u^{1/2} du = \frac{1}{2} \int_9^{25} \left(u^{3/2} - 9 u^{1/2} \right) du = \left[\frac{1}{5} u^{5/2} - 3 u^{3/2} \right]_9^{25} = 282.4$

Question 9: Let $u = \cos x$. Then $du = -\sin x \, dx$ and the integral becomes $\int \left(u - \frac{3}{2u} - 4\right)(-du) = -\frac{u^2}{2} + \frac{3}{2}\ln u + 4u + c = -\frac{\cos^2 x}{2} + \frac{3}{2}\ln|\cos x| + 4\cos x + c$.

Exercise 230 on page 175
Question 3: $3\sqrt{2}$
Question 4 (a) $8\frac{1}{2}$; (c) $\int_{-1}^{0} (x^3 - x^2 - 2x)\,dx + \int_{0}^{2} (-x^3 + x^2 + 2x)\,dx = \frac{37}{12}$.

Exercise 232 on page 179
Question 1 (b): $\int_{-2}^{1} x^3 dx = \left[\frac{x^4}{4}\right]_{-2}^{1} = -\frac{15}{4}$. Therefore we need c between -2 and 1 such that $3c^3 = -\frac{15}{4}$. Take $c = -\left(\frac{5}{4}\right)^{1/3}$.

(e): $\int_{0}^{3\pi/2} \cos x \, dx = -1$. We need c between 0 and $\frac{3\pi}{2}$ such that $\frac{3\pi}{2}\cos c = -1$. Take $\cos^{-1}\left(-\frac{2}{3\pi}\right)$.

Exercise 237 on page 188, Question 1c: Volume is

$$\int_{0}^{\sqrt{2}} 2\pi x \left(\sqrt{4 - x^2} - x\right)\,dx = \frac{8\pi}{3}\left(2 - \sqrt{2}\right).$$

Exercise 240 on page 191
Question 1: $f'(x) = x(x^2 + 2)^{1/2}$, therefore $1 + [f'(x)]^2 = 1 + x^2(x^2 + 2) = x^4 + 2x^2 + 1 = (x^2 + 1)^2$. Required length is $\int_{0}^{2} \sqrt{(x^2 + 1)^2}\,dx = \int_{0}^{2} (x^2 + 1)\,dx = \left[\frac{x^3}{3} + x\right]_{0}^{2} = \frac{14}{3}$.

Question 3: $f'(x) = \tan x$, and $1 + [f'(x)]^2 = 1 + \tan^2 x = \sec^2 x$. The required length is $\int_{0}^{\pi/3} \sec x \, dx = \left[\ln|\sec x + \tan x|\right]_{0}^{\frac{\pi}{3}} = \ln\left(2 + \sqrt{3}\right)$.

Question 5: The length is $\int_{0}^{\sqrt{3}} \left(1 + [f'(x)]^2\right)\,dx = \int_{0}^{\sqrt{3}} (1 + x^2)\,dx$. To integrate this, use the substituting $x = \tan\theta$ followed by the reduction formula (4.13) on page 162. The result is

$$\int_{0}^{\pi/3} \sec^3\theta \, d\theta = \left[\frac{1}{2}\left(\sec\theta\tan\theta + \ln|\sec\theta + \tan\theta|\right)\right]_{0}^{\frac{\pi}{3}} = \sqrt{3} + \frac{1}{2}\ln\left|2 + \sqrt{3}\right|.$$

Exercise 257 on page 206
Question 1 (a): $\int_{-R}^{0} e^x dx = \left[e^x\right]_{-R}^{0} = 1 - \frac{1}{e^R}$, thus $\int_{-\infty}^{0} e^x dx = \lim_{R \to \infty}\left(1 - \frac{1}{e^R}\right) = 1$.

(b) $\int_{1}^{R} \frac{1}{x^p}dx = \left(\frac{1}{R^{p-1}} - 1\right)\left(\frac{1}{1-p}\right)$. $\int_{1}^{\infty} \frac{1}{x^p}dx = \lim_{R \to \infty}\int_{1}^{R}\frac{1}{x^p}dx = \frac{1}{p-1}$.

Question 5. The substitution $u = \sqrt{x}$ gives $\int \frac{1}{\sqrt{x}(x+1)}dx = 2\int\frac{1}{1+u^2}du = 2\arctan u = 2\arctan\sqrt{x}$, (no need for a constant of integration). Therefore $\int_{1}^{\infty} \frac{1}{\sqrt{x}(x+1)}dx = \lim_{R \to \infty} 2\left(\arctan\sqrt{R} - \frac{\pi}{4}\right) = \frac{\pi}{2}$.

Exercise 260 on page 211. In each case, let $f(x)$ be the integrand.
Question 1:

$$\frac{1}{8}\left[f\left(\frac{1}{16}\right) + f\left(\frac{3}{16}\right) + f\left(\frac{5}{16}\right) + f\left(\frac{7}{16}\right) + f\left(\frac{9}{16}\right) + f\left(\frac{11}{16}\right) + f\left(\frac{13}{16}\right) + f\left(\frac{15}{16}\right)\right] = 0.87$$

Question 3: $\frac{1}{5}\left[f\left(0\right)+f\left(\frac{\pi}{10}\right)+f\left(\frac{2\pi}{10}\right)+f\left(\frac{3\pi}{10}\right)+f\left(\frac{4\pi}{10}\right)\right]=0.45$

Question 5 :

$$\frac{1}{24}\left\{f(0)+2\left[f\left(\frac{2}{8}\right)+f\left(\frac{4}{8}\right)+f\left(\frac{6}{8}\right)\right]+4\left[f\left(\frac{1}{8}\right)+f\left(\frac{3}{8}\right)+f\left(\frac{5}{8}\right)+f\left(\frac{7}{8}\right)\right]+f(1)\right\}$$

$$= 1.46$$

Exercise 263 on page 214

Question 1a: $\frac{1}{6}\left[f\left(0\right)+f\left(\frac{1}{6}\right)+f\left(\frac{2}{6}\right)+f\left(\frac{3}{6}\right)+f\left(\frac{4}{6}\right)+f\left(\frac{5}{6}\right)\right]=0.8778$. The value of $|f'(x)|=|-2x\sin x^2|$ on the interval $[0,1]$ does not exceed 2, therefore the error in the above estimate does not exceed $2\left(1\right)/2\left(6\right)=0.17$.

Question 1c: $\frac{1}{12}\left\{f(0)+2\left[f\left(\frac{1}{6}\right)+f\left(\frac{2}{6}\right)+f\left(\frac{3}{6}\right)+f\left(\frac{4}{6}\right)+f\left(\frac{5}{6}\right)\right]+f(1)\right\}=0.8395$. The value of $|f''(x)|=|-4x^2\cos x^2-2\sin x^2|$ on the interval $[0,1]$ does not exceed 6. Therefore error in the estimate does not exceed $\frac{6(1)}{12(36)}=0.014$.

Exercise 268 on page 217

Question 1: (a) 31500, (b) $30000+1500n$, There is a common difference equal to 1500.

Question 3: $34-\frac{9}{8}\left(n-1\right)$.

Exercise 270 on page 219

Question 3: If the first term is a and the common ratio is r then the $(n-1)$th term is $a_{n-1}=ar^{n-2}$, the nth term is $a_n=ar^{n-1}$ and the $(n+1)$th term is $a_{n+1}=ar^n$. Clearly $\sqrt{a_{n-1}a_{n+1}}=\sqrt{(ar^{n-2})(ar^n)}=\sqrt{a^2r^{2n-2}}=ar^{n-1}=a_n$

Exercise 279 on page 223

Question 1: (a) 3, 3^2, 3^3, 3^4, 3^5; (c) 2, 2, $\frac{4}{3}$, $\frac{2}{3}$, $\frac{4}{15}$; (e) $\frac{4}{3}$, $\frac{4}{9}$, $\frac{4}{27}$, $\frac{4}{81}$, $\frac{4}{243}$

Question 3: $|a_n-0|=|ar^{n-1}|$. Let $\varepsilon>0$ be given. Show that the inequality $|ar^{n-1}|<\varepsilon$ holds if $n-1>\frac{\log\varepsilon-\log|a|}{\log|r|}$ then complete the proof.

Exercise 290 on page 226

Question 1: We need positive integers n such that $-0.03<\frac{3}{2n}<0.03$. Since $-0.03<\frac{3}{2n}$ for all positive integers, it suffices to find integers n such that $\frac{3}{2n}<0.03$. Solve to get $n>50$. Therefore any $N>50$ will do.

Question 3: $\sqrt{3}$, $\sqrt{3}$, 0, $-\sqrt{3}$, $-\sqrt{3}$, 0. Since $\left|\sin\frac{n\pi}{3}\right|\leq1$, it follows that $\left|2\sin\frac{n\pi}{3}\right|\leq2$, therefore the sequence is bounded.

Question 4: By the Binomial Theorem, (see (3.10) on page 111), $|r|^n=1+nb+\frac{1}{2}n(n-1)b^2+\cdots+b^n$. Since all the numbers in this sum are positive, it follows that $|r|^n>1+nb$, hence $\{|r|^n\}$ is unbounded.

Question 6: (a) 3, (c) $-\frac{2}{5}$, (e) -2.

Question 7: If $\{ca_n\}$ were to converge then $\{\frac{1}{c}ca_n\}$ would also converge. But $\{\frac{1}{c}ca_n\}=\{a_n\}$ which we know diverges. The assumption that $\{ca_n\}$ converges leads to a contradictory conclusion, so it must be false. Therefore $\{ca_n\}$ diverges.

Exercise 301 on page 236

Question 2: $\frac{12}{(3n-1)(3n+2)}=\frac{4}{3n-1}-\frac{4}{3n+2}$. $s_k=2-\frac{4}{3k+2}$, hence $\sum\frac{12}{(3n-1)(3n+2)}=2$

Question 6: $\sum_{n=1}^\infty\left(-\frac{2}{3}\right)^n=\sum_{n=1}^\infty\left(-\frac{2}{3}\right)\left(-\frac{2}{3}\right)^{n-1}=\frac{-2/3}{1+2/3}=-\frac{2}{5}$.

Question 8: $\frac{12/100}{1-1/100} = \frac{4}{33}$

Question 10: (a) $s_k = ka$. $\{s_k\}$ diverges, therefore series diverges.

(b) $s_k = \begin{cases} -a & \text{if } k \text{ is odd} \\ 0 & \text{if } k \text{ is even} \end{cases}$. $\{s_k\}$ diverges, therefore the series diverges.

Exercise 319 on page 245

Question 1: (a) Converges. $\frac{2n}{3n^3+4} < \frac{2n}{3n^3} = \frac{2}{3n^2}$. (c) Diverges. Limit comparison test. Compare to the harmonic series. (e) Converges. $\frac{1}{n+2^n} < \frac{1}{2^n} = \left(\frac{1}{2}\right)^n$. (g) Converges. Alternating series test. (i) Converges. Alternating series test. (k) Diverges. p-series with $p = \frac{1}{3} < 1$. (m) Converges. Comparison test. $0 < \frac{\cos^2 n}{n^2} \le \frac{1}{n^2}$. (o) Converges. $\frac{2^{-n}}{n} < \frac{1}{2^n} = \left(\frac{1}{2}\right)^n$.

Exercise 331 on page 247

Question 1: $\lim\limits_{n\to\infty} \left(\frac{1}{n^2}\right)^{1/n} = \lim\limits_{n\to\infty} \frac{1}{\left(n^{1/n}\right)^2} = \frac{1}{1} = 1$. The series diverges because its terms do not shrink to 0.

Question 3: Since $\lim\limits_{n\to\infty} \frac{1}{n^2} = 0$ and $\lim\limits_{x\to 0} \frac{\sin x}{x} = 1$, it follows that $\lim\limits_{n\to\infty} \frac{\sin\left(\frac{1}{n^2}\right)}{1/n^2} = 1$. Since $\sum_{n=1}^{\infty} \frac{1}{n^2}$ converges, the limit comparison test implies that $\sum_{n=1}^{\infty} \sin\left(\frac{1}{n^2}\right)$ converges.

Question 5: $\frac{3^{n+1}(n+1)^{n+1}}{((n+1)!)^2} \cdot \frac{(n!)^2}{3^n n^n}$ simplifies to $3\left(1 + \frac{1}{n}\right)^n \cdot \frac{1}{n+1}$ which has limit $3e \cdot 0 = 0$. By the ratio test, the series converges.

Question 8: (a) Diverges. Ratio test. (c) Converges. Limit comparison test. (e) Diverges. Limit comparison test. Compare to the harmonic series. (g) Converges. A Geometric progression $\sum_{n=1}^{\infty} ar^n$ with $|r| < 1$. (i) Diverges. Terms do not shrink to 0. (k) Converges. Root test.

Exercise 333 on page 252

Question 1: (a) $f^{(n)}(x) = \frac{(-1)^n(n-1)!}{x^n}$. Series is $\sum_{n=1}^{\infty} \frac{(-1)^n}{n}(x-1)^n$.

(b) $f^{(n)}\left(\frac{\pi}{2}\right) = 0$ if n is even. However, $f'\left(\frac{\pi}{2}\right) = -1$, $f^{(3)}\left(\frac{\pi}{2}\right) = 1$, $f^{(5)}\left(\frac{\pi}{2}\right) = -1$, $f^{(7)}\left(\frac{\pi}{2}\right) = 1$, ... Series is $\sum_{n=1}^{\infty} \frac{(-1)^n\left(x-\frac{\pi}{2}\right)^{2n-1}}{(2n-1)!}$.

Question 4 (c) Differentiate both sides of $\sum_{n=0}^{\infty} nx^n = \frac{x}{(1-x)^2}$ then multiply by x to get $\sum_{n=1}^{\infty} n^2 x^n = \frac{x(1+x)}{(1-x)^3}$.

Question 6: Integrating both sides of $\frac{1}{1+y^2} = \sum_{n=0}^{\infty} (-1)^n y^{2n}$ On $[0, x]$ gives $\arctan x = \sum_{n=0}^{\infty} \frac{(-1)^n x^{2n+1}}{2n+1}$.

A list of Common Integrals

Where they appear, a and b are non-zero constants.

1. If $n \neq -1$ then $\int x^n dx = \frac{x^{n+1}}{n+1} + c$, and $\int x^{-1} dx = \int \frac{1}{x} dx = \ln |x| + c$

2. $\int e^x dx = e^x + c$. In general, $\int e^{ax} dx = \dfrac{e^{ax}}{a} + c$

3. $\int \sin x dx = \cos x + c$. In general, $\int \sin ax dx = \frac{\cos ax}{a} + c$

4. $\int \cos x dx = -\sin x + c$. In general, $\int \cos ax dx = -\frac{\sin ax}{a} + c$

5. $\int \tan x dx = \ln |\sec x| + c$

6. $\int \cot x dx = \ln |\sin x| + c$

7. $\int \sec x dx = \ln |\sec x + \tan x| + c$

8. $\int \csc x dx = -\ln |\csc x + \cot x| + c$

9. $\int \frac{1}{\sqrt{1-x^2}} dx = \arcsin x + c$, and $\int \frac{1}{\sqrt{a^2 - b^2 x^2}} dx = \frac{1}{b} \arcsin \left(\frac{bx}{a} \right) + c$

10. $\int \frac{1}{1+x^2} dx = \arctan x + c$, and $\int \frac{1}{a^2 + b^2 x^2} dx = \frac{1}{ab} \arctan \left(\frac{bx}{a} \right) + c$

11. $\int \sin^2 x dx = \frac{1}{2} x - \frac{1}{4} \sin 2x + c$.   12. $\int \cos^2 x dx = \frac{1}{2} x + \frac{1}{4} \sin 2x + c$.

The next 4 are reduction formulas

13. $\int \sin^n x dx = -\frac{\sin^{n-1} x \cos x}{n} + \frac{n-1}{n} \int \sin^{n-2} x dx$

14. $\int \cos^n x dx = \frac{\cos^{n-1} x \sin x}{n} + \frac{n-1}{n} \int \cos^{n-2} x dx$

15. $\int \tan^n x dx = \frac{\tan^{n-1} x}{n-1} + \int \tan^{n-2} x dx$

16. $\int \sec^n x dx = \frac{\sec^{n-2} x \tan x}{n-1} + \frac{n-2}{n-1} \int \sec^{n-2} x dx$

Hyperbolic functions

17. $\int \sinh x dx = \cosh x + c$ and   18. $\int \cosh x dx = \sinh x + c$

The integration by parts formula

19. $\int \left(f(x) \frac{dg}{dx} \right) dx = f(x)g(x) - \int \left(g(x) \frac{df}{dx} \right) dx$

Index

294

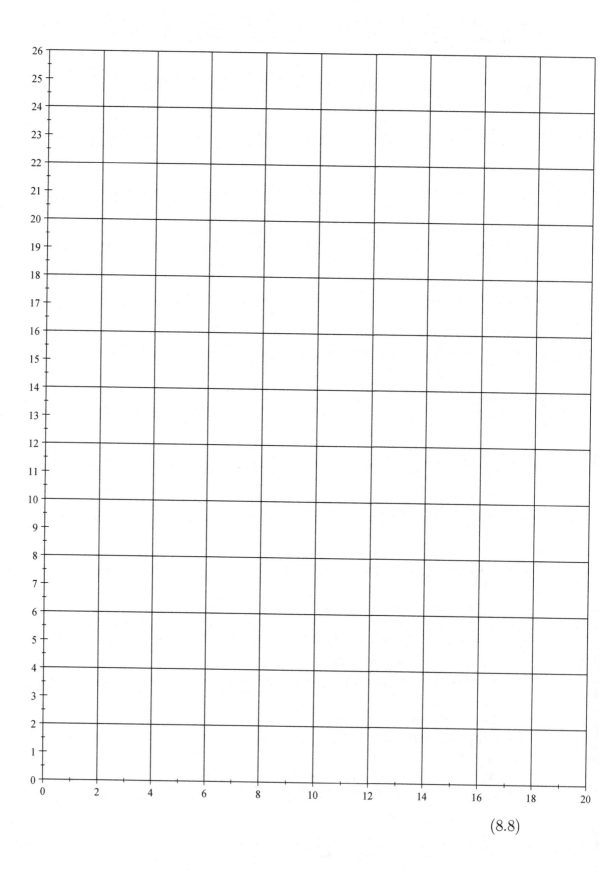

(8.8)